W9-CRB-638

PHYSICS OF
SUSTAINABLE ENERGY

To learn more about AIP Conference Proceedings,
including the Conference Proceedings Series, please visit the webpage
http://proceedings.aip.org/proceedings

PHYSICS OF SUSTAINABLE ENERGY

Using Energy Efficiently and Producing It Renewably

Berkeley, California   1 - 2 March 2008

EDITORS

David Hafemeister
California Polytechnic State University
San Luis Obispo, CA

Barbara G. Levi
Physics Today
Santa Barbara, CA

Mark D. Levine
Lawrence Berkeley National Laboratory
Berkeley, CA

Peter Schwartz
California Polytechnic State University
San Luis Obispo, CA

SPONSORING ORGANIZATION
APS Forum on Physics and Society

Melville, New York, 2008
AIP CONFERENCE PROCEEDINGS ■ 1044

Editors

Phys
Sep/oe

David Hafemeister
Physics Department
California Polytechnic State University
San Luis Obispo, CA 93405
E-mail: dhafemei@calpoly.edu

Barbara G. Levi
Physics Today
Santa Barbara, CA 93109
E-mail: bglevi@msn.com

Mark D. Levine
Environmental Energy Technology Division
Lawrence Berkeley National Laboratory
Berkeley, CA 94720
E-mail: mdlevine@lbl.gov

Peter Schwartz
Physics Department
California Polytechnic State University
San Luis Obispo, CA 93405
E-mail: pschwartz@calpoly.edu

L.C. Catalog Card No. 2008934526

ISBN 978-0-7354-0572-1
ISSN 0094-243X

Printed in the United States of America

CONTENTS

Preface

The public is awakening to the reality that anthropogenic additions of greenhouse gases to the atmosphere are affecting Earth's climate. They are focusing renewed attention to the need to reduce the world's dependence on fossil fuels for its energy sources. Rising prices of oil have further motivated public interest in alternatives to fossil fuels, adding to existing concerns in the US about its dependence on foreign sources of oil and urban air pollution.

As members of the American Physical Society's Forum on Physics and Society, we are concerned with the need to produce and use energy more wisely. Our hope is to assist in the education of fellow physicists, especially those who teach in our colleges and universities about the technical details of some of the more promising techniques for efficient and renewable energy.

To that end, we organized a short course on the *Physics of Sustainable Energy: Using Energy Efficiently and Producing It Renewably.* The short course was intended to give physicists the in-depth technical background needed either to teach about energy options or to become involved in energy research. The short course was held on the campus of the University of California, Berkeley on March 1-2, 2008. It attracted not only physics professors but also a large number of graduate and post-doctoral students who wanted to learn about wide aspects of energy physics, in some cases to redirect their careers. The chapters in this book are written versions of the talks presented at the short course, with a few additional chapters to supplement the topic of *sustainable energy.* Enthusiasm was extremely high among the 260 attendees of the *Woodstock of Sustainable Energy*, which was limited to the size of 10 Evans Hall.

This book is the third in a series of books resulting from APS-sponsored conferences on energy, all of which have been published by the American Institute of Physics. The first was the 1975 *American Institute of Physics Conference Proceedings 25*, titled *Efficient Use of Energy*. It resulted when APS, in the wake of the 1973-74 oil embargo, sponsored a meeting to study enhanced end-use energy, realizing that it is easier to save a kilowatt-hour than it is to produce a kilowatt-hour. AIP25 launched the energy careers of such physicists as Art Rosenfeld, Rob Socolow, Marc Ross and Dave Claridge, and spurred the establishment of energy programs at Lawrence Berkeley National Laboratory and at Princeton. The second energy book appeared twenty years ago, when the *APS Forum on Physics and Society* organized a short course, *Energy Sources: Conservation and Renewables*, in Washington, DC. The 700–page book, *AIP135,* resulting from that short course, became a valuable reference in many physics libraries, where applied topics are often scarce.

The current book is organized into four sections: (A) policy, (B) buildings, appliances and industry, (C) automobiles, and (D) alternative electricity production. The PowerPoint presentations of the talks presented at the short course are available online at http://rael.berkeley.edu/files/apsenergy. In addition, AIP25 and AIP135, and this volume can be obtained http://proceedings.aip.org/proceedings/cpreissue.jsp.

David Hafemeister   Barbara Levi   Mark Levine   Peter Schwartz
CalPolyU, San Luis   *Physics Today*   LBL-China Group   CalPolyU, San Luis

This book is dedicated to Alex Farrell who gave much to the ideas and issues in this book. His chapter on *The Race for Fuels in the 21st century* is a key part of these conference proceedings. We recall him as a compassionate person who knew energy issues and could address them without bias.   [Photos courtesy of Richard Cohen]

Physics of Sustainable Energy: Using Energy Efficiently and Producing It Renewably
APS Forum on Physics and Society, UC-Berkeley, March 1–2, 2008

SESSION A

POLICIES FOR SUSTAINABLE ENERGY

Opportunities in the Building Sector: Managing Climate Change

Arthur H. Rosenfeld and Patrick McAuliffe

California Energy Commission
1516 Ninth Street
Sacramento, CA 95814

Abstract. We review the data documenting the value of energy efficiency measures enacted in California, largely in the buildings and appliance sectors. We compare energy savings in the state to those achieved in the US as a whole. We also compare the cost and energy savings possible with efficiency standards enacted in China with the cost and quantity of energy expected from the construction of the Three Gorges Dam there.

US AND CALIFORNIA ENERGY SITUATIONS

Faced with increasing concentrations of atmospheric carbon dioxide, many countries are aggressively implementing measures to reduce these emissions. Although the United States has not yet committed to reducing its carbon dioxide emissions, the State of California is moving forward with its efforts to reduce carbon emissions to 1990 levels by the year 2020. The specifics of how California will proceed are under development. Full implementation is expected in 2012, with some earlier measures prior to that date. In this paper, we will provide an overview of energy consumption in the United States and in California with particular emphasis on efforts that California has made to increase the efficiency of its energy use. Also, we will discuss and describe cost curves for carbon reduction and contend that much of the reduction needed to modulate global warming could be achieved at negative costs.

In 1974, the California Energy Commission was formed to develop and implement the first energy efficiency standards for buildings and appliances in the United States as well as assess supply and demand conditions, and site new thermal power stations. Over the years, the Commission also has developed capabilities and funding for research and development (R&D) efforts related to energy and environmental issues. Currently, funding in the R&D area amounts to $80 million dollars per year with about half of this focused on energy efficiency and demand response.

A common measure of energy efficiency is energy intensity E/GDP, defined as the quantity of primary energy E consumed per unit of gross domestic product (GDP). Energy intensity in the United States has declined at five times the historical rate since

CP1044, *Physics of Sustainable Energy, Using Energy Efficiently and Producing It Renewably*
edited by D. Hafemeister, B. Levi, M. Levine, and P. Schwartz
© 2008 American Institute of Physics 978-0-7354-0572-1/08/$23.00

the 1973-74 oil crisis raised, not only, the price of energy but, along with it, an awareness of energy consumption and an appreciation for energy efficiency.

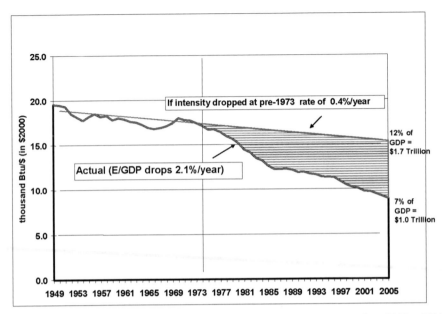

FIGURE 1. US energy Intensity in thousands of BTU per 2000-dollar per person from 1949 to 2005.

Figure 1 illustrates the decline in energy intensity in the US, especially since 1973. The impact of this improvement on primary energy demand is illustrated in Figure 2. If, instead of the actual 2.1 percent decline per year experienced since 1973, the United State's energy intensity had decreased by only the business-as-usual pre-1973 rate of 0.4 percent per year, energy use in the country would have risen by an additional 70 quadrillion Btus (quads) in 2005. Even with this improvement, primary energy use still climbed by 25 quads during these three decades. The monetary savings, associated with improvements in energy intensity in the US, amount to about $700 billion in 2005 as a result of reducing primary energy demand by about 70 quads, compared to what it could have been if pre-1973 energy intensity levels had remained unchanged through the subsequent three decades.

Improvements in energy intensity arise from many factors: improved technology, customers facing higher energy prices, consumer awareness and others. These improvements occur throughout the economy. We estimate that the $700 billion in foregone energy expenditures in the United States (in 2005 compared to what we would have spent if the energy intensity of the U.S. economy had improved at only 0.4% per year) was 1/3 due to major structural changes in the economy (less heavy industry and more high tech); 1/3 due to improvements in transportation (Corporate Average Fuel Efficiency, or CAFE, standards); and 1/3 from improvements in buildings and industry (compact fluorescent lightbulbs, better motors, building and appliance standards, etc.)

4

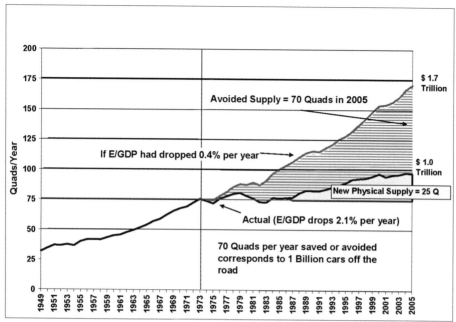

FIGURE 2. US energy consumption in quads per year from 1949 to 1974 oil embargo to 2005. The monetary savings associated with improvements in energy intensity in the United States amount to about $700 billion in 2005 as a result of reducing primary energy.

Next we address a comparison between California (34 million people) and the US (300 million people, including California). But figures 1 and 2 included transportation fuel, which in turn depends on US Federal policies and standards, which "pre-empt" California from adopting more stringent standards. Hence, we focus on electricity where California controls its own destiny.

Annual use of electricity in kWh per person from 1960 to 2005 with forecasts through 2008 in California and in the US is illustrated in Figure 3. Use in California is currently about 40 percent less than in the US as a whole, even though use was nearly the same in the 1960s. The lines start to diverge in the mid-1970s when the US experienced its first energy crisis. At times, petroleum was rationed and energy prices increased rapidly. For example, the price of electricity to residential customers in California and throughout the US nearly doubled (in nominal dollars) from the early 1970s to the later 1970s. In addition, in the late–1970s California began its building and appliance efficiency standards, which contributed to keeping per capita electricity use in California nearly flat since 1975. Of course, compared to the entire US, other factors such as a different mix of industries and differences in climate contribute.[i] Although not depicted on this slide, other policies also have led to electricity savings in California. For example, California standards allow electric water heating in homes only when it is cost effective: which is seldom the case. This has resulted in only limited electricity use for this purpose in California.

Thus, for a variety of reasons -- some policy and others due to climate or economic variables, electricity use per capita has been flat in California and should decrease slightly as California expands programs aimed at efficiency improvements.

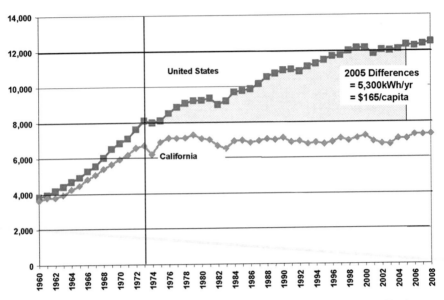

FIGURE 3. Per Capita Electricity Consumption in the United States and California.

Energy Efficient Appliances

In combination with technological improvement due to "naturally occurring" innovation, California beginning in the late 1970s introduced efficiency standards for some new appliances and buildings. In Figure 4, we provide examples of three appliance standards that were initially formulated by the state and later became US federal standards on gas furnaces, central air conditioning, and refrigerators. The trends are similar for all three but the magnitude of improvement in efficiency differs.

The amount of energy consumed in a year by the average new appliance sold in California from 1972 to 2006 (estimated) is illustrated in Figure 4. For each appliance, use is indexed to the year 1972, i.e., scaled to a value of 100. Arrows indicate when new standards took effect or will take effect. White arrows indicate state standards, which were first put in place in 1976 in response to the first oil crisis and generally rising fuel costs. US federal government standards are shown as black arrows. These did not begin until the early 1990s.

Energy use intensity by new appliances was greatly reduced by the early 2000s:

- Energy use by new gas furnaces declined by 25 percent (100% -> 75%)
- Energy use by new central air conditioners went down by 50 percent
- Refrigerators have shown the most improvement, with more than a 75% reduction in energy use.

6

Theses are just three examples. Many other appliances as well as building characteristics, such as insulation and windows, are regulated and these regulations are upgraded every few years as technological advancements continue to improve appliance efficiency.[ii] During development of these new regulations, industry representatives play an active and important role.

Index (1972 = 100)

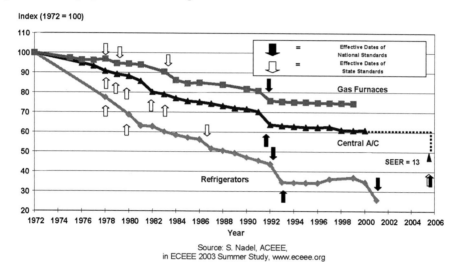

Source: S. Nadel, ACEEE,
in ECEEE 2003 Summer Study, www.eceee.org
FIGURE 4. The impact of efficiency standards for three appliances (1972–2006).

The most effective path toward energy efficiency has been to set standards for autos, buildings, appliances, equipment, etc. Figure 5 shows the remarkable gains in refrigerators. The smoothly rising curve shows that the average refrigerator has increased in size. Despite that size increase, and despite the elimination of chlorofluorocarbon use, the unit energy use has decreased dramatically since 1975. Beginning in that year, refrigeration labels and standards have improved efficiency 5 % per year for 25 straight years. In the US, improvements in refrigerators has saved enough energy to avoid the construction of 40, 1-GW power plants,. Through all of this, the price for refrigerators has declined when viewed in constant dollars even as both energy efficiency and size have improved.

Continuing with the impressive gains in refrigerator efficiency, we now compare the quantity of energy saved due to these improvements with various sources of electrical generation in the US. The refrigerator data assume that all refrigerators in use meet the current standard (which of course they do not yet, but eventually will as old units are replaced with new units). In Figure 6, the comparison is based on electricity saved or generated compared to the case in which the refrigerator efficiency was frozen at 1976 levels. Using this as a basis of comparison, refrigerators save about one-third of the amount of energy that the entire nuclear fleet in the United States generates. The data are for the year 2005.

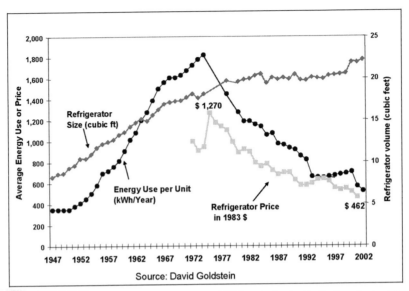

FIGURE 5. New US Refrigerators: Electricity use (kWh/year), size (cubic feet), and price (1983$).

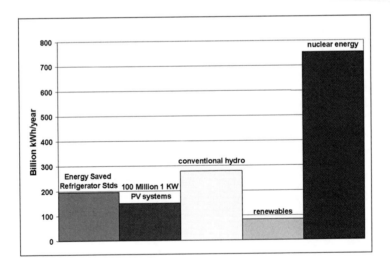

FIGURE 6. Annual US energy saved in billions of kWh/year from refrigerators vs. several sources of generation, 2005.

In the next image, Figure 7 presents a similar comparison to that in Figure 6, but here we value the electricity at the wholesale price (3 cents/kWh) for conventional hydro, renewables, and nuclear) and at the retail price (8.5 cents/kWh) for energy saved and photovoltaic systems. Using the value of the power as the metric, energy saved due to refrigerator standards has a value of nearly twice all the hydropower in the United States and about 75 percent of all electricity generated by the United States nuclear power stations. Again, we assume all refrigerators operate at the current standards for efficiency.

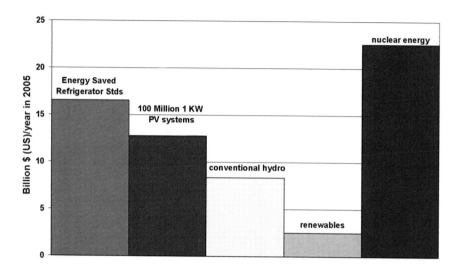

FIGURE 7. The value (billions of 2005 dollars) of electricity saved versus electricity produced in the US in 2005.

Of course, energy efficiency is not limited to the US. In Figure 8, we compare the energy production from the Three Gorges Dam in China to various efficiency standards. Figure 8 is divided into two parts:

- On the left is electricity generation from Three Gorges Dam compared to savings from China standards for refrigerators and air conditioners;
- On the right is a comparison of the dollar value of the electricity generated at the Three Gorges Dam to that of the electricity saved due to the efficiency standards illustrated on the left.

Figure 8 shows the energy saved due to efficiency standards put into effect in 2000 and in 2005 as well as the additional energy that could have been saved if the 2005 standards adopted in China had been equivalent to the current Energy Star standards in the US. Generation or savings depicted on the left side are in TWh/year, with expected generation of 100 TWh/year from the Three Gorges Dam and savings totaling nearly 90 TWh/year. These savings are calculated 10 years after the standards take effect to account for time for consumer to buy and install this equipment.

On the right side of the figure -- The value of generation from Three Gorges was calculated using wholesale electricity prices of 3.6 cents/kWh while the value of electricity saved through the standards was priced at the average cost to the consumer at 7.2 cents/kWh. The value of electricity saved is almost twice the value of that produced at Three Gorges, a somewhat startling discovery given the cost of Three

Gorges versus the cost of the standards and the incremental cost of more efficient refrigerators and air conditioners.

Comparison of 3 Gorges to Refrigerator and AC Efficiency Improvements

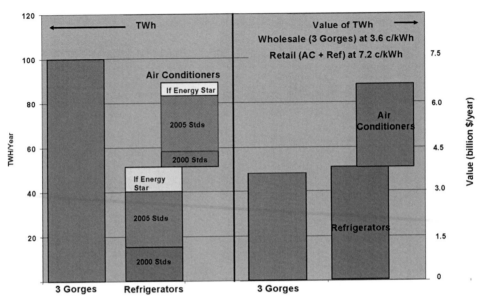

Savings calculated 10 years after standard takes effect. Calculations provided by David Fridley, LBNL

FIGURE 8. Electricity generation (TWh/year, left-hand panel) and cost of electricity (cents/kWh, right-hand panel) of the Three Gorges Dam in China compared to energy savings brought about by efficiency gains in refrigerators and air conditioners.

EFFICIENCY PROGRAMS AND STANDARDS

California's efforts to encourage efficiency through building and appliance standards provide an interesting example that is directly applicable to the issue of reducing greenhouse gas emissions. In the mid-1970s, in response to a rise in fuel prices, occasional limitations in fuel supply, concerns regarding environmental impacts of electricity production and other factors, California began to set building and appliance standards, and initiated utility programs aimed at reducing electricity use. We estimate that the current impact of these programs reduces electricity demand in California by about 40 TWh, or 15 percent. Figure 9 provides an illustration of these savings. They amount to a reduction of about 1,000 kWh per person currently.

Each year, the cost of conservation programs, public interest R&D, and standards adds about one percent to electric bills, but cuts one-half percent off the bill. So an investment of $1 in, say 1990, saves $0.50 per year for 10 to 20 years. The simple payback time is 2 years. We arrive at this by comparing the initial investment ($1) to a savings in each year of ($.50). So in two years we have paid off the initial investment, but savings continue for many more years.

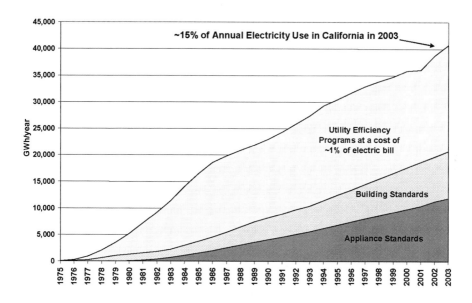

FIGURE 9. Annual Electricity Savings from Efficiency Programs and Standards in California.

However, to implement this extensive effort for utility efficiency programs, California had to put in place a number of policies. In Figure 10, we show the annual funding levels for investment in energy efficiency by California's investor-owned utilities.[iii] As the graph indicates, funding levels have fluctuated considerably since 1976. The state has now placed energy efficiency as its most preferred resource and has committed to fund these efforts aggressively for the next few years, as the figure illustrates. The figure also highlights a number of important policy decisions that the state made over this time period. These include:

- 1982 -- Decoupling utility profits from sales to eliminate the negative incentives associated with reduced sales
- 1990 -- Providing performance incentives to utilities that meet or exceed efficiency savings
- 2001 -- Including efficiency as a part of Integrated Resource Planning (IRP) and directly comparing savings to other options of meeting future load and load growth, including other policy considerations.

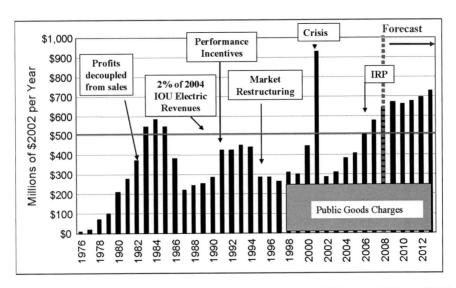

FIGURE 10. California Investor-Owned Utility Investment in Energy Efficiency. [Millions of 2002 dollars per year between 1976 and 2013]

GREENHOUSE GAS ABATEMENT

Figure 9 showed that by increasing energy efficiency in the electric sector, California currently saves about 40,000 GWh per year. We estimate that this results in an annual reduction of carbon dioxide emissions in California by 20 million metric tonnes, based on marginal generation from natural gas plants with emission rates of one-half tonne of CO_2 per MWH. California currently produces about 500 million metric tonnes of CO_2 per year.

Various estimates of the costs and methods to reduce greenhouse gas emissions are currently under discussion. Concerns abound regarding how costly it may be to reduce CO_2 emissions to acceptable levels to reduce the impact of global warming. In Figure 11, we reproduce a copy of a cost curve for greenhouse gas reductions, or abatements, prepared by McKinsey & Company (Per-Anders Enkvist, Tomas Nauclér, and Jerker Rosander[iv]) in collaboration with the Swedish utility Valtenfall. Note that in such plots, area is proportional to net annual euros saved (if area is below the x-axis) or expended (if above the x-axis). In more detail, the y-axis measures net cost of abatement in euros/tonne while the x-axis measures the size of the abatement in tonnes per year. The product (area) is the cost in euros per year. All data are for a single year – in this case the year is 2030. Total savings or costs per measure depend on the longevity of the measure. In Figure 11, considerable amount of emission abatement can be accomplished at a negative cost – that is, at a savings compared to business as usual practices. Most of these involve improving the efficiency of energy use:

- Increased building insulation
- Improved fuel efficiency in vehicles
- Improved air-conditioning system and water heating

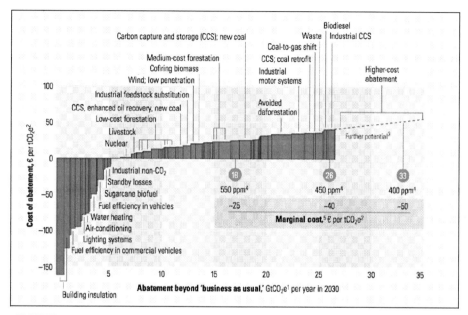

FIGURE 11.   Cost Curve of Greenhouse Gas Abatement, Worldwide [McKinsey & Company].

We have estimated the area below the x-axis in this figure at ~450 Billion Euros per year, mainly from efficiency measures. Interestingly, the area above the x-axis, mainly for renewable supply, is roughly of the same magnitude. If we can implement these at savings and costs illustrated above, there would be no net cost of getting to 450 ppm of CO_2.

American readers will want to read the US study now on the McKinsey web site (http://www.mckinsey.com/clientservice/ccsi/pdf/US_ghg_final_report.pdf). Many other examples of such costs curves can be found and, generally, they show that energy efficiency measures not only reduce greenhouse gas emissions but actually save money. However, just as California had to struggle to convince others that building and appliance standards were not only a good idea but highly cost-effective, we think the same problems will arise as we try to convince others that energy efficiency is an important tool in our effort to stem the ever rising tide of global warming.

13

REFERENCES

[i] For a thorough discussion of these factors, see Anant Sudarshan and James Sweeney, "Deconstructing the 'Rosenfeld Curve'", Stanford University, to be published in the *Energy Journal*.

[ii] Mark Ellis, *Experience with Energy Efficiency Regulations for Electrical Equipment*, International Energy Agency, Paris, March 2007.

[iii] These utilities provide service to about 75% of the state's population. The remainder is served by municipal utilities and other public agencies.

[iv] http://www.mckinseyquarterly.com/Energy_Resources_Materials/A_cost_curve_for_greenhouse_gas_reduction_abstract

Energy Efficiency in China: Glorious History, Uncertain Future

Environmental Energy Technologies Division
Lawrence Berkeley National Laboratory
Berkeley, CA 94720

Abstract. China's rapid economic growth of 10% per year has been accompanied by an annual energy growth rate of greater than 10% from 2001-2005. This in turn has led to the construction of 1 to 2 GWe of electrical generating capacity per week over the period, with the vast majority of the power plants using coal. Because of the energy growth, China has equaled the carbon consumption rate of United States at 6 billion tonnes/year in 2006, far sooner than was expected. This paper discusses the periods of energy growth and efficiency policy in China. This includes "Soviet Style" Energy Policy (1949-1980); Deng's Initial Reforms (1981-1992); Transition Period (1993 to 2001); Energy Crisis in China: 2001 to 2006, a repeat of much earlier "inglorious history; and the present efforts to return to an earlier period (Deng's Initial Reforms) in which energy growth may be more sustainable. Recommendations are given for policies to promote energy efficiency.

PART I: HISTORY OF ENERGY EFFICIENCY IN CHINA

It is well known that the United States, under capitalism, was frivolous in its use of energy before the oil embargo of 1973–74. US refrigerators required 1800 kWh/year; today, larger refrigerators with more features consume only 25% the electricity (450 kWh/year) of the 1980 refrigerator. There are similar stories to be told in other areas on the US; neglect of end-use efficiency with improved results over time. In a similar vein, China's first three decades under communism failed to take into account the wastefulness in end-uses of energy. The "Soviet Style" energy policy (1949-1979) had the single objective of growing industry as rapidly as possible to create an advanced economy. The effect of the underlying policies was to create a system of rapid energy supply growth. Energy prices were heavily subsidized. The central allocation system provided energy primarily to heavy industry to quickly transform China into a competitive power. As a consequence, there was little attention paid to end-use efficiency and little to no attention paid to the environment. China rapidly created one of the world's least efficient (and fastest growing) energy systems.

CP1044, *Physics of Sustainable Energy, Using Energy Efficiently and Producing It Renewably*
edited by D. Hafemeister, B. Levi, M. Levine, and P. Schwartz
© 2008 American Institute of Physics 978-0-7354-0572-1/08/$23.00

China's energy use rose by 500% between 1957 and 1980, while its GDP rose by 350%. Coal contributed over 80% of China's energy use in 1980.

"Soviet Style" Energy Policy (1949-1979)

- Single objective was rapid energy supply growth
- Energy prices greatly subsidized
- Central allocation system provided energy primarily to heavy industry
- No attention to environment
- **Result:** one of the world's least efficient (and fastest growing) energy systems

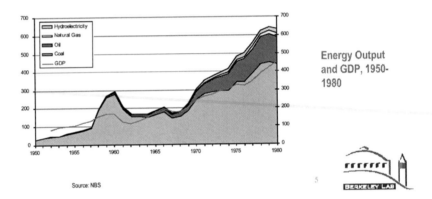

Energy Output and GDP, 1950-1980

Source: NBS

FIGURE 1. China's energy use rose by 500% between 1957 and 1980, while its GDP rose by 350%. Coal contributed over 80% of China's energy use in 1980.

Deng Xiaoping came to power in 1980. A group of academics challenged the energy policy that China had pursued from 1949 to the late 1970's. Key meetings among more than 100 academic energy experts were held in 1979 and 1980, which concluded that China energy's crisis was not severe with the endemic shortages it faced. Rather, the true energy crisis was the failure of its energy policy. This indicated an urgent need for major reform. Two areas needing reform were highlighted: energy price reform and serious attention to energy efficiency.

Deng responded to the unsolicited advice of the energy researchers. He was unable to institute significant pricing reforms at the time. However, he was able to create a system that strongly promoted energy efficiency in a variety of ways. The Chinese government quickly implemented reforms in the Sixth Five-Year Plan (1981-1985), including:

- Energy Management:
  - factory energy consumption quotas,
  - factory energy conservation monitoring,

–promotion of energy-efficient technology,
–closing of inefficient facilities, and
–controls on oil use.

• Financial Incentives:
    –low interest rate loans for energy-efficiency projects,
    –reduced taxes on sale of energy-efficient products,
    –incentives to develop new energy-efficient products,
    –monetary awards to energy-efficient enterprises.

The Chinese government initiated an investment fund for energy efficiency and created an office to administer the fund. This office later because the China Energy Conservation Investment Corporation, which, along with branch offices throughout China, became the arm of the government in selecting projects for loans and overseeing their progress. Funding for energy efficiency investments grew rapidly, reaching an authorization of 13% of total energy investment in the first years of the program. The share of total energy investment declined from 13% in 1983 to 7% in 1998 (Figure 2), but the total amount remained roughly constant.

Energy efficiency investment is stable, but declining as share of total investment

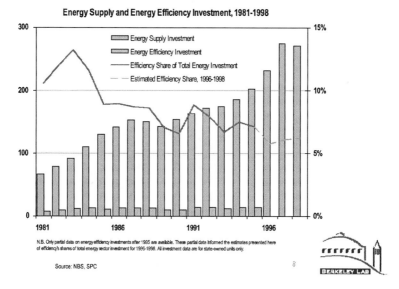

Energy Supply and Energy Efficiency Investment, 1981-1998

N.B. Only partial data on energy efficiency investments after 1985 are available. These partial data informed the estimates presented here of efficiency's shares of total energy sector investment for 1996-1998. All investment data are for state-owned units only.

Source: NBS, SPC

FIGURE 2. Investment in China's energy efficiency has been relatively constant, while the fraction of total energy funding dropped from 13% in 1983 to 7% in 1998.

As a result of these very strong programs to support energy efficiency, China's energy intensity (ratio of energy consumed to GDP) declined during 1980 to 1990 (Figure 3). This declining intensity continued through 2000 (Figure 4).

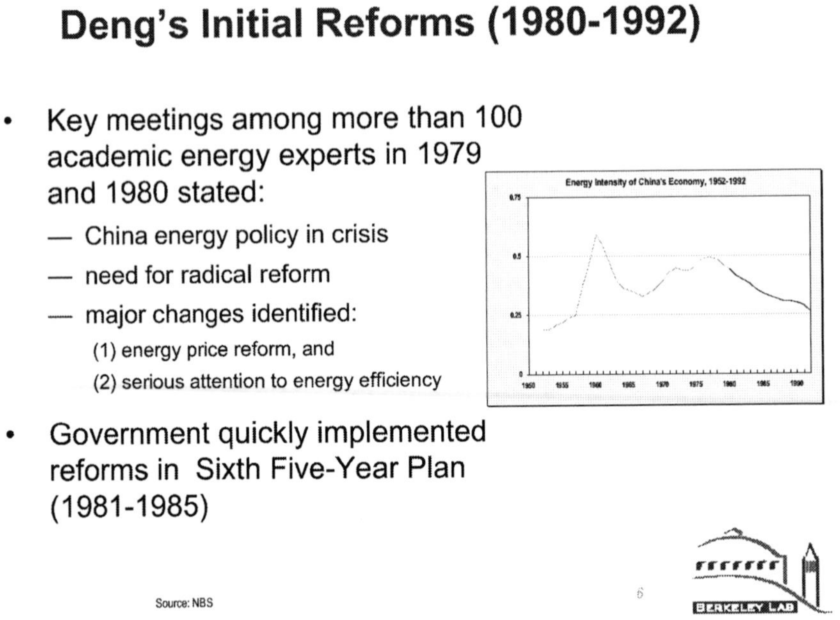
Deng's Initial Reforms (1980-1992)
• Key meetings among more than 100 academic energy experts in 1979 and 1980 stated:
— China energy policy in crisis
— need for radical reform
— major changes identified:
(1) energy price reform, and
(2) serious attention to energy efficiency
• Government quickly implemented reforms in Sixth Five-Year Plan (1981-1985)
Energy Intensity of China's Economy, 1952-1992
0.75
0.5
0.25
0
1950
1955
1960
1965
1970
1975
1980
1985
1990
Source: NBS
BERKELEY LAB

FIGURE 3. China's energy intensity (ratio of energy consumed to GDP) declined during 1980–1990.

However, in the middle 1990's, a period of transition between a government-controlled and a market economy, the energy efficiency apparatus in government began its decline. By 2000, as the market became king, energy demand took off. This occurred in spite of energy price reforms of the late 1990's and enterprise reforms that were supposed to increase price sensitivity. The past successes in improving energy efficiency were based on mechanisms that were disappearing: energy quotas for industrial enterprises disappeared; record-keeping of energy use by factors fell into disuse; energy efficiency loans from the government declined and eventually disappeared as the China Energy Conservation Investment Corporation was given goals appropriate for a private sector entity; and the new tax code of 1994 eliminated tax incentives for efficiency.

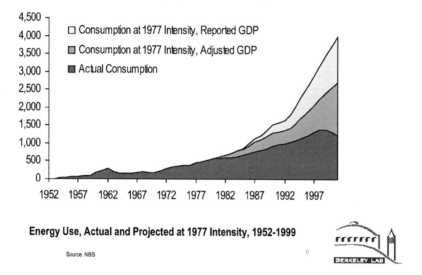

Investment in energy efficiency and other policies greatly reduced China's energy intensity (1980-2000)

- □ Consumption at 1977 Intensity, Reported GDP
- ▣ Consumption at 1977 Intensity, Adjusted GDP
- ■ Actual Consumption

Energy Use, Actual and Projected at 1977 Intensity, 1952-1999

Source: NBS

FIGURE 4. Energy use, actual and projected at 1997 intensity, 1952-1999.

Production of consumer goods, such as refrigerators, air conditioners and automobiles, grew dramatically between 1980 and 1999 (Figure 5). The increased consumption since 2001 caused energy consumption to rise faster than GDP (Figure 6). This is the first time since 1980 that energy demand had, in any single year, risen faster than GDP. Between 2001 and 2005, energy consumption grew by 61%, much more than the government target of 15%. During the same period, GDP increased by 42%, two thirds the rate of energy consumption. As is seen in Figure 7, by the year 2006 China equaled the US annual energy–related carbon emission rate. However, the US has emitted more than 3 times China's emissions on a long–term cumulative basis. (Figure 8).

Take-off of consumer goods highlights the need for efficiency standards

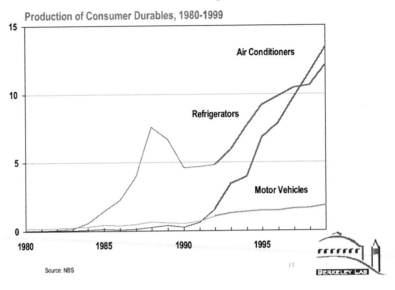

Production of Consumer Durables, 1980-1999

Source: NBS

FIGURE 5. The production of refrigerators, air conditioners and motor vehicles grew dramatically between 1980 and 1999 (in millions/year).

Since 2001, energy use has grown much faster than GDP, reversing patterns from 1980 to 2000

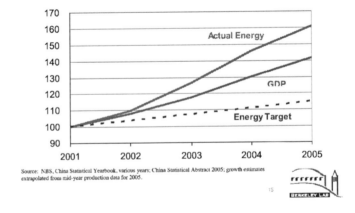

Source: NBS, China Statistical Yearbook, various years; China Statistical Abstract 2005; growth estimates extrapolated from mid-year production data for 2005.

FIGURE 6. Energy growth has outpaced GDP growth since 2001, reversing the trends of 1980–2000, normalized to 100 for 2001.

20

Annual energy-related carbon dioxide emissions, 1980-2006

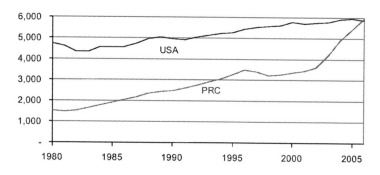

Source: US annual emissions amounts reported by US EIA in the 2006 Annual Energy Review and 2007 Flash Estimate; China emissions are derived from revised total energy consumption data published in the 2007 China Statistical Yearbook using revised 1996 IPCC carbon emission coefficients by LBNL.

FIGURE 7. Energy-related CO_2 emissions in millions of metric tons/year, US and China, 1980-2006.

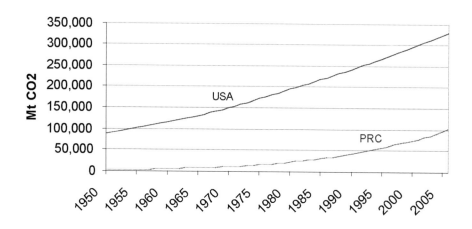

FIGURE 8. Cumulative energy–related CO_2 emissions in millions of metric tons, 1950–2006.

PART II: ENERGY CRISIS IN CHINA, 2001 TO PRESENT

China now faces a serious new energy crisis. Just as in 1979, the crisis is not in energy shortages (as many consumers see it) but rather in the lack of coherence and effectiveness of national energy policy. The Chinese government has lost the levers by which it restrained energy demand growth, and has also lost the ability to limit construction of new energy facilities. A failure to rein in energy demand growth has already had serious impacts, with more to come. China has rapidly increased its annual energy–related carbon dioxide emissions, as indicated in Figure 7. China equaled the US in annual energy–related carbon dioxide emissions at 6 billion tonnes/year in 2006. China's annual emissions doubled between 2001 and 2006, while the US annual emissions remained essentially constant over that period. However, US per-capita, energy related carbon dioxide emissions exceed Chinese per capita emissions by a factor of six (Figure 9). Rapid energy growth portends economic consequences of equal concern for China.

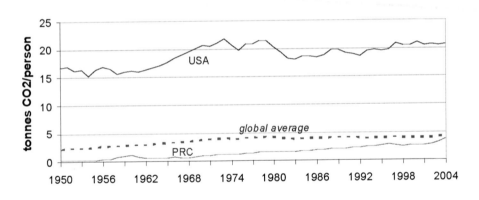

FIGURE 9. Global, Chinese and American per-capita, energy related carbon dioxide emissions in metric tons per person per year. U.S. per capita emissions exceed Chinese per capita emissions by a factor of six.

Energy demand is growing very, very fast. China is building the equivalent of three to four 500 MW coal–fired coal plants every week. The price of coal has soared. There are significant transportation bottlenecks for coal, causing large economic losses. There has been a surge in oil imports and the price of imported oil has risen considerably in the past two years. Oil is used to transport coal because inadequate rail capacity has led to widespread use of trucks to transport oil Since 2002 energy use has grown much faster than GDP, reversing patterns from 1980 to 2000. Growth in heavy industry has been extraordinary in the past five years. For these reasons, industrial efficiency is especially critical. China is the world leader in the production of many industrial commodities, such as cement, steel, and aluminum. China now produces about 50% of global cement, while the second largest producer, India, produces only 6% (Figure 10). Figure 11 shows the growth in cement and steel production from 1990 to 2007. Energy intensities within many industrial sub-sectors have declined by about 50% between 1995 and 2003 (Figure 12).

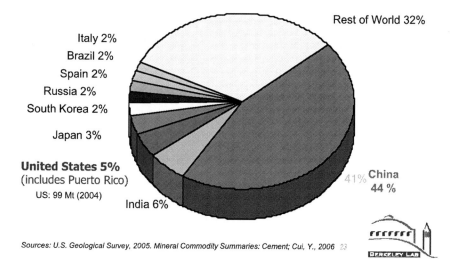

Cement Production Worldwide: 2004

Rest of World 32%

Italy 2%
Brazil 2%
Spain 2%
Russia 2%
South Korea 2%

Japan 3%

United States 5%
(includes Puerto Rico)
US: 99 Mt (2004)

India 6%

41% China 44 %

Sources: U.S. Geological Survey, 2005. Mineral Commodity Summaries: Cement; Cui, Y., 2006 23

FIGURE 10. China produces 50% of global cement, while the second largest producer, the United States, produces only 4%.

23

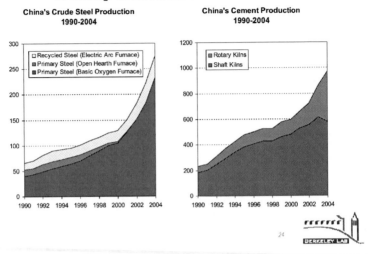

FIGURE 11. China's steel production increased by a factor of four between 2000 and 2007. China's cement production increased by almost a factor of three in this period.

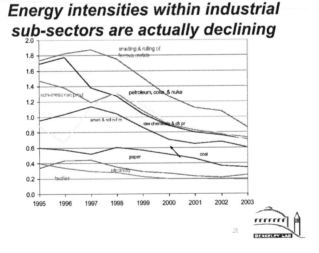

FIGURE 12. Energy intensities within many industrial subsectors have declined by about 50% between 1995 and 2003.

PART III: THE FUTURE

What might happen? What is to be done to end the crisis? The energy situation in China could get worse. Because of the unsustainability of China's energy system, there is a limit to how much more the situation will deteriorate. Or, China could continue on its present path energy path for a time. Or, things could get better.

China's National Energy Strategy has declared that "Energy development and *efficiency* have equal roles (emphasis on efficiency)." But supply investments are on the order of a factor of 20 greater than energy–efficiency investments. In light of the growing pressure that the rapid growth in energy use is placing on sources of energy supply and on the environment, as well as a part of its effort to move toward a more sustainable energy future, the Politburo of the Communist Party, in a statement released in November, 2006, called on the nation to reduce energy/GDP (energy intensity) by 20%. The 20% energy intensity reduction target was reiterated by Premier Wen Jiabao and passed into law by the Peoples Congress shortly after the Communist Party declaration. The National Develop and Reform Commission immediately began to plan the program, and was authorized to expand staff capabilities (through the use of contractors) for this purpose.

Is 20% energy intensity reduction by 2010 possible? Figure 13 presents an assessment of the impacts of different measures in reducing energy intensity. Assuming an average annual growth rate (AAGR) of GDP of 7.5%, energy consumption at constant energy intensity would rise by 43% in 2010. The second bar below, labeled "business as usual" has energy growing two-thirds as fast as GDP, or 11% below 2005 level energy intensity. This is a somewhat optimistic "business as usual" case; from 2001 until 2006, when the 2020 target was announced, energy has grown faster than GDP.[1] Aggressive efficiency policies could drop the AAGR to 3.9% (energy intensity 15% below 2005 level). These findings are displayed for industrial energy efficiency, appliances, and electricity production, transmission and distribution. Closing inefficient plants could lower AAGR to 3.55% (17% below the energy intensity in 2005).

This analysis suggests that it will be very difficult to meet the 2010 objective, and that achieving it will likely require measures beyond energy efficiency and plant closings. The only other feasible approach in the time period is to constrain the growth of energy-intensive industry, so that the structure of GDP changes over time.

Much of this seems possible if done over a longer period of time. There is good reason to believe that China, by dint of a strong effort to promote policies that reduce energy intensity, will achieve a substantial fraction of their 2020 goal. Even more important will be the execution of policies following the first five-year period. Chinese policy makers are already giving thought to using the efforts of the first five years as the foundation of a longer-term program.

---

[1] In 2006, energy demand grew 1.3% slower than GDP. In 2007, energy demand growth was 3.3% below that of GDP.

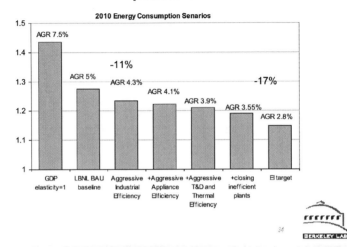

Is 20% energy intensity reduction by 2010 possible?

FIGURE 13. Analysis of measures to reduce energy intensity in China, 2006-2010.

RECOMMENDATIONS

Microeconomic policies are needed to establish targets for energy efficiency for specific industries. Accurate reporting of progress, along with incentives and penalties, and especially upgrading of technical expertise will be needed for all sectors and activities in the economy.

The provisions for information and technical guidance are of great importance. The network of energy conservation service centers and capabilities of the centers' staff should be strengthened.

A strong national program would have the following elements:

- enhanced enforcement of tighter building energy standards,
- improved compliance with tighter appliance efficiency standards,
- strict enforcement and strengthening of auto fuel economy standards,
- initiation of demand-side management programs of scale, either through electric utilities or through other administrative structures,
- investment in mass transit alternatives, especially rapid bus transit,
- revisiting energy prices and taxes to better reflect costs and to better achieve social/political objectives,

- large expansion of investments in energy efficiency through government incentives,
- policies that result in additional closings of energy inefficient factories, and
- macroeconomic policies can lead to structural changes, yielding reduced growth of energy-intensive industries.

Stabilization Wedges and Climate Change[1]

Robert H. Socolow

Department of Mechanical and Aerospace Engineering
Princeton Environmental Institute
Princeton University, Princeton, NJ 08542
Socolow@princeton.edu

Abstract. An informal pedagogical tour provides quantitative views of the magnitude of the challenge of mitigating climate change and many of the energy technologies available to address this challenge. The importance of energy efficiency is emphasized, as is the societal context for technological change.

Past, Present and Potential Future Carbon Levels in the Atmosphere

The atmosphere can be thought of as a bathtub (Figure 1). There's a certain amount of carbon in our Earth's atmosphere today, 800 billion tons. 200 years ago, the Earth's atmosphere contained 600 billion tons of carbon. In the depth of the ice age, approximately 20,000 years ago, it contained about 400 billion tons of carbon.

The ice core records are a marvelous piece of science. When we drill an ice core in the Antarctic, it's just like drilling into a tree to examine the tree rings; the deeper you go the further back in the past you are. Bubbles trapped in the ice tell us about the atmosphere when they were trapped. These records show that the quantity of carbon in the CO_2 in the atmosphere has gone back and forth between 400 and 600 billion tons

---

[1] This manuscript is similar to one being published in the New York State Bar Association's *Journal of Government Law and Policy*, produced by Albany Law School, in a special issue on climate change, (Kevin Healy, guest editor). Both manuscripts are based on a keynote talk I gave at Albany Law School, Albany, New York, on July 19, 2007, at the invitation of Judge Eleanor Stein, Administrative Law Judge for the New York State Public Service Commission, at a meeting of the Public Service Commission. Both manuscripts are the result of an overhauling and rewriting of a transcript of that talk. I appreciate having received permission from the *Journal of Government Law and Policy* to allow the publication of this second reworking of the same talk.

CP1044, *Physics of Sustainable Energy, Using Energy Efficiently and Producing It Renewably*
edited by D. Hafemeister, B. Levi, M. Levine, and P. Schwartz
© 2008 American Institute of Physics 978-0-7354-0572-1/08/$23.00

of carbon in about 100,000 years cycles, which are the ice-age cycles. For at least six cycles back, ice cores drilled into the Antarctic ice sheet provide such data.

Six hundred billion tons is the reference number for the pre-industrial quantity, and people talk about doubling or tripling it. When they just say "doubling," they mean 1,200 billion tons of carbon in the atmosphere. From the numbers in Figure 1, you can see that at the present time we are both as far above the pre-industrial level as the depths of the ice ages were below and one-third of the way to doubling.

Another unit is the fraction of the molecules in the atmosphere that are CO_2 molecules. It is now 380 out of every million (380 ppm). The concentration was about 285 ppm in the pre-industrial period. The connection is that 2.1 billion tons of carbon is one part per million.

A third unit is tons of carbon dioxide. A ton of carbon is contained in 3.67 tons of carbon dioxide, since the atomic numbers of carbon and oxygen and 12 and 16, respectively. Most of the prices used in the discussions of the economics of carbon are in dollars per ton of carbon dioxide, not dollars per ton of carbon. However, of these three units, here I'm going to use the unit "tons of carbon."

Past, present, and potential future levels of carbon in the atmosphere

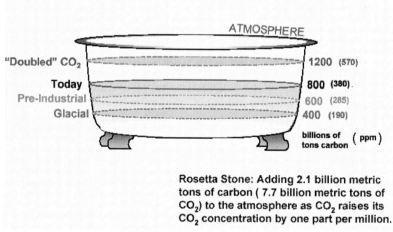

Rosetta Stone: Adding 2.1 billion metric tons of carbon (7.7 billion metric tons of CO_2) to the atmosphere as CO_2 raises its CO_2 concentration by one part per million.

FIGURE 1. Carbon in the atmosphere.

About half of the carbon we burn stays in the atmosphere for centuries

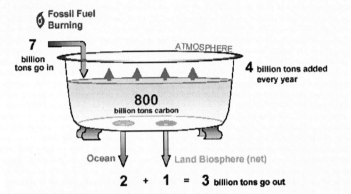

FIGURE 2. Three of seven billion tons/year of carbon do not enter the atmosphere.

Carbon Removal Mechanisms

Now let's look at what's going into that bathtub and what's going out (Figure 2). Each year, seven billion tons of carbon come out of the ground. Approximately the same amount of carbon is going into the atmosphere, because not long after it's taken out of the ground, typically months, it will be burned and become CO_2.

The atmosphere does not grow each year by seven billion tons of carbon, but by something less. That's because there are two removal mechanisms, drains in the bathtub. One is that at the surface of the ocean. If there's extra CO_2 in the atmosphere, some of it goes into the ocean and dissolves in the ocean. About two billion tons out of the seven get removed in this way. There are impacts on the ocean when this happens; notably, the ocean becomes more acidic.

The size of the other removal mechanism is found, in fact, by subtraction. It's hard to measure, and no one can model it terribly well. On average there's a net movement of CO_2 *into* terrestrial plants. This flow outward from the atmosphere happens in spite of deforestation, whose representation in Figure 2 on its own would be shown by the "land" arrow pointing *up*. Deforestation brings between 1 and 2 billion tons of carbon *into* the atmosphere each year. However, the net exchange between the biosphere and the atmosphere, including deforestation, is one unit going *out* of the atmosphere, a land arrow that goes down.

When the measurement of CO_2 in the atmosphere started at Mauna Loa, Hawaii, in 1958, there were less than 700 billion tons of carbon in the air, and in the 50 years since, that number has climbed to 800 (Figure 3).

Mauna Loa CO$_2$ data, 1958-2004

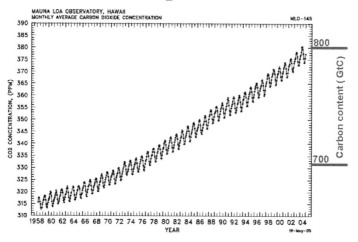

FIGURE 3. Carbon dioxide concentration in ppm at Mauna Loa, Hawaii, 1958–2004. The carbon content of the atmosphere rose from 700 to 800 gigatonnes between 1975 and 2005. [Source: Oak Ridge National Laboratory, Carbon Dioxide Information Analysis Center http://cdiac.ornl.gov/trends/co2/graphics/mlo145e_thrudc04.pdf]

The oscillation in Figure 3 is a result of an exchange of CO$_2$ between the forests and the atmosphere on an annual basis. When the forests grow, CO$_2$ comes out of the atmosphere into the leaves. When the leaves decay on the forest floor, the CO$_2$ goes back where it came from. The rise in the curve is because we're burning fossil fuels and to a lesser extent deforesting. The climb would be twice as steep, were it not for those two sinks in Figure 2.

Climate Change Impacts

Sea-level rise will strongly affect those places that are near sea level and flat, such as Florida. Much of southern Florida will disappear if sea level is eight meters higher than it is right now. Will it get eight meters higher? The answer used to be: "We don't have to worry about that for a long time." In the last couple of years, it's: "Well, maybe we do have to worry about that, even now."

There are two ice masses on the planet that are secure for the moment. One is the glaciers of Greenland and the other is what's called the West Antarctic Ice Sheet, which juts northward toward Argentina and Chile. Each of those, if it were to melt, would be worth about six to eight meters of sea level. You just melt the mass of the ice, spread the water over the surface of the ocean, which is two-thirds of the surface of the planet, and that's how much sea-level rise you get.

Which of the impacts of Climate Change are the ones that are going to be politically salient? Is it going to be sea level rise -- with a lot of uncertainty about whether it's something we have to be concerned about? Suppose we were told that there is a 10% chance that sea level will rise by 10 meters over the next 1,000 years if we do not address climate change, and that only after 100 years will we know whether this is the track we're on? Would that be enough to engender political action?

What about hurricanes instead of sea-level rise (both sometimes affecting the same territory, e.g., southern Louisiana)? Will salience adhere to the impacts of rare events becoming more frequent? If a bell curve describes the occurrence of intense storms, droughts, very hot days, and other unwanted environmental phenomena, and climate change simply shifts these bell curves to the right, enriching the upper tail, then there's bigger change for extreme events than a focus on average values would suggest. Is that what's going to drive people to action?

We can think of our response to climate change as buying insurance. My colleague, Stephen Pacala, calls these dangers "the monsters behind the door." There are quite a few monsters. As we learn more, we find out about more monsters. Every once in a while, we discover that a monster is not as fearful as we thought it was. There was a lot of concern about the shutting down of the Gulf Stream five years ago, and that was a monster. This outcome may not be as likely as people thought it was. Not everything is getting more scary. But usually new knowledge reveals more ways by which our adding CO_2 to a complex climate system brings problems for us. Yes, for other species too, but clearly, primarily, for us.

The Stabilization Triangle and the Size of the Job

Pacala and I tried to make sense of what all this had to do with energy and policy. Start with Figure 4 and look back in the past. Fifty years ago the global emissions rate was less than one-third of what it is today. In 1955 it was 2 billion tons of carbon per year and now it is about 7 billion tons of carbon per year.

The first question we asked is: "If the world does not care about carbon for the next 50 years, what will the emissions rate be?" Suppose, for example, that everyone were to buy into Senator Inhofe's view that climate change is a hoax being perpetrated on the American people. What would be the global emissions rate in 2055? There are thousands of papers answering this question, generally written by people called econometricians. They use the past as a guide to the future, try to develop what the rate of increase of the Gross National Product will be and how fast new technology will come in. They come up with lots and lots of answers, a wide band of answers.

The other question we asked is: "If the world really cares about the climate problem and works very hard, what should the goal be for fifty years from now?" Another thousand papers exist on that topic.

Because the many papers, in aggregate, produced so much noise and so little signal for those of us who are onlookers, Pacala and I asked: "Can't we cut through this?" And we drew Figure 4. This picture says that about double the carbon extraction rate, 14 billion tons of carbon a year, fifty years from now, is where the world is heading if we ignore climate change. Of course, you can make cases for higher or lower numbers, but we needed to make a single choice. Pacala and I tried to be in the middle of what is out there. The picture also says that if we humans could keep global carbon emissions to today's level for 50 years, we should be very pleased. Most students reading this are going to be around in 2055; I'd like to endow a party that they could throw themselves if the interim goal is achieved.

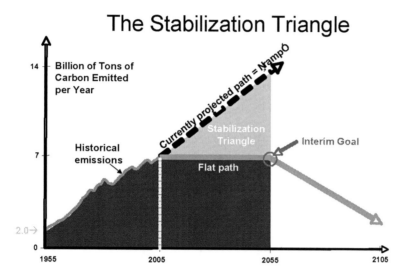

FIGURE 4. Fifty years is projected to double carbon in the atmosphere.

I circled one point on Figure 4 and called it our "interim goal": 50 years from now, the same global CO_2 emissions rate as today. I am optimistic that we can meet this interim goal, for three reasons. First, we have a terribly energy-inefficient energy system. Second, most of what will be the world's capital stock in 50 years is not yet built. Third, we are just beginning to put a price on carbon – so far, only in a few markets, notably in the European Trading System. These are the three reasons why I find it possible to imagine achieving all of the savings in the stabilization triangle in Figure 4.

Most of the criticism of Figure 4 since its publication asserts that it underestimates the job ahead. The rising line isn't rising steeply enough to capture what "Business As Usual" will bring, and the flat line is too timid a course of action to avoid climate change. Keep these criticisms in mind, because to the extent that these criticisms are valid, addressing climate change adequately means doing even more of what we'll be talking about here.

Some of you know that the language of "two degrees" and "three degrees" is another way of talking about climate change goals. These are proposed values for targets expressed in terms of the maximum rise of the average surface temperature of the planet, compared to its -industrial value (in Celsius degrees). We're 0.6°C (one degree Fahrenheit) above the pre-industrial temperature already. Figure 4 can be restated in this language. We're on track for a 3°C temperature rise if we follow the flat path, and for perhaps a 5°C rise if we follow the rising path. Many argue today that 3°C is too much, and that we should aim for 2°C. To do so requires, roughly, cutting the global emissions rate by half in fifty years, a much tougher job than keeping it constant.

Yet another way to illuminate the interim goal is to note that over the next 50 years the average global population will be about seven billion people. So our share as individuals over this period is a ton of carbon per year, taking it out of the ground, putting it in the atmosphere. In later sections of this paper, I'll relate 1 ton of carbon per year to other things.

The Wedge Model

Pacala and I divided the stabilization triangle into seven equal pieces and named these pieces "wedges," creating a unit of discussion for the subject (Figure 5). A wedge is a campaign or a strategy that leads to one billion tons of carbon per year not being emitted on the planet fifty years from now. It could be a campaign of various kinds, and so you can compare campaigns.

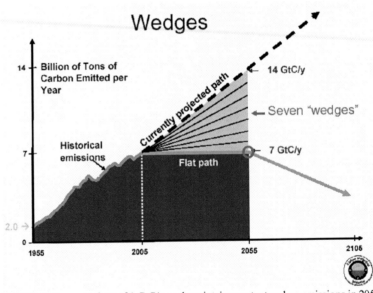

FIGURE 5. Seven wedges of 1 GtC/y each maintain constant carbon emissions in 2055.

Our definition of a carbon wedge is a triangle of carbon emissions reduction over 50 years, which attains 1 GtC/yr in 2055 (Figure 6). So, 25 billion tons of carbon are not added to the atmosphere over the fifty years. Figure 6 also introduces a price for carbon emissions, $100 per ton of carbon (about $30 per ton of CO_2). This price, in my view, is the approximate price one ought to have in mind as required to deal with climate change. It's not cheap; I'll say more a little later about how expensive it is. This price makes a wedge a $2.5 trillion enterprise. That's a lot of jobs around the world.

What is a "Wedge"?

A "wedge" is a strategy to reduce carbon emissions that grows in 50 years from zero to 1.0 GtC/yr. The strategy has already been commercialized at scale somewhere.

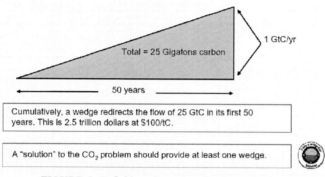

FIGURE 6. Definition of a wedge of carbon reduction.

CO$_2$ Emissions by Sector and Fuel

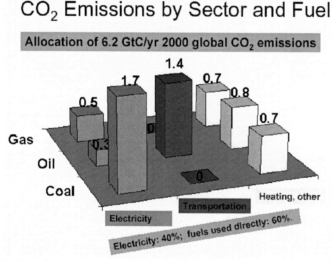

FIGURE 7. Gas, oil and coal consumption by the electricity, transportation and building/industry sectors.

Now, let's go on a hunt for wedges. First, let's find out where the seven billion tons of carbon emissions are originating right now. Take Figure 7 as a starting point. The three-by-three set of skyscrapers shows how emissions are split between gas, oil and coal. These are the three forms of carbon that come out of the Earth. The slide also shows the split between power, mobile applications and stationary applications that are not in the form of electricity but use fuels directly.

The two tallest skyscrapers are about equally high, and between them they add up to half of the total, which was six billion tons of carbon in 2000 (but seven when Pacala and I wrote the "wedges" paper). The two tallest are coal-to-power and oil-to-transport, as you might expect. At the right, you find natural gas and fuel oil going to buildings, gas going to the glass industry, oil going to petrochemicals, and coal going to metallurgy. Focusing on the electricity column, you find that it's 40% of global emissions. Also for the U.S., power plants are responsible for close to 40% of total emissions.

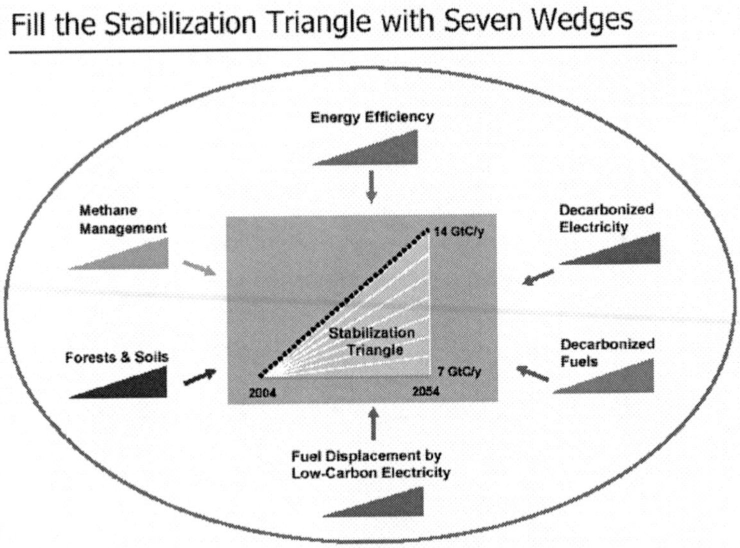

Fill the Stabilization Triangle with Seven Wedges

FIGURE 8. Seven wedges (energy efficiency is two wedges) can save 7 GtC/year by 2055.

We seek broad categories for sorting out the wedge strategies (Figure 8). Energy efficiency is at 12 o'clock, because that's where I think it belongs, right at the top. It can provide three wedges or more. At 2:00 and 4:00, we recognize that both power use and fuels use must be decarbonized; because of electricity's 40% share (noted above), neither electricity nor fuels can be ignored. At 6:00, we acknowledge that it's harder to decarbonize the use of fuels than to decarbonize electricity. At least that's our current wisdom. So when there's a price on carbon and the economy tilts away from emitting carbon, there'll be a shift toward electricity and away from direct fossil applications. An example is the plug-in hybrid car, where much of the energy for driving is coming by way of a battery charged from an electric grid. Another example is the electric heat pump for space heating. To be sure, the plug-in hybrid and the electric heat pump are only carbon-saving strategies when power comes from a low-carbon grid.

At 8:00 are forests and soils, deliberately manipulated to store additional carbon. Planting trees stores carbon.

Methane management is at 10:00, reminding us that CO_2 is not the whole story, that there are other important greenhouse gases. Methane is less well-understood and harder to address than CO_2.

Pacala and I wrote two papers, in *Science* in 2004[2] and in *Scientific American* in 2006.[3] Both have the same list of fifteen wedges:

1. Increase fuel economy of 2 billion cars from 30 to 60 mpg.
2. Reduce annual mileage of 2 billion cars from 10,000 miles/yr to 5,000 miles/yr at 30 mpg.
3. Cut electricity use in buildings by 25%.
4. Raise efficiency of 1,400 large (1000 MW) coal–fired plants from 40% to 60%.
5. Replace 1,400 large coal–fired plants with gas-fired plants.
6. Install carbon capture and storage (CCS) at 800 large coal-fired plants.
7. Install CCS at coal plants that produce hydrogen for 1.5 billion cars.
8. Install CCS at coal–to–syngas plants.
9. Add twice today's nuclear output to displace coal.
10. Increase wind power 40–fold to displace coal.
11. Increase solar power 700–fold to displace coal.
12. Increase wind power 80–fold to make hydrogen for cars.
13. Drive 2 billion cars on ethanol using one sixth of world cropland.
14. Stop all deforestation.
15. Expand conservation tillage to 100% of cropland.

People say, "Well, here's one that's not on your list. It must not be important." Read our papers. We said that there are wedges not on our list that are important. Four examples, quite different from one another, are industrial energy efficiency, "upstream" emissions, concentrated solar power, and population.

Industrial energy efficiency didn't happen to be on our list. We included buildings efficiency and vehicle efficiency, but not industrial efficiency, which of course is important. Industrial efficiency is more easily internalized by the decision-makers, who will pay more attention to any carbon price that comes along. Carbon efficiency emerges naturally for many businesses, especially when carbon emissions costs become a significant fraction of the total cost. In businesses where carbon costs are small, the business becomes more like a building.

"Upstream" investments are the oil and gas and coal industries' own emissions of carbon during extraction and conversion, as they prepare their product for market. Examples are the emissions come from flaring and venting at oil and gas fields; methane releases at coal mines; energy expenditures to transport coal, oil, and gas; CO_2 that comes out of the ground as a component of natural gas. Reducing such emissions is often a relatively low-cost mitigation opportunity.

---

[2] "Stabilization Wedges: Solving the climate problem for the next 50 years with current technologies," *Science*, (August 13, 2004).

[3] "A Plan to Keep Carbon in Check," *Scientific American*, (Sept. 2006).

Concentrated solar power (CSP) belongs with wind and photovoltaics. The most intriguing version of CSP is an array of troughs in the desert focusing sunlight onto long tubes to produce high-temperature fluids that can run engines.

Lastly, population. For more than two decades, linking population with environment has been out of fashion. But in the 1970s the link was strong. The best textbook of that period, by Paul and Anne Ehrlich and John Holdren, was called *Ecoscience: Population, Resources, Environment* (Freeman, San Francisco, 1977).. I tell my students that the choice ahead of them that will make the largest impact on the environment is how many children to have. They tell me no one has said this to them before. .

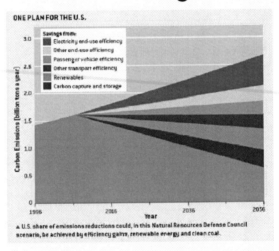

U.S. Wedges

Source: Lashof and Hawkins, NRDC, *in* Socolow and Pacala, *Scientific American*, September 2006, p. 57

FIGURE 9. US wedges could cut deeper. [D. Lashof and D. Hawkins]

"The Wedge Model is the iPod of climate change. You fill it with your favorite things." Thus says David Hawkins of the Natural Resources Defense Council, who produced Figure 9 with his colleague, Dan Lashof. Figure 9 shows U.S. wedges in a world consistent with the Princeton global wedges. Al Gore shows this image in *Inconvenient Truth* to convey the U.S. role in the global story. Compare Figure 4 and Figure 9: if global emissions remain constant, Hawkins and Lashof (and you too, reader, yes?) expect U.S. emissions to fall.

Figure 9 shows how Hawkins and Lashof would fill their iPod. There is no nuclear power, because NRDC doesn't like nuclear power; but there are four efficiency wedges, one renewables wedge, and one carbon capture and storage wedge.

Every wedge strategy can be implemented well or poorly. These are not miracles. In fact, they're dangerous. For example, nuclear power can be done well, but we're nowhere near doing it well. We certainly don't want to trade climate change for nuclear war.

Other examples: Conservation can lead to too much regimentation: how much can you intrude on the way people use energy indirectly and directly? Renewables can be done badly by not paying attention to the competing uses of land. "Clean" coal, a phrase widely used, generally refers only to burning coal well, with minimal emissions, including emissions of CO_2. But "clean" should only be used when coal is handled cleanly upstream too: mining, land reclamation, worker safety all count.

In short, one must assume that any solution to climate change can be done badly. How will it get screwed up? Ask that question at the front end.

Efficiency Wedges

Let's turn to specific wedges. I'm going to discuss only two classes of wedges here: wedges of efficiency and wedges related to substitutes for conventional coal power plants. These two classes of wedges are, I think, the most urgent ones for the next decade or so.

When we search for efficiency wedges, we address the consumers, those people on the planet who already have some means, the members of post-industrial society. They have appliances in their homes, and their vehicles dominate the scene. The significance of consumption for the global environment is relatively new, as you saw in earlier slides. Globally, 60% of oil is used in vehicles and 60% of electricity is used in buildings. In the U.S., 70% of electricity is used in buildings. The CO_2 mitigation challenge is a challenge to both energy supply systems and energy use systems, but here we'll consider the use systems.

Here's a carbon number: If your car gets thirty miles per gallon and goes 10,000 miles per year, you're going to put a ton of carbon into the atmosphere. That was your quota as a global citizen, if you remember, for *all* of your carbon. That one part of your footprint is the global average. Some of you are driving a sixty-mile-per-gallon (60 mpg) car 10,000 miles a year, and some of you are driving a 30 mpg car 5,000 miles a year. If either case, you're putting half a ton of carbon in the atmosphere.

The first wedge calculation concerns auto CO_2 emissions. The auto industry believes there will be 2 billion vehicles on the planet in 2057, about three times as many as we have right now. If they are the reference vehicles that I just referred to, 2 billion tons of carbon will go into the atmosphere. If, instead, by deliberate policy driven by climate concerns, these are 60 mpg vehicles on average, we'll have a wedge from energy efficiency in vehicles. Alternatively, if we have restructured our cities and commute less, and if we are using video-conferencing and drive less on the job, we might actually have a wedge in a different way, 30 mpg cars driven an average of 5,000 miles per year. Or we could do both, and we would have a wedge and a half.

As for efficiency in electricity use, if 40% of CO_2 will continue coming from power plants, and 70% of that power will be used in buildings, and 14 billion tons of carbon is our baseline, then cutting out one quarter of electricity use in buildings will be a wedge. Cutting out half would be two wedges. These are promising and exciting wedges. Obviously, if we're decarbonizing the power system at the same time, we're doing better still. And if we're recarbonizing, moving to coal, these are even more important wedges.

One example of an efficient electricity-using device is the variable-speed-drive motor, Another example is the compact fluorescent bulb. An example of an efficient electricity facility is the cogeneration plant, which uses both electricity and the byproduct heat. The Public Utilities Regulatory Policy Act (PURPA) of 1978 enabled a significant expansion of cogeneration. It forced utilities to allow non-utility generators to sell their electricity to the grid. In particular, industries requiring substantial heat sources were able to build cogeneration plants that sold electricity to the grid while producing the required heat. PURPA was one of the most important carbon related initiatives from the 1970s.

Which policy innovations will the next generation of energy analysts produce, the equivalents of PURPA, that the energy policy community will talk about with admiration twenty years from now?

Five ways to cut 1 tonC/yr by half

	1 ton carbon/yr	Cut in half	How?
a) Drive	10,000 mi/yr 30 mpg	60 mpg	Lighter, less power(?)
b) Drive	10,000 mi/yr, 30 mpg	5,000 miles/yr	Live closer to work
c) Fly	10,000 miles/yr	5,000 miles/yr	Video-conference
d) Heat home	Nat. gas, av. house, av. climate	Insulate, double-pane windows, fewer leaks, condensing furnace,	
e) Appliances	300 kWh/month when all-coal power (600 kWh/month, NJ)	Permanently replace twenty 60W incandescent bulbs, lit 6 hrs/day, with compact fluorescents.	

FIGURE 10. Five ways to cut carbon.

To be concrete about energy efficiency, consider Figure 10. I list activities that emit a ton of carbon per year and how to cut them in half. The first two are from our already discussed reference car, which we can drive less or exchange for a car with better fuel efficiency.

The third is about air travel. A mile flying in a commercial aircraft has about the same associated CO_2 emissions as a mile of driving alone in our reference car. The carbon footprints of only a small fraction of the people on this planet are dominated by air travel, but it's an awkward and common situation among analysts who work on energy efficiency.

The fourth item addresses residential heating. The carbon emissions that accompany the natural gas used to heat of my own home in Princeton, which is not a

McMansion, are very close to a ton of carbon a year. (I split that with my wife, so that's a half a ton of carbon for each of us.) Finding this out was not easy. My gas bill is in therms. A therm is a unit of energy, one hundred thousand Btus, or 105.5 MJ. Approximating natural gas by pure methane, looking up that the heat of combustion of methane (higher heating value) is 55.51 MJ, and assuming that methane is 75% carbon by weight, we find that using a therm of natural gas will send 1.42 kg of carbon into the atmosphere as CO_2.

Not exactly a layman's calculation! But the gas company could do the work for us, giving us this information on our monthly bills, showing carbon emissions histograms that reveal the carbon footprint of our home. The bills could present annual numbers, compare these numbers with past values, compare them to a reference group, and lots more.

For electricity consumption, the final item in Figure 10, the story is slightly more complicated than with gas. We need extra information from the electric utilities. With gas, you can go from therms to carbon without further information, and so you can work out your emissions directly from the consumption data on your bill. But for electricity calculations, you need another number: the carbon intensity of the electricity supplied by your particular utility for some particular time period. What exactly were the energy sources that produced the electricity you bought last month? That's known inside each utility, but it's not known by consumers today.

If someone uses 300 kilowatt hours per month (about a third of my own actual electric bill) his or her carbon footprint will be a ton of carbon a year – provided that all the electricity is coal-based. But New Jersey is about half as carbon intensive as that, so for someone living in New Jersey, 300 kilowatt hours would be associated with half a ton of carbon. The carbon footprint for electric power is geographically dependent, even within a state, because the key conversion factor depends on the mix of hydropower, nuclear power, natural gas, and coal

Lessons Learned about Efficiency from the 1970s and 1980s.

I was one of the researchers learning about energy efficiency in the 1970s and 1980s. Here's a summary of what we need to convey to our successors:

Measure, measure, measure. Don't give prizes for the design of a building before it's built. Too often there is a large shortfall in performance when people move in. President Reagan said, "Trust, but verify." That principle sums up the most important lesson we learned about efficiency the first time around.

For existing buildings, go building by building. They're all different. In the 1970s and 80s, trained workers were going building by building, sometimes working for the gas and electric utilities, which had put these costs in their rate base. My own research group at Princeton developed diagnostic tools using an infrared camera and equipment to pressurize a building, so that trained personnel could understand energy efficiency opportunities, which were numerous and were usually related to deficiencies in building design and construction.

For new buildings, anticipate the undoing of good intentions. My own group monitored nominally low-energy buildings that were designed so that daylight would penetrate deep into the interior. The designer imagined that the perimeter office would be occupied by an executive who would be perfectly happy to have a glass interior wall. But, alas, he wasn't, or she wasn't. The executive valued privacy and used a curtain. As a result, daylight did not go to the interior

Nominally "low-energy" or "low-carbon" buildings generally assume a low demand for the energy that enables the occupants to do whatever they choose to do inside. But this assumption is often wrong. The interior decorator in one building we studied thought that there should be oil paintings on the walls and that they should be lit by task lighting. So to save energy in buildings, we must get the interior decorators into the electricity efficiency business. So far, they've not been told that saving energy is what their client wants them to do. The same can be said of lighting specialists, who could find lighting solutions using less energy if asked to do so.

Performance standards. These clearly have great impact. They determine appliance efficiency, interior temperature, and light levels. Buildings researchers in the 1970s learned that lighting standards were captives of the lighting industry, which found ways to justify the need for great amounts of interior light in order to do various tasks. We asked for evidence that you need the extra light in order to do some particular task. We asked whether there might be a concept called over-lit?

Bounty. Decades ago, California authorities were paying people to give up their old, inefficient refrigerators, and trucks would come to your house to pick them up. Some of these inefficient units had been put in the basement when a person bought a new refrigerator; they were often running while hardly being used. California was doing same thing for old cars.

Time of day pricing and congestion charges. Adam Smith can help conserve energy.

Lifeline rates. One of the arguments against efficiency improvements that should always be challenged is that such improvements inevitably hurt the poor. This never needs to happen, because one can always implement lifeline rates, where the first block of consumption is less expensive than the next block of consumption. It's a progressive policy idea. If the overall result of some policy is that retail electricity or retail gas gets more expensive on the average, there's nothing conceptually difficult about protecting the first block of kilowatt hours or therms from a price increase. The richer consumers then carry a bit more of the total burden. Any governing body can use lifeline rates as much as it wishes, perhaps after a political fight.

Decouple profits from sales. This is a goal Amory Lovins, in particular, has been articulating for as long as I have been in this game. The regulatory body sets utility revenue rules that create incentives to sell not raw kilowatt-hours but the services that power produces. With such an arrangement, an investment in energy efficiency that reduces kWh sold is rewarded, not penalized.

Anticipate increases in kWh consumption via shifts from fuel to power As I already noted, strong carbon policy is likely to add kilowatt-hours to sales as a result of shifts to heat pumps and hybrid vehicles. You don't want to set electricity production goals that result in fighting these shifts. A goal of simply reducing kilowatt hours is not sufficiently subtle.

Wedges of Energy Supply

In the United States, the electricity sector is becoming more carbon-intensive, which is not good news from a climate change perspective. This development reverses a trend of a very long period, fifty years or more, when the nationally averaged carbon emissions per kilowatt-hour produced fell steadily. The surprise of the last five years has been that natural gas is turning out to be a less competitive electricity source for incremental power in most of the country, relative to coal, than had been expected earlier. This is bad news from a climate perspective.

Another useful carbon number is this one: seven hundred 1,000-megawatt power plants (big ones), running on coal, will put a billion tons of carbon into the atmosphere a year. So not building those plants is a wedge.

The International Energy Agency said in 2005 that we're going to put the equivalent of 1,400 new coal plants of 1000-megawatt capacity into place globally, a lot of that in China, but some of it here, by 2030. So we have a tremendous challenge to build a different plant than the kind we're heading for. And because coal plants run for many decades, carbon policy makers lock in lots of future emissions when they procrastinate.

Carbon bookkeeping today keeps track of expenditures this year, but not expenditures in future years incurred as a result of capital investments made this year. Carbon analysts do only one-column bookkeeping, and they could be doing two-column bookkeeping. The new column would report future carbon emissions commitments resulting from investments. Private industry does such double bookkeeping all the time: expenditures and investments. Firms routinely estimate future obligations when they build something. Carbon analysts don't yet do that.

Of course, an additional assumption is required, before one can make estimates of future committed emissions, namely how long is the thing going to be around? I might argue that a coal plant is going to generate electricity for 60 years, and someone else might argue for 45 years; this would need to be settled. To institutionalize "commitment accounting," a government would have to debate these additional assumptions and then embed its choices in its reporting methodology. Commitment accounting could include the lifetime fuel consumption of not only new power plants, but also new residences and commercial building.

The case for commitment accounting is implicit in Figure 2, where the atmosphere is a bathtub. From the perspective of long-term climate impact, it doesn't matter if CO_2 enters the atmosphere next year or twenty years from now. Carbon is around for so long that we really can sum over future years and learn something meaningful.

Figure 11 shows when the currently operating U.S. power plants were built. The bottoms of each bar are the coal plants and the light parts of the bars in the 1970s and 80s are the nuclear plants. We have many power plants that are thirty to forty years old. As a result, industry and government are confronting relicensing, grandfathering, retirement, and "scrap and build." Grandfathering means exempting old plants from new rules. Scrap and build means tear down the current plant, stay at the same site, and build something new and spiffy, a process with considerable virtue from an environmental perspective.

U.S. Power Plant Capacity, by Vintage

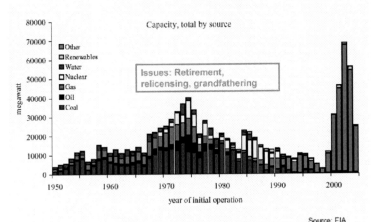

FIGURE 11. Age and type of US Powerplants.

Note the remarkable lemming-like behavior at the far right. Less than a decade ago, companies built an extraordinary amount of natural gas power, when many investors persuaded themselves that this was a brilliant thing to do. It may have been brilliant if each of them had been the only builder on the scene, but it was not brilliant when many others were doing the same thing. The price of natural gas went way up with all this new demand (and for other reasons), with the result that many of the plants on the right in Figure 11 are today either mothballed or running many fewer hours a year than they were expected to. Several firms went bankrupt. It is sobering that very few years ago, a large number of investors made a collectively wrong decision.

What can we build instead of conventional coal plants? Wind power is one answer. You need a huge amount of wind. To replace 700,000 megawatts of coal requires about 2 million megawatts of wind. (The reason the two numbers don't match is because the watt that we're talking about in both cases is a peak watt, and the intermittency of wind costs you about a factor of three when you compare wind to coal.) Wind is growing 30% per year globally. It's growing substantially but fitfully in the U.S.

Decentralized electricity production is another option. Every roof is a potential energy collector. It's not obvious exactly how to count decentralized kilowatt-hours versus centralized kilowatt-hours, because of the cost of intermittency and the benefit of no transmission and distribution. "Net metering," where a single meter runs forward when a consumer buys from the utility and backward when the consumer sells to the utility, provides a simple, but far from perfect, measure of the value of decentralized energy.

Nuclear energy is another option. Today's nuclear reactors have a 40-year operating license. As Figure 11 shows, many key issues over the next ten years will involve relicensing.

Last of the alternatives to coal-as-we-know-it is the option where coal plants capture their own CO_2 emissions so that they don't reach the atmosphere. Coupled to such "capture" plants must be "storage" facilities, typically places where CO_2 is injected deep underground. This half-a-loaf strategy is matched to a world that is unwilling to shut down coal power plants and, moreover, continues building new ones. I described carbon capture and storage (CCS) in an article I wrote in the August 2005 *Scientific American*, called "Can We Bury Global Warming?"

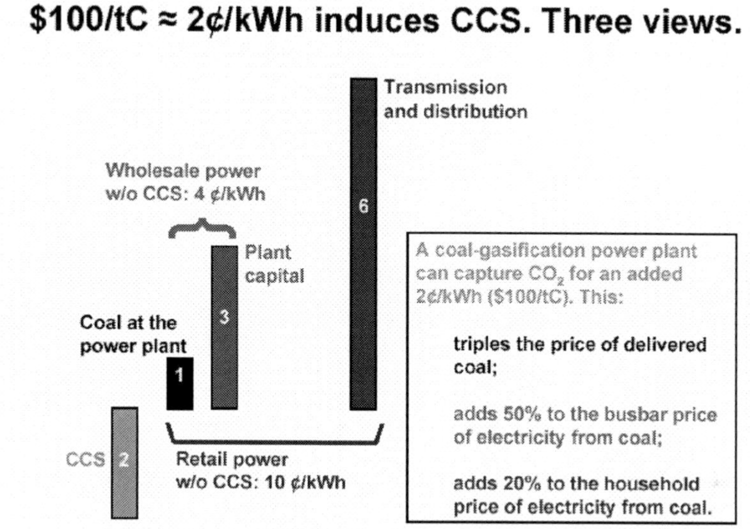

$100/tC ≈ 2¢/kWh induces CCS. Three views.

FIGURE 12. Costs to produce electricity with carbon capture and storage.

Carbon management is going to increase the price of electricity. Figure 12 presents three ways of thinking about this increase. Suppose the cost of adding CCS to a power plant is two cents per kilowatt-hour. The cost of power from such a plant will be about the same as the cost of power for the same coal plant without CCS when its CO_2 emissions are charged at a rate of $100 a ton of carbon (about $30/tonCO_2$) – called the breakeven price. To what can we compare two cents per kilowatt-hour? There are three interesting answers: We can compare it to the cost of coal, to the wholesale cost of power, or to the retail customer's cost of power.

We'll use ballpark numbers, all in cents per kWh. It's about one cent for the coal burned, three cents more for paying off the capital costs for building the plant, and another six cents for the transmission, distribution, and retail handling costs between the power plant and the residential consumer. Those numbers aren't exactly right for any specific situation, but they get us to the important insights. If you're in the coal industry, you're looking at a tripling of the cost of your product, and you could be losing out in your competition with natural gas. If there is a $30/ton CO_2 carbon tax, it will triple the cost of your coal when delivered to the utility. By contrast, the utility is looking at a 50% increase in its plant-gate ("busbar") costs, and the residential customer is looking at a 20% increase in the costs on the bill. How hard people will fight a carbon policy that leads to a $30/tonCO$_2$ emissions price (whether they will tie you up in court, for example) is implicit in the numbers in Figure 12.

To be sure, Figure 12 assumes that the extra CO$_2$ cost gets passed from one transaction to the next with neither overheads being charged nor costs being absorbed. Legislation could assure this outcome; similar legislation governs the pass-through of fuel escalation costs in electricity markets. Without such legislation, all along the value chain from coal mine to retail customer, percent overheads could be charged, whereupon two cents per kilowatt-hour on the coal could turn it into seven at your home. Not a good outcome.

The utilities have the opposite concern. They want to make sure they can recover the full two cents. The policy maker should assure that the cost of carbon mitigation moves through all the transactions right in the middle of the fairway.

Avoid Mitigation Lite

Carbon emission charges in the neighborhood of $30/tCO$_2$ can enable scale-up of most of the wedges, if supplemented with sectoral policy to facilitate transition.

Form of Energy	Equivalent to $100/tC or $30/tCO$_2$
Natural gas	$1.50/1000 scf
Crude oil	$12/barrel
Coal	$65/U.S. ton
Gasoline	25¢/gallon (ethanol subsidy: 50¢/gallon)
Electricity from coal	2.2¢/kWh (wind and nuclear subsidies: 1.8 ¢/kWh)
Electricity from natural gas	1.0¢/kWh

$100/tC was the approximate EU trading price for a year ending April 2006, when it fell sharply.

FIGURE 13. Fuel Equivalency Cost of $30/tCO$_2$

How can we understand $30/tonCO$_2$? First, it's more than the emissions price usually being talked about in Washington today. It's far more than that the price that will come into being with the Regional Greenhouse Gas Initiative (the interstate initiative designed by northeastern U.S. states, scheduled to come into force in January 2009).

How much is $30/ton CO$_2$ in other energy units? It would help if more people could know the answers. Because there's a specific amount of carbon in any ton of fuel or gallon of fuel, these answers are well defined. See Figure 13

Natural gas is measured in the U.S. either in therms or in standard cubic feet. A value of $30/tonCO$_2$ is about fifteen cents per therm or $1.50 per thousand standard cubic feet. Wholesale natural gas prices, at the point where the gas enters our

46

interstate pipeline system, are about four times higher than that today, and for the residential consumer, often ten times higher. The corresponding price in the unit used for crude oil is twelve dollars per barrel. Coal prices are usually in tons, and $30/ton CO_2 is about $65 per ton of coal, approximately twice what many coal-burning utilities pays for coal.

Coal, oil, and gas are affected unequally by a price on CO_2 emissions, because the three feedstocks produce different amounts of CO_2 when they deliver the same amount of energy. Natural gas emits only a little more than half as much CO_2 as coal and about two thirds as much as oil. The underlying reason is a difference in the amount of hydrogen in each fuel, relative to the amount of carbon. Hydrogen burns to water and produces no CO_2. As a result, when more hydrogen is present for the same amount of carbon, more energy is produced for the same amount of CO_2. Natural gas has the highest hydrogen-to-carbon ratio of the three fuels. Accordingly, a $30/ton CO_2 price on CO_2 emissions to the atmosphere has a truly big impact on the competition between coal and natural gas for electric power (favoring natural gas) and the competition between fuel oil and natural gas for home heating fuel (again, favoring natural gas)..

Returning to Figure 13, you see that by the time the price of $30/ton CO_2 reaches the consumer, if it's a straight pass through, it's twenty-five cents per gallon of gasoline, a price that isn't likely to have a big effect on driving. It's two cents per kilowatt-hour for a customer who gets his or her electricity exclusively from coal power plants. It's one cent per kilowatt-hour for a customer whose power comes from natural gas. As noted above, it's also about one cent per kilowatt-hour for an average New Jersey resident, given the mix of the nuclear, coal, and gas power plants that produce our electricity.

Given the way these numbers work, I think you will see why it is important for U.S. carbon policy to levy the CO_2 emissions charge far "upstream," ideally, where the fossil carbon comes out of the ground or across our borders. The further upstream, the higher the percent impact on the price of the product for the same CO_2 charge, and also the fewer the emissions that escape notice. If one places the charge far downstream, where gasoline is purchased and electricity bills are paid, the result of the same CO_2 emissions charge is likely to be much less CO_2 emissions reduction. If there is a CO_2 tax, impose it on the fossil fuel producer and importer; if there is a CO_2 cap and trade system, cap the carbon flows of the same players. I think designers of CO_2 policy haven't focused enough on putting under the cap the largest possible fraction of the economy's carbon emissions.

If for societal buy-in you desire involvement of the downstream consumer (the retail consumer of gasoline and electricity) in carbon policy, you will need to supplement the CO_2 price signal with targeted policy. An example of targeted policy is CAFE, the corporate-average fuel economy standard that governs the new-car market.

We can't expect to arrive at $30/ton$CO_2$ instantly. We need a ramp. It seems in this particular discussion that it falls to academics to make options vivid, so, to be specific, I recommend a ramp that climbs to thirty dollars per ton of CO_2 in ten years – an increase of three dollars per ton of CO_2 every year throughout the decade? Five years into the policy, the price is fifteen dollars per ton of CO_2. The start date might be 2010.

If, instead, we lock in much lower CO_2 prices, we set up what I call "Mitigation Lite." I say: Avoid Mitigation Lite. Mitigation Lite has the right words and the wrong numbers. Advocates of Mitigation Lite argue that we can fix the numbers after we've gotten used to the right words. The trouble with this line of reasoning is that industry negotiators are saying, "We'll take anything you want to throw at us as long as you promise not to change it." Mitigation Lite is a poor option, if regulatory certainty for a decade or more is attached to it.

Can We Do It?

Finally, can we do it? People, *we*, are becoming increasingly determined to lower the risk that we and our children will experience major social dislocation and environmental havoc as a result of rising CO_2 in the atmosphere, and we are learning that there are many ways of changing how we live, what we buy, and how we spend our time that will make a difference.

We are in the midst of a discontinuity. What once seemed too hard has become what simply must be done. Precedents include abolishing child labor, addressing the needs of the disabled, and mitigating air pollution.

What once seemed too hard has become what simply must be done.

Acknowledgements.

I would like to thank Judge Eleanor Stein, Administrative Law Judge for the New York State Public Service Commission, for inviting me to give the keynote talk at the meeting of the New York Public Service Commission on July 19, 2007, and for taking the first steps to turn the transcript of my talk into a readable manuscript. I would like to thank Margaret Barry and Rose Mary Bailly for follow-up editing in Albany that will result in publication of a different and longer version of this article in the New York State Bar Association's *Journal of Government Law and Policy*, produced by Albany Law School, in a special issue on climate change, (Kevin Healy, guest editor). I wish also to thank Jim McMahon and Dave Hafemeister for their parallel editing of the Albany manuscript for inclusion with the talks presented at the Short Course on Energy held at the University of California at Berkeley, March 1–2, 2008. For both articles, I have done much rewriting. Responsibility for the final versions of these articles is entirely mine.

Additional Articles by the Author

S.W. Pacala and R.H. Socolow, "Stabilization wedges: Solving the climate problem for the next 50 years with current technologies," *Science*, Vol. 305, pp. 968-972 (2004), with supporting online material.

R.H. Socolow, "Can we bury global warming?" *Scientific American*, Vol. 293, pp. 49-55 (July 2005).

R.H. Socolow and S.W. Pacala, "A Plan to Keep Carbon in Check," *Scientific American*, Vol. 295, pp. 50-57 (September 2006).

R.H. Socolow and S.H. Lam, "Good enough tools for global warming policy making," *Philosophical. Transactions of the Royal Society* A, Vol. 365, pp. 897–934 (2007).

Science and Technology to Support a National Energy Strategy for the United States

Daniel M. Kammen[a, b, c]

Energy and Resources Group
Goldman School of Public Policy
Renewable and Appropriate Energy Laboratory
University of California, Berkeley, CA 94720-3050

Abstract. Over the next five decades progress to meaningfully address the risk of significant climate change will require an estimated 80%, or more, reduction in the global emissions of greenhouse gases. From the baseline in 2007 of over seven billion tons of greenhouse gas emissions, three-quarters of which comes from fossil fuel combustion (with the remainder largely from land conversion and forest burning), the reductions required are from a global emissions portfolio that is currently *increasing*. As the largest current emitter, at roughly 25% of the global total – but more importantly as the nation with the largest energy resource and research base to affect change -- the United States and its inaction on climate protection for the last several years is poised to play a, if not *the* critical role in our collective climate future.

INTROCUCTION: THE CLIMATE-ENERGY CHALLENGE

It is now very clear that through action or inaction, our collective climate future is strongly tied to what course the United States steers in the beginning of the 21st Century.

A range of technologies exist that can protect the environment and improve our economic and political security—in many cases not at a cost, but instead with political and economic benefits to the nation in the form of reasserted leadership both technologically and financially, through increased geopolitical stability and flexibility, and through job growth in the 'clean energy' sector (Kammen, Kapadia, and Fripp, 2004).

To accomplish these goals, not only will a comprehensive strategy be needed (Augustine, 2005; Kammen and Nemet, 2005), but we must develop a balanced approach that recognizes that replacing the vast infrastructure and economic machinery developed to exploit fossil fuels will be a central challenge of the 21st Century, and one where the fundamental mindset of large-centralized energy monopolies will need to evolve to one of a decentralized clean energy marketplace.

CP1044, *Physics of Sustainable Energy, Using Energy Efficiently and Producing It Renewably*
edited by D. Hafemeister, B. Levi, M. Levine, and P. Schwartz
© 2008 American Institute of Physics 978-0-7354-0572-1/08/$23.00

This is the issue where --more than any set of technologies or economic incentives-- climate change causes the most uncertainty, and in some cases fear and 'pushback'.

DEVELOPING A SCIENCE-BASED ROADMAP FOR ACTION

Developing a balanced portfolio of energy research, development, and deployment projects (RD&D) is central to meeting the challenge of climate change, but it is equally clear that 'technology push' projects must be accompanied by 'demand pull measures'. Among the most important demand-pull – or market creating or enabling -- options available to us today are:

- A national commitment to saving money and energy through energy efficiency measures at every step of the economic value chain (some states, including California are fully 40% more efficient than the national average);
- The pursuit and steady increase of renewable energy portfolio standards as a baseline, and in the cities, states and regions with mandate to pursue more aggressive policies, the addition of 'feed-in' laws to diversify and expand the number and type of clean energy producers;
- Low-carbon fuel standards that evolve in time into sustainable fuel standards (LCFS parts 1 and 2; Kammen, 2007);
- The use of carbon taxes or 'cap and trade' systems under which carbon emission rights are limited;
- Developing and using for business, industrial, municipal and – critically – personal purchases carbon footprint analyses; and
- International collaborations, and public-private partnerships designed to commercialize, or at least open 'market space' for clean energy and energy efficient technologies.

This is a remarkably simple list, but one that has enough teeth, and economic opportunities, to truly harness the innovative power of the Superpower economy. It also happens to be a simple enough plan that a suitably committed presidential candidate, or president, could put it into action.

Despite a *great deal* of sound and furry, it is critical to recognize that we currently *do not* have an energy plan. In the United States arguably there has not been anything even remotely resembling an 'energy plan' since the efforts by Presidents Ford and Carter three decades ago.

Recently, however, integrated planning on climate and energy has begun to emerge, although largely at the state and regional level. The precedent for this changing the course of national energy policy is, however, a strong one. Supreme Court Justice Louis D. Brandeis wrote in 1932 that:

> ... a single courageous state may, if its citizens choose, serve as a laboratory; and try novel economic and social experiments....

Conservative and liberal justices have quoted this line over 30 times in subsequent Supreme Court Opinions. Courageous experiments are now taking place in a number of U. S. states, and can form the basis of needed federal legislation and leadership. The Global Warming Solutions Act of 2006 (AB32) in California, as well as the regional Greenhouse Gas Initiative (RGGI, http://www.rggi.org/) in the Northeast and Mid-Atlantic States are such examples. By contrast, the U. S. Federal government's current target will require only a slight change from the business as usual case (Figure 1) (EPA 2005). More relevant to the climate problem, reaching this target would actually allow emissions to grow by 12 to 16%. This target would thus represent a larger increase than the 10% increase that occurred in the previous decade. If we are to be serious about meeting the climate challenge we need to set a goal consistent with the U. S. Department of Energy's Climate Change Technology Plan (CCTP) objective of moving in the long term (e.g. ~ 2050) toward 80% reductions in net emissions. In fact, the CCTP actually mentions a zeroing of net emissions at some time after mid-century.

The California climate change protection plan is one to carefully consider in developing a comprehensive climate plan. The Governor of California's five decade GHG emissions targets of 80% below 1990 levels (EE 3-05) and the 25% GHG reductions adopted via AB32 (signed on September 27, 2006) (Kammen, 2006) include both near-term and longer-term goals – including market-based cap and trade mechanisms -- that delineate a path of emissions *reductions* toward climate stabilization. Congress should act to set a series of targets that show a clear path to meaningful emissions reductions.

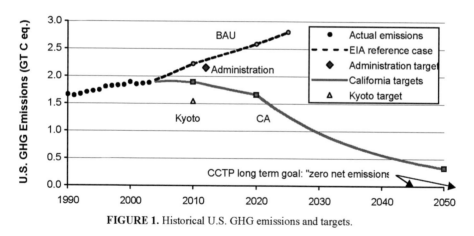

FIGURE 1. Historical U.S. GHG emissions and targets.

Figure 1 shows the actual U.S. GHG emissions from 1990 through 2003 (EPA 2005) in giga-tons of carbon equivalent. Four future paths for future U.S. emissions are shown; circles show the business-as-usual (BAU), or "reference case," as calculated by the Energy Information Agency (EIA). The diamond shows the Administration's GHG intensity target for 2012 of 18% below 2002 level in tons of carbon per unit of GDP, or a 3.6% reduction in emissions from BAU. The squares show U.S. emissions if the nation were to meet the percentage reductions that have been announced in California for 2010, 2020, and 2050 (California Executive Order 3-05, and California AB32). The triangle shows the U.S.'s target for 2010 under the Kyoto Protocol. Arrows indicate the levels required to meet the U. S. Department of Energy's Climate Change Technology Plan (CCTP) long-term goal of "levels that are low or near zero".

What is needed is a sustained commitment to emissions reductions and a time scale that conveys to the country the urgency of the need for future options. The California plan, for example, does not start or end with AB32, but includes a set of mutually reinforcing laws and executive orders. The most recent of which, the Low Carbon Fuel Standard (EE 1-07) makes a significant advance in our regulatory power to discriminate between the full range of liquid (petroleum or fossil-fuel based) fuels or electricity to power plug-in hybrid vehicles.

A Self-Consistent Energy Plan: Recent California Energy and Climate Legislation

California Renewable Energy Portfolio Standard (RPS)
Renewables to constitute 20% by 2010 (& 33% by 2020)
AB 1493 (Pavley)
30% reduction in automobile GHG emissions (MY2016)
Executive Order S-3-05
Statewide GHG emission reduction targets (~25% in 2020)
AB 32 (Pavley/Nuñez – The California Climate Solutions Act of 2006)
25% GHG reduction from stationary sources/statewide plan
CPUC action further requires that electricity sold into California meet a carbon standard based, today, on the current generation of natural gas-fired power plants. Further reductions will proceed as CA meets
AB 1007 (Pavley 2)
"develop a comprehensive strategy…alternative fuels"
and measure the Cleanenergy jobs dividend
Executive Order 06-06
Statewide biofuels production targets (40% in 2020)
Executive Order 1-07
California Low-Carbon Biofuel Standard (& State of the State address, January 2007)

The California plan represents only one such path to a low-carbon society, but it embodies the key features that are required in federal legislation: an integrated, consistent approach that both initiates early action *and* clarifies the long-term roadmap to a decarbonized future.

The U. S. has under-invested in energy research, development, and deployment for decades, and sadly the FY2008 budget request is no exception. This history is shown in Figure 2: federal energy research and development investment is today back at *pre-OPEC* levels – despite a panoply of reasons why energy dependence and *in-security,* and climatic impact from our energy economy are dominating local economics, geopolitics, and environmental degradation.

As an example of the 'commitment' to clean energy, consider the U. S. federal energy budget. At $2.7 billion for energy research, the overall federal energy research and development budget request for 2008 request is $685 million higher than the 2006 appropriated budget. Half of that increased request is accounted for by increases in fission, and the rest is in moderate increases in funding for biofuels, solar, FutureGen, and $147 million increase for fusion research. However, the National Renewable Energy Laboratory's (NREL) budget is to be cut precisely at a time when concerns over energy security and climate change are at their highest level.

The larger issue, however, is that as a nation we invest *less* in energy research, development, and deployment than do a few large biotechnology firms in their own, private R&D budgets. This is unacceptable on many fronts. The least of which is that we *know* that investments in energy research pay off at both the national and private sector levels.

A series of papers (Margolis and Kammen, 1999; Kammen and Nemet, 2005; Nemet and Kammen, 2007; all these papers are available on the website of my laboratory, http://rael.berkeley.edu) my students and I have documented a disturbing trend away from investment in energy technology—both by the federal government and the private sector, which largely follows the federal lead. The U.S. invests about $1 billion less in energy R&D today than it did a decade ago. This trend is remarkable, first because the levels in the mid-1990s had already been identified as dangerously low, and second because, as our analysis indicates, the decline is pervasive—across almost every energy technology category, in both the public and private sectors, and at

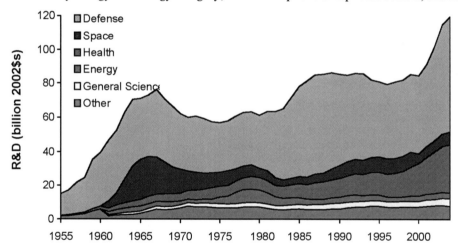

FIGURE 2. Overall federal investment in science and technology, with energy highlighted as the third sliver from the bottom. Note the comparison with the life sciences R&D budget, directly over the energy component. The federal health R&D budget experienced a doubling from the mid-1980s to today, and at the same time, private sector health investment increased by a factor of 15. Source: *Margolis, R. and Kammen, D. M. (1999) "Underinvestment: The energy technology and R&D policy challenge", Science, 285, 690 - 692.*

multiple stages in the innovation process. In each of these areas investment has been either been stagnant or declining. Moreover, the decline in investment in energy has occurred while overall U.S. R&D has grown by 6% per year, and federal R&D investments in health and defense have grown by 10 to 15% per year, respectively.

Figure 2 shows all U.S. federal R&D programs since 1955. Notice the thin strip showing the small energy R&D program relative to other sectors. The current budgets for energy R&D would continue this situation, or even reduce R&D investment (Kammen and Nemet, 2005). This is not in the best interests of the nation.

We are now in a moment – perhaps a first – where a growing view exists that energy and climate could be a *front burner* issues for candidates and voters. The time is right to focus on the energy system we want, not on the one we had, and sadly, still have.

ACKNOWLEDGMENTS

It is pleasure to acknowledge support of the Energy Foundation, and the Karsten Family Foundation's support of the Renewable and Appropriate Energy Laboratory.

REFERENCES

N.R. Augustine, *Rising Above The Gathering Storm: Energizing and Employing America for a Brighter Economic Future*. Washington, DC, National Academies Press (2005).

A.E. Farrell, Plevin, R. J. Turner, B. T., Jones, A. D. O'Hare, M. and D.M. Kammen, "Ethanol can contribute to energy and environmental goals", *Science*, vol. 311, 506 – 508 (2006).

G. Gruener and D.M. Kammen, "How to save the planet? You decide", *The Los Angeles Times*, (31 January 2007).

N. Hultman, Koomey, J. G. and D.M. Kammen, "What can history teach us about costs of future nuclear power?" *Environmental Science & Technology (ES&T)*, vol. 40, 2088 – 2093 (1 April 2007).

D.M. Kammen, "September 27, 2006 – A day to remember", *San Francisco Chronicle* (27 September 2006).

D.M. Kammen, "Transportation's Next Big Thing is Already Here", May, *GreenBiz.com, Climate Wise*. URL: http://www.greenbiz.com/news/columns_third.cfm?NewsID=35189 (2007).

D.M. Kammen, Kapadia, K. and M. Fripp, *Putting Renewables to Work: How Many Jobs Can the Clean Energy Industry Generate?* A Report of the Renewable and Appropriate Energy Laboratory, University of California, Berkeley (2004). Available at: http://socrates.berkeley.edu/~rael/papers.html#econdev

D.M. Kammen and G. F. Nemet, "Reversing the Incredible Shrinking Energy R&D Budget" *Issues in Science and Technology*, vol. 22, 84 - 88 (2005).

R. Margolis and D.M. Kammen, "Underinvestment: The energy technology and R&D policy challenge", *Science*, vol. 285, 690 - 692 (1999).

G.F. Nemet, "Beyond the learning curve: factors influencing cost reductions in photovoltaics, *Energy Policy*, vol. 34(17), 3218 - 3232 (2006).

G.F. Nemet and D.M. Kammen, "U.S. energy research and development: Declining investment, increasing need, and the feasibility of expansion", *Energy Policy*, vol. 35 (2007) 746–755 (2007).

R.N. Schock, W. Fulkerson, *et al*, "How much is Energy Research and Development Worth as Insurance?" *Annual Review of Energy and Environment*, vol. 24, 487 - 512 (1999).

Water Security: A Growing Crisis and the Link to Energy

Allan R. Hoffman

Office of Energy Efficiency and Renewable Energy
U.S. Department of Energy
1000 Independence Avenue
Washington, DC 20585

Abstract. This paper explores the global situation with respect to fresh water availability and the linkage between water and energy.

INTRODUCTION

Water has always been mankind's most precious resource. There are no substitutes and the struggle to control water resources has shaped human political and economic history. As the U.N. Committee on Economic, Cultural and Social Rights stated in 2002: "The human right to water is indispensable for leading a healthy life in human dignity. It is a prerequisite to the realization of all other human rights."

Lack of water security is a serious and growing global crisis. Water security can be defined as the ability to access sufficient quantities of clean water to maintain adequate standards of food and goods production, sanitation and health. Many parts of the developing world already face significant water shortages, with severe implications. In coming years, the problem will become more widespread, extending to the United States and other developed countries.

Complicating this crisis is the inextricable linkage between water and energy. To address water security issues one must have the energy to extract water from underground aquifers, transport water through canals and pipes, manage and treat impaired water for reuse, and desalinate brackish and sea water to provide new fresh water supplies. But just as access to clean water that people can afford to buy is limited, so too, is access to affordable energy.

Energy can play a key role in delivering clean water. Fossil-fuel powered desalination already plays an important role in desalination of sea water in several countries. Nuclear power plants can serve a similar role. In addition, direct and indirect forms of solar energy, a plentiful resource, have the potential to provide significant new sources of water where it is most needed.

CP1044, *Physics of Sustainable Energy, Using Energy Efficiently and Producing It Renewably*
edited by D. Hafemeister, B. Levi, M. Levine, and P. Schwartz
2008 American Institute of Physics 978-0-7354-0572-1/08/$23.00

WATER AND ENERGY LINKED

The linkage between water and energy extends beyond the need for energy to extract, transport, treat and desalinate water to create fresh water supplies. Many forms of energy production depend on the availability of water — for example, the production of electricity at hydropower sites, the cooling of thermal power plant exhaust streams, the processing of crude oil, tar sands and oil shales, the growing of biomass, the transportation of coal slurries, and the release of hydrogen from water by electrolysis or high temperature dissociation.

Other, indirect, linkages between water and energy exist as well. The production and conversion of energy resources, and the use of energy in end use applications, produce waste products that can contaminate surface and underground water supplies. There is now incontrovertible evidence that carbon dioxide released through combustion of fossil fuels contributes to global warming and associated climate change, with significant implications for global precipitation patterns. If competing water uses limit use of waterways for transport of goods, rail and truck will require more energy to move those goods. Water and energy are also critical to sustainable economic development, a principal goal of U.S. economic and foreign policy. Without access to both, economies cannot grow, jobs cannot be created and poor people cannot move out of poverty.

Water and energy are linked in yet another way. Energy, in absolute terms, is not in short supply in the world. The world's total annual consumption of commercial energy is about 450 quads (the U.S. share is 100 quads), whereas the sun pours 6 million quads a year into the earth's atmosphere. While approximately 30% of this energy is reflected back into space, the majority is absorbed and reflected in the earth's energy balance in various direct and indirect forms. What is in short supply is energy that people can afford to buy. The same can be said about water. The earth is a water-rich planet and annual consumption of fresh water is much less than 1 percent of the planet's total supply. What is in short supply is affordable clean water.

Energy policy and water policy can also be expressed in similar terms. If one recognizes that energy is a means to an end and not an end in itself – i.e., energy is important only as it allows us to provide the services that are important to human welfare (heating, cooling, illumination, communication, etc.) - it follows that energy security rests in part on using the least amount of energy to provide a given service. It also rests on access to technologies providing a diverse supply of reliable, affordable and environmentally benign energy sources. The first priority of energy policy must then be the wise, efficient use of whatever energy supplies are available. The same is true of water. Only after ensuring the wise, efficient use of existing resources must we focus on harvesting new energy and water supplies that meet sustainability and environmental requirements.

Demand Grows for Scarce Water

The earth's total water supply is estimated to be 329 million cubic miles (1,371 cubic kilometers), with each cubic mile containing more than 1 trillion gallons (3.75 trillion liters). The problem is that 317 million cubic miles (1,321 cubic kilometers), or just over 96 percent, is found in the oceans and is saline, on average 35,000 parts per million of dissolved salts. Another 7 million cubic miles (29 cubic kilometers) is tied up in icecaps and glaciers, and 3.1 million cubic miles (13 cubic kilometers) in the earth's atmosphere. Ground water, fresh water lakes and rivers account for just over 2 million cubic miles (8.3 cubic kilometers) of fresh water. The net result is that 99.7 percent of all the water on earth is not available for human and animal consumption. Of the remaining 0.3 percent of water, the vast majority is stored in ground water.

An important feature of the earth's supply of fresh water is its non-uniform distribution around the globe. Water has been a source of tension wherever water resources are shared by neighboring peoples. Globally, there are 215 international rivers and 300 ground water basins and aquifers shared by two or more countries. Water-related tensions around the world can have significant implications for U.S. national security. In the Middle East, for example, water is a source of conflict not only between Israel and its Arab neighbors, but also between Egypt and Sudan, and among Turkey, Syria, and Iraq. Many have forgotten that the progression towards the 1967 War, whose impact lingers to this day, was triggered by the water dispute between Israel and Syria over control of the Jordan River. Water conflicts add to the instability of a region on which the U.S. depends heavily for oil. Such conflicts could extend to Central Asia as new fossil fuel supplies become available there as well.

Global demand for water has more than tripled over the past half century. Global water withdrawal in 2000 is estimated to be 1,000 cubic miles (4.168 cubic kilometers), about 30 percent of the world's total accessible fresh water supply. By 2025, that fraction may reach 70 percent. Over-pumping of ground water by the world's farmers (including in the U.S., India and China) already exceeds natural replenishment by more than 38 cubic miles (158 cubic kilometers), roughly 4 percent of total withdrawals.

How Serious is the Situation Today?

The World Health Organization estimates that, globally, more than 1 billion people lack access to clean water supplies and more than 2 billion lack access to basic sanitation. The amount of water deemed necessary to satisfy basic human needs is 1,000 cubic meters per capita annually. In 1995, 166 million people in 18 countries lived below that level. By 2050, experts project that the availability of potable water will fall below that level for 1.7 billion people in 39 countries. Water shortages plague almost every country in North Africa and the Middle East.

These shortages have significant health effects. Water-borne diseases account for roughly 80 percent of infections in the developing world. Nearly 4 billion cases of diarrhea occur each year, with diarrheal diseases killing millions of children. Another 60 million children are stunted in their development as a result of recurrent diarrheal episodes. In addition, 200 million people in 74 countries are infected with the parasitic disease schistosomiasis, intestinal worms infect about 10% of the population in the developing world, and an estimated 6 million people are blind from trachoma, with an at-risk population of 500 million..

The gender implications of fresh water shortages are also serious. Women head one-third of the world's families (in parts of Latin America families headed by women are the majority) and frequently are the financial mainstays of and principal water providers for their families. They are responsible for half of the world's food production, and produce between 60 and 80% of the food in most developing countries. To produce adequate sanitation and food they must first 'produce' water. As the principal water providers women and girls in developing countries spend up to 8 hours daily finding, collecting, storing and purifying water. This reduces significantly the time they might otherwise use for education, community involvement and cottage industries. If safe and reliable water sources do not exist nearby they are forced to pay exorbitant prices to street vendors or rely on unsafe local water resources. This has major implications for hygiene and the spread of diseases among poor women and their families. Finally, poor women's access to water is less than that of poor men because decisions are most likely made by men and the water needs of women are often ignored or undervalued. This has led to a situation where women are among the poorest of the poor in most parts of the world, leading to a "feminization of poverty."

Responding to the Growing Crisis

A number of voices have sought to sound the alarm for more than a decade. Only recently, though, has broad world attention begun to focus on water. World Water Forums were held in 1997, 2000, 2003 and 2006. The U.N. Millennium Summit in 2000 identified water and energy availability as critical global issues, as did the 2002 World Summit on Sustainable Development. The U.N. declared 2003 the International Year of Freshwater, and designated the period 2005-2015 the U.N. Decade of Water. It is now widely recognized that access to clean water and sanitation is critical to the eradication of poverty and the achievement of sustainable economic development. Nevertheless, despite this growing awareness international support for water projects is marginal and declining.

At its 2000 Summit the United Nations adopted two Millennium Development Goals related to water and sanitation: to reduce by half, by 2015, the proportion of people without access to (a) safe drinking water, and (b) basic sanitation. Assuming a world population in 2015 of 7.2 billion, to meet these goals 1.6 billion more people will need to be supplied access to safe drinking water and an additional 2.2 billion access to basic sanitation. Even if the 2015 goals are reached, which is questionable, 600 million people in 2015 will still lack access to clean water and 1.5 billion to adequate sanitation.

THE U.S. WATER SITUATION

U.S. water withdrawals in 2000 are shown in the following figure.

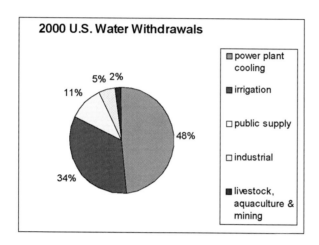

FIGURE 1. U.S. water withdrawals in 2000.

Power plant cooling is the largest user when total withdrawals (fresh plus saline) are counted. A 500 MW closed-loop power plant requires 7,000 gallons per minute (10.1 million gallons per day). Of the 195 million gallons per day used in 2000 for cooling thermal power plants, 70 percent was fresh water, and 30 percent saline (only about 3 percent of this water is actually consumed through evaporation). Nationally, power plant cooling and agricultural irrigation each accounted for 39 percent of fresh water use.

Sustainable withdrawal of fresh water is currently an issue in the U.S. The fast growing demand for clean water, coupled with the need to protect and enhance the environment, has already created shortages in some parts of the U.S. and will make other areas of the U.S. vulnerable to water shortages in the future. For example, California's allocation of Colorado River water has been reduced because competing urban, agricultural and environmental interests could not agree on a conservation plan. The Ogallala fossil water aquifer in the Central Plains is being depleted by agricultural and urban extraction (and contaminated by agricultural runoff) with no effective recharge. Lake Michigan sees increasing amounts of pollution from the area around Chicago. An increasing number of water disputes are taking place as well in the eastern U.S. - between Virginia and Maryland, Virginia and North Carolina, and among Georgia, Florida and Alabama. A brackish water desalination plant is being operated in Tampa, Florida, and other desalination plants are being planned for sites in California, Texas, Utah and Hawaii.

Competition for fresh water is already limiting energy production. For example, Georgia Power lost a bid to draw water from the Chattahoochee River, the Environmental Protection Agency ordered a Massachusetts power plant to reduce its water withdrawals, Idaho has denied water rights requests for several power plants, Duke Power warned Charlotte, NC to reduce its water use, and a Pennsylvania nuclear power plant is planning to use wastewater from coal mines. Other utilities are warning of a power crunch if water availability is reduced.

In response, the Electric Power Research Institute (EPRI), the research and development arm of the private electric utility sector, has initiated a major new research program that will address the connection between fresh water availability and economic sustainability. As a first step, EPRI, which has projected that the world will need 6-7,000 GW of additional electrical generation capacity by 2050 (today's total is about 4,000 GW), undertook a screening study aimed at characterizing the probable magnitude of the quantity of water demanded and supplied, as well as the quality of such water, in the U.S. for the next half century (2000-2050). This screening study, published in 2002, concluded that "...the water budget of the United States in the next 50 years is more uncertain than the currently available predictions suggest," that "...the cost of insufficient water availability over the next 50 years can be huge," and that "...water availability can severely constrain electricity growth."

An additional problem is that many U.S. federal agencies address water issues, but none at the water-energy nexus. No federal agency is responsible for water-related impacts on energy policy, water used by energy production, or energy used by water systems.

What Can Be Done to Improve Water Security?

In addition to providing adequate financial resources and training, we can build on lessons learned in recent years by the development community. First, for services to be sustainable, community members, both male and female, must participate in decisions about the design, management and maintenance of the services. Second, demand for water must drive our strategy, rather than supply. Third, solutions must be tailored to local conditions. Finally, we must recognize water as a scarce resource, with costs attached to its provision.

In his November 28, 2003 *Science* article, "Global Freshwater Resources: Soft-Path Solutions for the 21st Century," Peter Gleick states, "The most cited estimate of the cost of meeting future infrastructure needs for water is $180 billion per year to 2025 for water supply, sanitation, wastewater treatment, agriculture, and environmental protection." In this article, Gleick proposes a different, "soft path", approach to global water security, analogous to the soft-path approach for energy first proposed by Amory Lovins in the 1970s.

The soft path assumes that people's fundamental interest is in satisfying demands for water-related services such as food, fiber, waste disposal and sanitation. Thus, society's focus should be not on the use of water per se but on the services and benefits provided per unit of water used. Gleick estimates that if a soft-path approach is used, the cost to improve global water security could be in the range of $10 billion to $25 billion per year for the next two decades, a much more achievable level of investment.

A Role for Renewable Energy

Globally, energy used in delivering water is approximately 26 quads, 6 percent of world consumption. It is used in the following ways:

Lifting ground water: Power needed = (water flow rate) x (water density) x (head). For example, lifting water from a depth of 100 feet (30.5 meters) at a flow rate of 20 gallons (75.7 liters) per minute, and assuming an overall pump efficiency of 50 percent, requires 1 horsepower (hp)/0.75 kilowatts(kW).

Pumping water through pipes. Power needed = (water flow rate) x (water density) x (H+HL), where H is the lift of water from pump to outflow and HL is the effective head loss from water flow in the pipe. For example, moving water uphill 100 feet at 3 feet (30.5 meters at 0.9 meters) per second through a pipeline that is 1 mile long (1.6 kilometers) and 2 inches (5 centimeters) in diameter requires 4.8 hp/3.6 kW.

Energy needed to treat water. Average energy use for water treatment, according to Southern California studies, is 652 kilowatt-hours (kWh) per acre-foot (AF), where 1 AF = 325,853 gallons (1,233.5 kiloliters). (Note: in many remote parts of the world treatment must be more basic and less expensive than in developed regions).

Energy needed for desalination. Extensive use of desalination will be required to meet the needs of a growing world population. Energy costs are the principal barrier. The two most widely used desalination technologies are reverse osmosis (RO; 44%) and multi-stage flash distillation (MSF; 40%). Worldwide, more than 15,000 desalination units are producing approximately 37 million cubic meters of fresh water every day. Fifty two percent of this capacity is in the Middle East, largely in Saudi Arabia where 30 desalination plants meet 70% of the Kingdom's current drinking water needs and several new plants are under construction. North America has 16%, Asia 12%, Europe 13%, Africa 4%, Central America 3%, and Australia 0.3%. U.S. plants (Florida, California, Texas, Arizona) produce just under 6 million cubic meters per day. The energy required to produce a cubic meter of potable water, exclusive of energy required for pretreatment, brine disposal and water transport, is approximately 5 kWh via RO, and 25 kWh via MSF. Costs of desalinated water from large scale plants are typically in the range 60-90 cents per cubic meter.

If water security issues are to be addressed, where is the required energy to come from? Historically, the answer has been the grid in developed regions, and either human power (e.g., foot-peddle pumps) or diesel generators in remote regions. The use of diesel generators is neither cheap nor environmentally benign. Even in developed regions that draw power from the grid, reliability can be an issue, as recent major outages demonstrate.

Renewable energy can play a key role in meeting this challenge, both in developing and developed countries. Solar-powered water pumping can raise clean water from depth and transport it to where it is needed, and already does so in many locations. In addition, in many parts of the developing world, two plentiful resources are brackish water and solar energy. Use of solar energy to power local desalination of brackish water can provide significant new sources of water in areas that have few if any other potable water supply options. Jordan, Israel, the Palestinian Authority and the United Stages, in a joint effort, have undertaken such an activity in the Middle East. These results are easily replicated in many parts of the world.

Similarly, solar energy can be used in remote locations to power disinfection systems for contaminated water. WaterHealth International's UV Waterworks technology, for example, uses ultraviolet radiation in a 60-watt system to kill bacteria and viruses via DNA disruption. A single solar panel (or wind energy or low-head hydro) can power this system, which is capable of disinfecting 4 gallons per minute. Even less energy will be needed in the future for UV disinfection of contaminated water as UV LEDs become available, where the emitted radiation is narrowly focused on biologically active UV wavelengths.

In developed countries WorldWater & Power Corporation has demonstrated the value of solar energy in providing water services. The energy crisis in California, which led to limited power availability for the state's billion-dollar agriculture, winery and water utilities markets, presented an opportunity for WorldWater to provide solar-powered water-pumping systems (up to 600 hp) to these markets.

Other promising renewable technologies include solar thermal power systems that heat water to drive steam turbines. Such systems can be jointly operated as electricity and potable water providers. An ocean energy technology such as the open cycle version of OTEC, which taps the temperature differences at selected locations in the ocean to generate electricity, can also produce large quantities of distilled water. In remote island locations, the water produced is likely to prove more valuable than the electricity. Other ocean technologies (wave energy, tidal and ocean current energy) can also be used to provide electricity and/or water.

Concluding Thoughts

The problem of global water security is already serious and growing more serious each year. For example, six years of drought in Australia have taken a toll, reducing Australia's rice crop by 98%, leading to "skyrocketing" prices worldwide. "Many scientists believe it is among the earliest signs that a warming planet is starting to affect food production."

The United States is not immune. A new "...study by researchers at the Scripps Institution of Oceanography concludes that the growing demand for water in the West, combined with reduced runoff due to climate change, are causing a net deficit of nearly 1 million acre-feet of water per year in the Colorado River system..." and "...the researchers estimate a 50% chance that Lake Mead could drop too low for power production (at Hoover Dam) by 2017." The literature includes other frightening stories as well.

The bottom line is that energy (and food) issues cannot be separated from water issues, and we can no longer take water resources for granted if the U.S. and other countries are to achieve energy security in the decades ahead. No longer can U.S. and global water security be guaranteed without careful attention to related energy issues. The linkage between the two must be explicitly recognized and acted upon. In the U.S. this will require a new partnership between the federal government, which has primary responsibility for energy security, and state and local governments, where water issues have historically been addressed.

SESSION B

ENERGY USE IN BUILDINGS, APPLIANCES, AND INDUSTRY

Energy Savings by Treating Buildings as Systems

L.D. Danny Harvey

Department of Geography
University of Toronto
100 St George Street
Toronto, M5S 3G3, Canada

Abstract. This paper reviews the opportunities for dramatically reducing energy use in buildings by treating buildings as systems, rather than focusing on device efficiencies. Systems-level considerations are relevant for the operation of heat pumps (where the temperatures at which heat or coldness are distributed are particularly important); the joint or separate provision of heating, cooling, and ventilation; the joint or separate removal of sensible heat and moisture; and in the operation of fluid systems having pumps. Passive heating, cooling, and ventilation, as well as daylighting (use of sunlight for lighting purposes) also require consideration of buildings as systems. In order to achieve the significant (50-75%) energy savings that are possible through a systems approach, the design process itself has to involve a high degree of integration between the architect and various engineering disciplines (structural, mechanical, electrical), and requires the systematic examination and adjustment of alternative designs using computer simulation models.

INTRODUCTION

The energy use of buildings depends to a significant extent on how the various energy-using devices (pumps, motors, fans, heaters, chillers, and so on) are put together as systems, rather than depending on the efficiencies of the individual devices. The savings opportunities at the system level are generally many times what can be achieved at the device level, and these system-level savings can often be achieved at a net investment- cost savings. Thus, by failing to analyze buildings as systems, the energy savings potential in the building sector will be vastly underestimated, costs will be overestimated, and, as a result, less stringent policies with regard to energy use and greenhouse gas emissions from this sector will be adopted than if policymakers are fully aware of the systems-level savings potential. At the same time, the systems-level analysis presents a greater intellectual challenge to building designers than analyses at the device level, but the large efficiencies achievable at the system level provide a greater motivation for work at the systems

CP1044, *Physics of Sustainable Energy, Using Energy Efficiently and Producing It Renewably*
edited by D. Hafemeister, B. Levi, M. Levine, and P. Schwartz
© 2008 American Institute of Physics 978-0-7354-0572-1/08/$23.00

level. The purpose of this paper is to provide examples of how system-level considerations lead to large energy savings opportunities in the buildings sector.

SYSTEM-LEVEL CONSIDERATIONS FOR MECHANICAL SYSTEMS

Heat Pumps, Operating Principles

Heat pumps provide a simple example of the difference between system and device efficiency. A heat pump transfers heat from cold to warm (against the macro temperature gradient), although at each point in the system, heat flow is from warm to cold. It relies on the fact that a liquid cools when it evaporates, and the cooling effect is greater the lower the pressure of evaporation, while a gas releases latent heat as it condenses and is warmed to a greater temperature the greater the pressure. The key energy-using device in a heat pump is a compressor, which increases the pressure of the working fluid on the discharge side and creates low pressure on the suction side. The cycle during heating mode is illustrated in Figure 1a. As the working fluid is compressed, it is heated to a temperature in excess of the indoor air temperature. This allows heat to be transferred to the indoor air stream in an indoor heat exchanger, thereby extracting heat from (and condensing) the refrigerant. The more the gas is compressed (i.e., the greater the pressure), the more it warms up. The liquid refrigerant travels through an expansion valve to a heat exchanger that is connected to the suction side of the compressor. The low pressure there induces evaporation and hence cooling of the refrigerant. The lower the pressure, the greater the cooling that occurs. The refrigerant must by cooled to below the temperature of the outdoor air in order to absorb heat from the outside air. The cool, low-pressure refrigerant, now in the gaseous state, returns to the compressor, where the cycle is repeated.

By simply reversing the direction of fluid flow, a heat pump can act either as a heating unit (transferring heat from the outside to inside) or as an air conditioner (transferring heat from the inside to outside). This is illustrated in Figure 1b. An air conditioner is a heat pump that operates in only one direction.

The efficiency of a heat pump is represented by its coefficient of performance (COP), which is the ratio of heat delivered to energy input. The difference between the temperatures of the evaporator and condenser is called the temperature lift. The maximum possible COP (called the Carnot cycle COP) pertains to an ideal (fully reversible) heat pump, and is related to the temperature lift as follows:

$$COP_{cooling,ideal} = \frac{T_L}{T_H - T_L} \qquad (1)$$

and

$$COP_{heating,ideal} = \frac{T_L}{T_H - T_L} + 1.0 \qquad (2)$$

where T_L is the evaporator (lower) temperature and T_H is the condenser (higher) temperature. The Carnot COP for heating is equal to the Carnot COP for cooling plus 1.0 because the energy input to the heat pump is ultimately dissipated as heat and so adds to the heat that can be supplied.

(a) Heating Mode

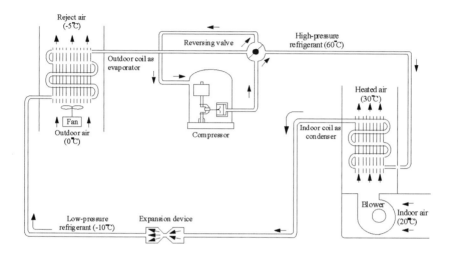

(b) Cooling Mode

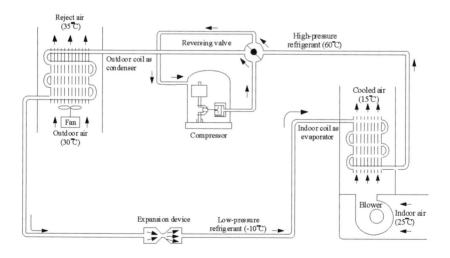

FIGURE 1. Refrigerant flow in a heat pump operating in heating mode (upper panel) and cooling mode (lower panel). Source: (1).

The actual COP is given by

$$COP_{cooling,real} = \eta_c \left(\frac{T_L}{T_H - T_L} \right) \tag{3}$$

where η_c is the Carnot efficiency. A typical Carnot efficiency is around 65%. If one focuses on the device efficiency, one might imagine increasing the efficiency to 75%, which would reduce the energy use by only 14%. However, if one can reduce the temperature lift required for given heating or cooling load, then vastly greater reductions in energy use are possible.

The evaporator must be colder than the heat source in order to draw heat from it, and the condenser must be warmer than the heat sink in order to supply heat to heat. The relationships are shown in Figure 2 for the case of heat source and sink temperatures of 16°C to hot water at 30°C, with temperature differentials of 10 K at both the evaporator and condenser. The apparent lift, based on the heat source and sink temperatures, is 14 K, and the apparent Carnot COP is 20.6. However, the real lift and real Carnot COP are 34 K and 8.5, respectively. The real COP, given a Carnot efficiency of 0.65, would be 5.53. However, if both temperature differentials are cut in half, the COP would increase to 7.75, a 40% improvement.

There are two ways to increase the real COP: to reduce the apparent lift by reducing the difference between the heat source and sink temperatures, and to further reduce the real lift by reducing the temperature differentials. The temperature differentials can be reduced by (i) reducing the required heat flux, (ii) building larger heat exchangers, or (iii) designing the heat exchangers to be more effective in transferring heat for a given area and temperature difference. The temperature differential is the driving force for the heat flow, so with smaller required heat flows, the temperature differentials do not need to be as large. In heating mode, the heat flow from the condenser must balance the heat loss from the building, so a high-performance thermal envelope (high levels of insulation, windows with low heat loss) will permit more efficient operation of the heat pump by permitting a smaller real temperature lift. Similarly, measures that reduce heat gains in a building (a high-performance thermal envelope and efficient lighting and equipment to minimize internal heat gains) will permit a more efficient cooling heat pump (or more efficient air conditioners or chillers).

The apparent temperature lift can be reduced by distributing heat at the coolest possible temperature or distributing coldness at the warmest possible temperature. In buildings with hydronic cooling (that is, where coldness is distributed via cold water), it is common to distribute coldness at a temperature of 6-8°C. This would require an evaporator temperature of 0°C or less, while the condenser might be at a temperature of 50°C, so as to discharge sufficient heat to the surrounding air. The cooling units usually involve a small coil with a fan in the rooms that need to be cooled. However, temperatures as warm as 20°C can provide adequate cooling if the entire ceiling is cooled by circulating the cooling water through panels mounted on the ceiling (illustrated in Figure 3) or inside the concrete core of the ceiling. This permits a much warmer evaporator, a smaller temperature lift, and hence a larger COP. Similarly, distributing heat at the coolest possible temperature minimizes the temperature lift required in heating mode. Low heating temperatures are possible if the entire ceiling or floor is heated and serves as a radiator.

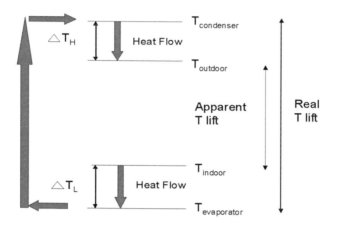

FIGURE 2. Relationships between real and apparent temperature lifts, temperature differentials (ΔT_L and ΔT_H), and heat flow in a heat pump.

FIGURE 3. A chilled ceiling panel. Source: www.advancedbuildings.org.

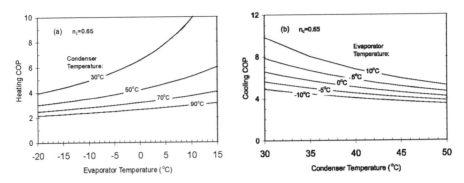

FIGURE 4. Variation in the COP of a heat pump in heating mode and in cooling mode for various evaporator-temperature combinations, assuming a Carnot efficiency of 0.64. Source: (1).

The other way to reduce the apparent temperature lift is to draw heat from the warmest possible temperature source in heating mode (such as the ground or exhaust airflow, rather than from cold outside air), and to reject the heat to the coolest possible heat sink in air conditioning mode (such as to evaporatively cooled water, the ground, or lake, river, or sea water rather than to the hot outside air). Figure 4 shows the variation in the COP of a heat pump in heating mode and in cooling for various evaporator-temperature combinations. There can frequently be a factor of two difference in the heat pump COP between best-case and worst-case combinations of evaporator and condenser temperature.

Heating, Ventilation, and Air Conditioning (HVAC) Systems

The combined heating, ventilation, and air conditioning systems (HVAC systems) in buildings present many opportunities for large reductions in energy use through better integration of the system components. These components might be air conditioners or chillers (which produce cold water), pumps, fans, cooling towers, and various devices for exchanging heat and possibly moisture from one airstream to another.

Separation of ventilation from heating & cooling functions

In many buildings, air is circulated both for ventilation purposes (to deliver fresh air) and for heating and cooling. However, the airflows usually required in order to deliver adequate heat or coldness (without unduly warm or cold supply air) are usually several times greater than the airflows required for ventilation purposes. If heat or coldness are instead delivered by circulating warm or cool water, the required airflows can be greatly reduced. This in turn leads to a large savings in fan+pump energy use, because delivering heat or coldness by moving water is inherently more efficient than by moving air. Further large reductions in fan energy use are possible if the ventilation airflow can be reduced through more effective ventilation, and if ventilation is avoided or greatly when it is not needed (which is possible if heating and cooling are independent of ventilation).

The power that must be imparted to a moving fluid in order to sustain its motion is given by

$$P_{fluid} = \Delta P \times Q \tag{4}$$

where ΔP is the pressure head created by the fan or pump (and is equal to the sum of all the pressure losses through the flow circuit) and Q is the volumetric flow rate. The rate at which heat is given off (in Watts) by an air or water flow, Q_H, is given by

$$Q_H = \rho c_p Q (T_{sup\,ply} - T_{return}) = \rho c_p Q \Delta T \tag{5}$$

where ρ and c_p are the fluid density and specific heat, respectively. From Eqs. (4) and (5), the ratio of energy supplied to keep a fluid moving to heat removed or supplied by the fluid flow is given by $\Delta P / \rho c_p \Delta T$. Given typical ρ, c_p, and ΔT values for hydronic and air systems, it turns out that the energy used to deliver a given amount of heat or coldness is about 25 times less using water than using air (1, Section 7.1.2).

The electric power that must be supplied to the fan or pump used to push the fluid is given by

$$P_{electric} = \frac{\Delta P Q}{\eta_m \eta_p} \tag{6}$$

where η_m and η_p are the motor and pump (or fan) efficiencies, respectively. As ΔP α Q^2 for turbulent flow,

$$P_{fluid} \; \alpha \; Q^3 \tag{7}$$

Thus, cutting the required rate of flow in half reduces the required rate at which energy needs to be supplied to the flow by a factor of 8. Motor and pump efficiencies decline slightly at lower flow rates, but the electrical energy requirement still decreases by a factor of 6-7. For this reason, it is far more effective to deliver heat and coldness by circulating warm or cold water, with the airflow reduced to the much smaller flow rates needed for ventilation purposes alone. Further savings are possible if the ventilation airflow requirements are themselves reduced through more effective delivery of ventilation air to the breathing zone, as discussed next.

Displacement ventilation

Ventilation air flow requirements can be reduced by a factor of two by using displacement ventilation (DV) rather than ceiling-based mixing ventilation, while improving air quality and reducing total heating loads on the chillers. DV involves supplying slightly cooled ventilation air at floor level (or from small, scattered diffusers in the floor itself), then relying on internal heat gains to heat the ventilation air, causing it to gradually rise and displace the room air, and then exit through ceiling vents. This is in contrast to conventional systems, where fresh air is supplied at ceiling level, and reliance is placed on diluting rather than displacing the stale room air through turbulent mixing. DV is inherently more efficient than mixing ventilation, and saves energy by permitting a factor-of-two reduction in ventilation air flow and by permitting the supply of air at warmer temperatures (due in part to the fact that heat that accumulates in the ceiling area is directly removed from the room rather than mixed it into the room air). When combined with chilled ceiling cooling (described below), a savings of 40-60% in cooling energy use occurs in various US cities compared to a standard mixing ventilation system with variable airflow (2-5).

Chilled ceiling and cooling towers for direct evaporative cooling

The natural complement to DV is chilled ceiling (CC) cooling, in which cooling water at 16-20°C is circulated through large ceiling panels. Conventional systems provide chilled water at a temperature as low as 6-8°C (and generally no warmer than 14°C). There are a number of energy-saving synergies with CC cooling:

- the room temperature can be 2 K warmer with the same perceived temperature, which will reduce heat gains due to conduction and leakage through the building envelope; and
- by providing cooling at 16-20°C, the chiller efficiency will be much larger (as noted earlier) or the chiller can be bypassed altogether if the cooling-tower water is used for chilling (as discussed below).

However, if a chilled ceiling is combined with an otherwise conventional system, both of these energy savings can be lost. In particular,

- if dehumidification is accomplished by over-cooling the ventilation air and then reheating it, more energy will be needed for reheating because the ventilation air is supplied at a warmer temperature in a CC system;
- if ice thermal storage is used in order to shift some of the cooling load from daytime to night-time, then the COP benefit of a warmer chilled-water temperature and the greater opportunity for directly cooling the chilling water using the cooling tower are lost, because initial cooling down to freezing occurs (a partial solution for storing coldness is to use materials, such as eutectic salts, with a warmer freezing point).

Thus, the magnitude (and even the sign) of the change in energy use from a given design change can depend on the other parts of the system.

In most HVAC systems, a cooling tower (shown schematically in Figure 5) is used to provide water at 20°C or less through evaporative cooling. This evaporatively-cooled

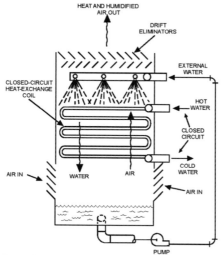

FIGURE 5. Schematic illustration of a cooling tower. Source: (1).

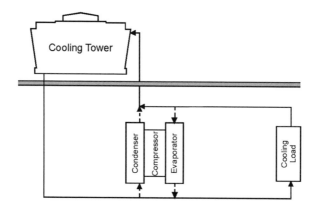

FIGURE 6. Cooling tower used as an evaporative chiller. Under normal operation, the cooling tower water would circulate through the condenser only (removing heat from the condenser) and building cooling water would circulate through the evaporator (as shown by dashed lines). In free cooling mode, the condenser and evaporator would be bypassed altogether and the compressor shut off. Modified from (1).

water is used for removing heat from the chiller condenser, permitting a lower condenser temperature than if it were cooled with ambient air. However, when cooling is provided with 18-20°C water in ceiling panels, the cooling tower water will often be cool enough to be directly used for cooling purposes, bypassing the chiller altogether and greatly reducing the required energy. This arrangement is illustrated in Figure 6. Assuming the chilling water to be supplied at 18°C, a cooling tower could directly meet cooling requirements 97% of the time in Dublin and 67% of the time in Milan according to (6). If chilling water at 20°C is adequate, then evaporative cooling in a cooling tower is sufficient 99% of the time in Dublin and 78% of the time in Milan.

Dedicated outdoor air systems

As noted above, heat and coldness are delivered in many systems HVAC systems solely by circulating warm or cool air, but the amount of air that is circulated is many times that needed for ventilation purposes alone. So as to reduce the volume of outside air that needs to be heated and humidified (in winter), or cooled and dehumidified (in summer), it is common to recirculate, say, 80% of the indoor air on each circuit, while exhausting only 20% and replacing the exhausted air with fresh outside air. However, this means that 80% of the internal heat gains picked up by the moving air will have to be removed by the chiller. However, if heating/cooling and ventilation functions are separated, then the amount of air that is circulated can be reduced to that needed for ventilation alone. All of the air will then have to be directly exhausted to the outside, producing a Dedicated Outdoor Air Supply (DOAS) system. If a DOAS system is combined with DV, then internal heat gains from lighting, and warm plumes that rise to the ceiling from office equipment, will be directly vented to the outside, reducing the cooling load on the chillers by up to 30% (7). Ventilation rates can be reduced to near zero when the building is not occupied (because ventilation is not used for temperature control). This is referred to as demand-controlled ventilation (DCV), and can save 20-30% in total heating+cooling+ ventilation energy use on an annual basis (8).

Separating cooling and dehumidification functions

The conventional approach in dehumidifying air is to cool it to a temperature of 6-8°C, so as to condense out adequate water vapor, then to reheat the air to the desired supply-air temperature (typically 12-14°C). This is inefficient because (i) the chillers must operate with a lower evaporator temperature than if the air were directly cooled to the supply-air temperature, (ii) greater cooling is carried out, and (iii) subsequent reheating is required. The inefficiency would be greater in a DV system because the supply-air in such systems is 16-18°C, so greater reheating from the same cold temperature would be required. An alternative is to directly remove water vapor using passive or active desiccant wheels. Both consist of a rotating drum that contains a solid desiccant and rotates from the incoming airflow (picking up moisture) to the outgoing airflow (releasing moisture) and back. In a passive desiccant wheel, the dryness of the outgoing airflow is the driving force for driving moisture from the desiccant. In an active desiccant wheel, supplemental heating of the outgoing air stream is used to assist in driving moisture from the desiccant. The combination of active desiccant wheels with conventional (electric vapor-compression) cooling systems reduces electricity use both by shifting some of the cooling load (the latent portion) to the desiccant wheel, which in turn can be regenerated using solar thermal energy, and by permitting a higher chiller evaporator temperature and hence greater chiller COP. Total energy savings in cooling and dehumidification can vary from as little as 6% to almost 50%, depending on the way the components of the system are put together, and up to 75% if solar thermal energy is used to regenerate the desiccant (9-11).

Desiccant wheels can be used to over-dry the incoming ventilation air, which then permits cooling through evaporative cooling without producing air that is too humid. In effect, desiccant systems extend the applicability of evaporative cooling into the hot-humid regions of the world, where it otherwise cannot be used. Air quality problems associated with moist evaporators are eliminated too.

An interesting synergy between different system components arises from the fact that the COP of a solid-desiccant cooling system increases with increasing initial temperature and humidity even though the regeneration temperature must be increased in order to produce the same final temperature and humidity (this is explained in 1, Box 6.5). An important corollary is that the COP will be lower if the desiccant cycle is applied to outside air that has been mixed with recirculated indoor air (as in most conventional systems). This is another factor in favor of DOAS systems, but these require supplemental hydronic heating and cooling so as to avoid excessively large ventilation airflows.

A building envelope with less uncontrolled infiltration of outside air directly reduces both sensible and latent cooling loads, by reducing the amount of heat and moisture that enter the building. The reduced influx of outside moisture means that the ventilation air supply does not need to be dried as much, which in turn permits a lower desiccant regeneration temperature and more efficient operation of the desiccant system. If the desiccant is regenerated with solar heat, more solar heat can be collected and with fewer losses if can be supplied at a lower temperature. Thus, there are a number of interactions through which a better building envelope increases the efficiency of solar-desiccant systems.

Pump systems

A system as simple as a pump that circulates water through a circuit exhibits energy savings opportunities that are vastly greater than can be expected by improving the device (pump) efficiency. The rate of flow through a pump varies directly with the speed of the impeller's rotation and its diameter, while the pressure developed varies as the square of the impeller speed and the square of the impeller diameter. The pump would, ideally, be chosen such that the pressure head created at the desired flow rate exactly balances the pressure drop in the piping system. However, this does not happen in practice. This is because there are errors in the calculation of the pressure drop created by the piping network for a given flow rate, and multiple safety factors are routinely built into the calculations, so that the resulting flow is invariably greater than specified. Thus, after the pump has been installed, it has to be operated to determine the actual flow at the specified pump speed. The pump-piping system should then be "rebalanced" so that the desired flow is created (12). This is done as follows: A piping system will typically have many parallel circuits, and the relative flow errors will generally be different for each circuit. A control valve in each circuit will need to be slightly closed (thereby throttling the flow), but this should be done only to bring the ratio of actual flow to predicted flow down to the ratio found in the circuit with the smallest ratio (so no adjustment of the control valve is needed in this circuit). At that point, the pressure head developed by the pump can be reduced, thereby reducing the flow in each circuit and bringing them all simultaneously down to the design-flow. The net result is to achieve the desired flow in all circuits with a minimum of throttling.

The pressure head developed by the pump can be reduced by trimming the impellor inside the pump (which requires disassembling and re-assembling the pump), or by reducing the rotational speed of the impellor. However, even if excess flow is recognized, the pump is rarely adjusted to achieve the design-flow, especially if the system is operating smoothly.

To understand the implications for energy use by the pump, we need to examine the pressure-head/flow variation for both the pump and the piping system on a single diagram, as shown in Figure 7. The convex-upward curves show the variation of pressure head and flow rate for a variety of fixed impeller diameters (these are a characteristic of the particular pump). For a given impeller diameter, a greater flow rate (due to less resistance in the system to which the pump is connected) is associated with a smaller pressure head. The concave-upward curves show the variation of piping system pressure

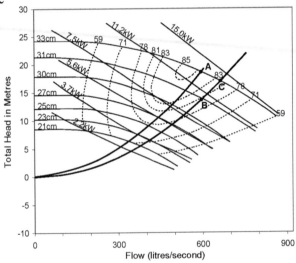

FIGURE 7. Pressure-flow relationship for a pump with different impellor diameters (convex upward curves) and for a piping system as originally estimated (upper concave up curve) and in reality (lower concave up curve). Point A: intended operating point. Point C: actual operating point without trimming the impeller – energy use is 6% larger. Pont B: operating point with impellor trim – energy use drops by 25%! Source: (1).

drop with flow rate; one curve is the calculated relationship, and the other is the actual relationship for that particular system. Also shown in Fig. 7, as dashed lines, are contours of pump efficiency. Point "A" is the expected operating point. However, as is typically the case, the pressure drop for the desired flow is less than the computed pressure drop (due to various safety factors). This is represented by point "B". Since the pump generates more pressure than needed, flow increases, the pump pressure head decreases, the system pressure drop increases, and the two come into balance at point "C", but with excess flow. The fluid power (P_{fluid}), equal to the product of pressure head and flow, is essentially unchanged in this example. However, the pump would have been chosen so as to have the greatest efficiency near the design flow, and is somewhat less at the actual operating point. The net result is a 5.6% increase in pump energy use, and most systems would be allowed to operate in this manner. However, at point B (where the pump efficiency is even lower), the required pump power is 24% less than at point C. This is not a negligible savings!

Summary for HVAC Systems

To sum up, to minimize the energy use in supplying and delivering heat or coldness and fresh air with mechanical systems, one should

- separate heating/cooling and ventilation functions by circulating hot or cold water for heating and cooling purposes;
- separate cooling and dehumidification functions by using desiccants for dehumidification;
- use chilled water at the warmest possible temperature for cooling (ideally, 20°C);
- use hot water for heating at the coolest possible temperature (ideally, 30°C);
- circulate only the amount of air required for ventilation purposes alone, varying with time based on actual needs, using displacement ventilation in a dedicated outdoor air supply system; and
- rebalance pumps and fans after cooling or heating circuits have been operated, so that they are attempting to do the least possible amount of work.

The minimization of temperatures used for heating requires large radiator surfaces (as in floor or ceiling radiant heating) and minimal heat loads, the latter requiring a high-performance envelope to minimize heat losses during the heating season. Similarly, maximization of temperatures used for cooling requires large radiator surfaces and minimal cooling loads, the requiring a high-performance thermal envelope and efficient equipment and lighting so as to minimize internal heat gains. Displacement ventilation (DV) works best if uncontrolled air flows (due to leaky building envelopes) are minimized. Thus, a high performance envelope saves energy both by directly reducing heating and cooling loads, by permitting alternative mechanical systems, and by permitting more efficient operation of conventional systems and equipment. Desiccant dehumidification avoids the need for overcooling and reheating of air for dehumidification, and will be most beneficial in combination with DV as one can take full advantage of the warmer supply-air temperature permitted with DV. Dedicated Outdoor Air Supply (DOAS) systems increase heating and cooling loads (and thus energy use) in a conventional all-air cooling system because large airflow rates are required in order to provide adequate heating and cooling without excessive supply temperatures, and all of this air is exhausted and replaced after one circuit in a DOAS system rather than partly recirculated. A system where air flow meets ventilation requirements only (with radiant heating and cooling for most of the heating and cooling loads) is, of necessity, a DOAS system and, when combined with DV, saves energy because waste heat from lights and other sources at ceiling level is directly vented to the outside.

PASSIVE VENTILATION, HEATING, AND COOLING

The focus thus far has been on mechanical HVAC systems. However, it is possible to design buildings to be passively heated, cooled, and ventilated. In this case, the building and its interaction with surrounding buildings through the outside airflow has to be understood and designed as an integrated fluid-dynamic and thermal system. In combination with a high-performance thermal element, the potential exists to completely eliminate the need for mechanical ventilation and cooling altogether, and to largely eliminate the need for heating even in cold climates. That is, energy savings can approach 100%, not through improved device efficiencies, but by drastically reducing what mechanical devices are required to do. Passive heating, cooling, and ventilation systems are extensively reviewed in (1, Chapters 4, 6, and 8), so only some of the key system-related elements and interactions are outlined here.

Passive ventilation can be achieved through building designs that permit cross-ventilation and create wind suction; or that exploit the "stack" effect (the natural tendency of warm air to rise) through solar chimneys, stairwells, and atria; or that make use of air-flow windows, double-skin façades, and cool towers. Solar chimneys are tall open columns that are heated by the sun, creating a rising plume of warm air that draws outside air into and through the building. An early example (built in 1982) is provided by the solar chimneys on the south façade of the British Research Establishment office building in Garston, illustrated in Figure 8. Passive ventilation not only reduces energy use, but can improve air quality (if the outdoor air is not overly polluted!) and gives people what they generally want (a connection to the outside).

Inasmuch as a building's internal temperature will tend to be several degrees warmer than the outside air, there will be times when cooling is called for and the outside air is cool enough to provide the required cooling. Passive ventilation, by inducing the flow of outside air through the building, will thus provide some or all of the required cooling. During winter, ventilation air may need to be preheated. This can be done if passive ventilation systems are designed to draw outside air through the gap in a double-skin façade or through an airflow window. The incoming air will be preheated by picking up heat that would otherwise be lost to the outside and also (during the daytime) by picking up solar heat. Airflow windows can be designed to operate in different modes in summer and winter, as illustrated in Figure 9.

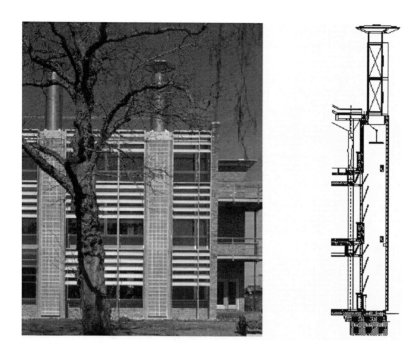

FIGURE 8. View of two solar chimneys and intervening glazed section with external shading louvres on the south façade of the British Research Establishment office building in Garston (UK) (left), cross section of the solar chimney in the British Research Establishment offices (middle) and of the glazed portion of the façade between the solar chimneys (right). Photographer: Dennis Gilbert, London. Source of diagram: (13).

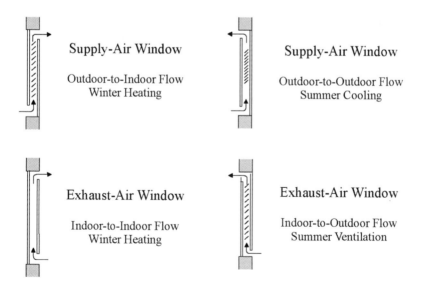

Supply-Air Window

Outdoor-to-Indoor Flow
Winter Heating

Supply-Air Window

Outdoor-to-Outdoor Flow
Summer Cooling

Exhaust-Air Window

Indoor-to-Indoor Flow
Winter Heating

Exhaust-Air Window

Indoor-to-Outdoor Flow
Summer Ventilation

FIGURE 9. An airflow window that functions as a counterflow heat exchanger. Source: (1).

Alternatively, ventilation air can be pre-warmed (in winter, and pre-cooled in summer) by drawing it through underground pipes. Note that the thermal driving force for passive ventilation is related to the difference between the interior and ambient temperatures, which will be largest in winter.

In hot climates, thermal insulation alone is of little benefit in reducing cooling requirements, especially if applied internally, because it can inhibit heat loss at night. Thermal mass reduces the rate of temperature change, and so can greatly attenuate the amplitude of internal temperature variations associated with high frequency (i.e., diurnal) external temperature variations. However, thermal mass is of little use after an extended warm period. However, the combination of external insulation, internal thermal mass, and night ventilation to remove heat from the thermal mass, can be very effective in reducing cooling loads. External insulation in this case serves to inhibit the daytime penetration of heat from outside, but does not inhibit night-time heat rejection because the insulation is bypassed through direct ventilation of the building interior with relatively cool night air. Thus, in the absence of aggressive night ventilation and exterior insulation, high thermal mass in hot climates can be a liability in that it slows the cooling of the building at night (this is especially an issue for residential buildings), but in the presence of strong night ventilation, thermal mass is an asset because it minimizes the daytime temperature rise. Thermal mass is an asset during the winter in cold climates because it allows greater passive absorption of solar thermal energy without overheating, and the absorbed heat can be slowly released at night. Thermal mass is more effective if indoor temperatures are allowed to float, as this permits more effective charging and discharging of the thermal mass. This in turn requires adoption of an 'adaptive' thermal comfort standard, in which the target interior temperature is allowed to vary with the outdoor air temperature rather than being fixed during the day and year round (this issue is discussed further later).

LIGHTING SYSTEMS

Lighting systems provide another example where system-level considerations and especially human behavior can provide large savings, in this case comparable to the large (factor of two) savings that can be achieved improved energy-using devices (lamps and ballasts). System-level considerations involve: a mix of task and ambient lighting (lower background lighting levels, with individually-controlled, greater levels of lighting when and where it is needed); incorporation of daylighting with light sensors, occupancy sensors, and dimmable electric lighting so that electric lighting levels can actually be varied according to the sunlight contribution and automatically turned off when a space is unoccupied; and wiring of lighting controls to coincide with zones having different degrees of daylighting. These issues are extensively discussed in Harvey (2006, Chapter 9) and in many references therein.

Building form and glazing area can significantly affect the extent to which daylighting can replace electric lighting, but this also has implications for heating and cooling loads. Thus, lighting energy use should not be analyzed in isolation, but as part of an optimized system of overall energy use. Computer algorithms to automatically control adjustable shading devices can be designed to minimize the sum of lighting plus cooling energy use for a given building, occupancy schedule, and climate. If only the amount of daylight needed for a given task is allowed to enter a building, cooling energy requirements will be reduced compared to the use of electric lighting, adding to the energy savings from reduced electric lighting loads. Peak electrical and cooling loads are also reduced, allowing downsizing of cooling systems and electrical transformers, thereby reducing costs.

It appears that most people require less light at night and on cloudy days, because of less window glare but also because requirements are conditioned in part by expectations (14). This parallels the finding that the acceptable temperature range depends in part on expectations, which vary with outside conditions. An added benefit of task/ambient lighting, then, is that it allows the lighting level to be adjusted to changing preferences, as well as to differences between users.

RECIRCULATION-LOOP DOMESTIC HOT-WATER SYSTEMS

Recirculation-loop (RL) domestic hot-water systems provide another example of dramatic energy savings that can be achieved at the system level, with no change in the efficiency of individual devices (hot-water boilers and pumps in this case). In RL systems, water is heated and stored in a central tank, continuously circulated through a closed loop to all the points of use, and consumed as needed. This keeps the hot-water pipes warm, so that hot water is instantly available when the faucet is opened. Apart from convenience, this avoids wasting hot water by running the faucet until the pipes have warmed sufficiently to deliver hot water to the faucet. Since the purpose of recirculation is to keep the pipes warm, the required flow can be reduced by insulating the pipes well. Since pumping power varies with the flow rate to the third power, dramatic reductions in pump energy use are possible along with reduced heat loss. However, even with well-insulated piping, piping heat losses can constitute 40% to more than 50% of the total hot-water load (15, 16). An alternative is point-of-use water (POU) heaters, in which water heaters are located at or very close to the point of use.

Hiller et al. (17) monitored the energy use in a new (1997 opening) school in Tennessee using an RL system serving six points of use, and again after it was converted to a POU system with three water heaters and short piping. An impressive 91% savings in total (pump+water heater) energy use was achieved. The POU water heaters were operated continuously, but it is estimated that they could have been shut down at night, on weekends, and during school holidays, with a further energy savings of 40% (bringing the total savings to 94.3%).

HUMANS AS PART OF THE ENERGY-USING SYSTEM

Humans are a critical part of the building energy systems, especially where passive ventilation and daylighting are part of the building, and with regard to the control of HVAC systems. In particular, adoption of an adaptive thermal comfort standard – in which the indoor temperature is allowed to vary with the outdoor temperature – can save substantial amounts of energy. Fortunately, it turns out that the indoor temperature perceived as "comfortable" increases with increasing outdoor temperature. Thus, buildings do not need to be cooled to as low a temperature on the hottest days of the year as on other days. Furthermore, a large body of evidence indicates that the temperature and humidity set-points in general are significantly lower than necessary (18, 19). Increasing the thermostat from 24°C to 28°C in summer will reduce annual cooling energy use by more than a factor of three for a typical office building in Zurich and by more than a factor of two in Rome (20). The temperatures deemed to be acceptable on hot days are 1-2 K warmer in buildings with natural ventilation than in buildings with mechanical ventilation, thereby extending the range of ambient conditions under which mechanical cooling can be avoided in passively ventilated buildings.

IMPLICATIONS: THE NEED FOR AN INTEGRATED DESIGN PROCESS

This paper has illustrated some of the many ways in which dramatic energy savings can be achieved in buildings by treating buildings as systems – savings that are many times greater than what can be achieved through improved device efficiencies. However, the conventional process of designing a building is a linear process, in which the architect makes a number of design decisions without extensive or even any consideration of their energy implications, and then passes on the design to the engineers, who are supposed to make the building habitable through mechanical systems. The design of mechanical systems, however, is also largely a linear process with, in some cases, system components specified without yet having all of the information needed in order to design an efficient system (given the constraints imposed by the architect) (21).

What is required is an alternative design process, referred to as the Integrated Design Process (IDP). This is not to say that there is no integration or teamwork in the traditional design process, but rather, that the integration is not normally directed toward minimizing total energy use through an iterative modification of a number of alternative initial designs and concepts so as to optimize the design as a whole (rather than optimizing individual subsystems). Two steps in the design process can be said to be "integrated" if the effect of the first design decision on the second design decision is taken into account when making the first design decision. The greater the number of design decisions that are linked in this way, the more "integrated" is the design. Integrated design means, as a minimum, to consider the impact of a change in early design decisions such as building shape and orientation, envelope characteristics, and inclusion of heat recovery on subsequent decisions, such as the nature and sizing of mechanical systems.

A number of early alternative choices should be considered as part of the IDP, with the goal of minimizing some objective criterion such as lifecycle cost, net present value of capital and operating costs, or energy cost alone subject to some upper limit concerning the acceptable payback time. If this is complemented by the choice of the most efficient equipment available, optimization of the equipment operation, and commissioning to ensure that everything works as intended, then savings on the order of 35-50% can be achieved relative to conventional practice for new buildings. This is the most basic IDP, a simple "back-to-basics" approach that stresses a quality envelope, quality equipment, and quality sizing and operation of mechanical equipment. Buildings produced in this way are still conventional buildings, but computer simulation will be needed if one wants to optimize the design choices.

To push the savings beyond the 35-50% achievable through a "back-to-basics" approach, an increasing number of unconventional measures will need to be carefully combined – measures such as passive ventilation, heating, and cooling involving perhaps double-skin façades and airflow windows; thermal mass with night ventilation; chilled ceiling, displacement ventilation, and desiccant dehumidification; daylighting; and adaptive thermal comfort. Computer simulation involving simulation specialists who serve as a liaison between the architects and engineers is essential. The integrated design process entails two-way interactions between the client and design team, between the architect and engineers, and between the design team and the contractors. Once the design is complete, the design team must be available during construction to explain details that are not clear, because no matter how thorough the plans and specifications, some details that affect the energy use by the building will be overlooked. The integrated design process in a number of buildings in Europe and North America is discussed in some detail, along with lessons learned, in IEA (22, 23).

de Wilde and van der Voorden (24) present a strategy for the selection of energy-saving components during the building design process. This strategy draws upon principles of systems engineering and decision theory. The elements of the strategy are:

- To define an option space, consisting of different energy-savings combinations that are to be considered.
- To identify all the performance characteristics (dimensions) of all the design options, in order to determine a set of criteria for the selection of the preferred option.
- To specify objectives, constraints, and performance indicators.
- To predict the performance of the design options, primarily using computer tools but possibly also using experimental set-ups.
- To evaluate the predicted performance, based on how well each option performs in the various performance dimensions, with some weighting of the different performance dimensions.

Computer simulation is needed when designing a high-performance building in order to be able to assess, in advance, the impacts on energy use of alternative designs and in order to be able to determine the optimum design for a given building in a given climate. In conventional practice, simulation is used only for the final confirmation of the performance of the mechanical system, rather than as an integrated element of the design process. For building simulation software to lead to improved designs, it must be easy to use in a comparative mode, that is, to be able to systematically generate alternative designs by varying input parameters and to generate an n-dimensional design-performance space. Current simulation environments do not support this process (25). For further information on building energy simulation tools and their use in the design profession, the interested reader can consult the detailed review by Jacobs and Henderson (26). For information on 291 building software tools, see the US DOE Energy Efficiency and Renewable Energy website (www.eere.doe.gov/buildings/tools_directory).

OPPORTUNITIES FOR PHYSICISTS

There are a number of system-related considerations that may be of interest to physicists wanting to move into the building energy-efficiency field. These include:

- computational fluid dynamics (CFD) to simulate passive ventilation or hybrid passive-mechanical systems;
- optical properties of windows with regard to passive solar gain, minimization of cooling loads, maximization of daylighting opportunities;
- research pertaining to phase-change materials and thermal mass; and
- all elements of building energy simulation.

REFERENCES

1. L.D.D. Harvey, *A Handbook on Low-Energy Buildings and District-Energy Systems: Fundamentals, Techniques and Examples.* EarthScan, London, 701 pages (2006).
2. N. Bourassa, P. Haves, and J. Huang, "A computer simulation appraisal of non-residential low energy cooling systems in California," in *Proceedings of the 2002 ACEEE Summer Study on Energy Efficiency in Buildings*, vol. 3, American Council for an Energy Efficient Economy, Washington, pp. 41-53 (2002).
3. M. Howe, D. Holland, and A. Livchak, "Displacement ventilation—Smart way to deal with increased heat gains in the telecommunication equipment room," *ASHRAE Transactions* 109(Part 1), 323-327 (2003).
4. S.A. Mumma, "Ceiling panel cooling system," *ASHRAE Journal* vol. 43(11), 28-32 (2001).
5. F. Sodec, "Economic viability of cooling ceiling systems," *Energy and Buildings*, vol. 30, 195-201 (1999).
6. B. Costelloe and D. Finn, "Indirect evaporative cooling potential in air-water systems in temperate climates," *Energy and Buildings*, vol. 35, 573-591 (2003).
7. K.J. Loudermilk, "Underfloor air distribution solutions for open office applications," *ASHRAE Transactions* vol. 105 (Part 1), 605-613 (1999).
8. M.J. Brandemuehl and J.E. Braun, "The impact of demand-controlled and economizer ventilation strategies on energy use in buildings," *ASHRAE Transactions* vol. 105(Part 2), 39-50 (1999).
9. J.C. Fischer, J.R. Sand, B. Elkin, and K. Mescher, "Active desiccant, total energy recovery hybrid system optimizes humidity control, IAQ, and energy efficiency in an existing dormitory facility," *ASHRAE Transactions* vol. 108(Part 2), 537-545 (2002).
10. S.A. Mumma and K.M. Shank, "Achieving dry outside air in an energy-efficient manner," *ASHRAE Transactions* vol. 107(Part 1), 553-561 (2001).
11. J.L. Niu, L.Z. Zhang, and H.G. Zuo, "Energy savings potential of chilled-ceiling combined with desiccant cooling in hot and humid climates," *Energy and Buildings* vol. 34, 487-405 (2002).
12. A. Egan, "Reasons, results, and remedies for pump safety factors overuse," *ASHRAE Transactions*, vol. 107(Part 2), 559-565 (2001).

13. W. Gething, "The Environmental Building: The Building Research Establishment, Watford," in *Green Buildings Pay*, edited by B. Edwards, London, Spon Press, pp. 86-93 (2003).

14. P.A. Torcellini, M. Deru, B. Griffith, N. Long, S. Pless, R. Judkoff, and D.B. Crawley, "Lessons learned from field evaluation of six high-performance buildings," in *Proceedings of the 2004 ACEEE Summer Study on Energy Efficiency in Buildings*, vol. 3, American Council for an Energy Efficient Economy, Washington, pp. 325-337 (2004).

15. F.S. Goldner, "Control strategies for domestic hot water recirculation systems," *ASHRAE Transactions*, vol. 105(Part 1), 1030-1046 (1999).

16. J.D. Lutz, G. Klein, D. Springer, and B.D. Howard, "Residential hot water distribution systems: Roundtable session," in *Proceedings of the 2002 ACEEE Summer Study on Energy Efficiency in Buildings*, vol. 1, American Council for an Energy Efficient Economy, Washington, pp. 131-144 (2002).

17. C.C. Hiller, J. Miller, and D.R. Dinse, "Field test comparison of hot water recirculation loop vs. point-of-use water heaters in a high school," *ASHRAE Transactions*, vol. 108(Part 2), 771-779 (2002).

18. R.J. de Dear and G.S. Brager, "Developing an adaptive model of thermal comfort and preference," *ASHRAE Transactions*, vol. 104(Part 1), 145-167 (1998).

19. M.E. Fountain, E. Arens, T. Xu, F.S. Bauman, and M. Oguru, "An investigation of thermal comfort at high humidities," *ASHRAE Transactions*, vol. 105, 94-103 (1999).

20. P. Jaboyedoff, C.-A. Roulet, V. Dorer, A. Weber, and A. Pfeiffer, "Energy in air-handling units--results of the AIRLESS European project," *Energy and Buildings*, vol. 36, 391-399 (2004).

21. M. Lewis, "Integrated design for sustainable buildings," *Building for the Future, A Supplement to ASHRAE Journals*, vol. 46(9), 22-30 (2004).

22. IEA (International Energy Agency), *Examples of Integrated Design: Five Low Energy Buildings Created Through Integrated Design*, International Energy Agency, Solar Heating and Cooling Programme, Task 23, Paris (2000). Available from www.iea-shc.org/task23.

23. IEA (International Energy Agency), *The Integrated Design Process in Practice: Demonstration Projects Evaluated*, International Energy Agency, Solar Heating and Cooling Programme, Task 23, Paris (2002). Available from www.iea-shc.org/task23.

24. P. de Wilde and M. van der Voorden, "Providing computational support for the selection of energy saving building components," *Energy and Buildings*, vol. 36, 749-758 (2004).

25. A. Mahdavi and B. Gurtekin, "Computational support for the generation and exploration of the design-performance space," in *Seventh International IBPSA Conference*, Rio de Janeiro, Brazil, pp. 669-676 (13-15 August 2001). Available from www.ibpsa.org.

26. P. Jacobs and H. Henderson, *State-of-the-Art Review: Whole Building, Building Envelope, and HVAC Component and System Simulation and Design Tools*, prepared for the Air-Conditioning and Refrigeration Technology Institute, Arlington, Virginia (2002). Available from www.archenergy.com.

Physics of Energy Efficient Buildings

David Hafemeister

Physics Department
Cal Poly University
San Luis Obispo, CA 93407

Abstract. A scaling model shows that that the free temperature offset between inside and outside increases with the size of buildings. Reducing the lossiness in buildings offers three energy savings advantages: (1) reduce the multiplicative term, (2) increase the free–temperature subtractive term, and (3) cater to the degree-day distribution. Thermal storage can be made to transfer sufficient heat of the day into the coolth of the night. The daily passive solar input is modeled by integrating over the change in zenith angle during the day. Passive solar storage is sized to give a 12–hour time constant.

BASIC PHYSICS

"Heat goes by itself from the hotter body to the colder body."
"Insulate before you insolate."
"Glass plus mass prevents you from freezing...."

Energy use in buildings accounts for 40% of total energy use. Reducing this fraction could significantly stabilize national security, improve the environment and enhance the national economy. Residential and commercial buildings consumed energy totaling 37.5 quads/yr in 2001, at a cost of $300 billion/yr. Buildings dominate the use of electricity at 26 quads/yr, which is 67% of US electricity consumption of 40 quads/yr. Buildings built prior to the oil embargo of 1973–74 were often an energy disaster, built without insulation, wasting winter heat and summer air conditioning alike. New buildings now consume one-half their former level per square feet because energy intensity of big buildings has dropped from 270,000 to 100,000 Btu/ft^2-yr of primary energy.[1] But these gains are being countered by homes that have grown from an average of 1400 ft^2 in 1970 to today's 2225 ft^2. This increase is caused by more bathrooms and other design extras. In 2000, US home ownership stood at 66%, which means that for 1/3 of residences the landlord makes the energy decisions while the renter pays the bills.

Before the oil embargo, leaders of the expansive and burgeoning building industry were not engaged in coordinated energy research. The embargo changed this,

CP1044, *Physics of Sustainable Energy, Using Energy Efficiently and Producing It Renewably*
edited by D. Hafemeister, B. Levi, M. Levine, and P. Schwartz
© 2008 American Institute of Physics 978-0-7354-0572-1/08/$23.00

catalyzing serious research and development at the Lawrence Berkeley Laboratory and Princeton University on the use of energy in buildings. The State of California established the first energy standards for buildings in 1975. Many states and the federal government followed California's lead.

Linear Heat Transfer

Summer heat gains and winter heat losses are caused by temperature differences through walls, ceilings and floors (50–70%) and windows (15–25%), and by air infiltration (20–30%). *Thermal resistance*, or the *R-factor*, is defined in terms of a unit area of 1 m² (1 ft²) area. For material of thickness L, R is defined as $R = L/k$, where conductivity is k. For an electrical resistor, the resistance R takes into account the three physical dimensions of the device, but thermal resistance R is described by a thickness, a material type and a basic area unit. Thus a builder purchasing insulation may specify 100 ft² of R13 insulation for 2" x 4" stud walls, R19 for 2" x 6" stud walls, but an electrician does not specify area when buying a 100-ohm resistor. The *thermal transmittance U-factor*, is defined as the inverse of thermal resistance R, or $U = 1/R$. Heat flow through a slab of material is thus given by

$$dQ/dt = A\Delta T/R = UA\Delta T. \qquad (1)$$

SI and English units are as follows: Heat flow, watts (Btu/hour); area, m² (ft²); ΔT, °C (°F); R, m²-°C/W (ft²-°F-h/Btu) and U, W/(m²-°C) (Btu/(h-ft²-°F). R19 walls and U1 windows in the United States become R3.4 and U6 elsewhere. The conversion factors between the two cultures are

$$R_{SI} = R_{English}/5.67 \text{ and } U_{SI} = 5.67 \text{ x } U_{English}. \qquad (2)$$

Series and parallel heat paths

The R-value of materials layered in series is the sum of the *R*-values for the individual layers. For a wall with two layers of plasterboard and a layer of insulation,

$$R_{total} = R_1 + R_2 + R_3 = L_1/k_1 + L_2/k_2 + L_3/k_3 = 1/U_{total}. \qquad (3)$$

Buildings have inside-to-outside, parallel heat paths through walls, windows, roofs, floors and infiltration. Total heat flow is the sum of parallel heat flows,

$$dQ/dt_{total} = dQ/dt_1 + dQ/dt_2 + dQ/dt_3 = (U_1A_1 + U_2A_2 + U_3A_3)\Delta T. \qquad (4)$$

Two electrical resistors in parallel have an effective resistance of $R_{eff} = R_1R_2/(R_1 + R_2)$. If R_1 is much larger than R_2, the effective resistance is determined *from the smaller value*, or $R_{eff} \approx R_2$. When combining parallel thermal paths, one needs to take into account the relative area of each, so that $A_{Total}/R_{eff} = A_1/R_1 + A_2/R_2$, where $A_{Total} = A_1 + A_2$. If a homeowner tightens 75% of his R10 house with R1000 insulation, the effective R-value of the house does not become R-750, but $R_{eff} = 40$, only 4 times the initial value. If insulation paths are essentially closed with very large R-values, a point of diminishing returns is reached since infiltration will dominate.

U-Factor for Convection

Surfaces lose heat by radiation and convection. The rate of heat loss from a surface of area A, emissivity ε_1, and temperature T_1 is $dQ/dt = \sigma A \varepsilon_1 T_1^4$, where σ is the Stefan-Boltzman constant. The same surface will receive radiation from its surroundings, assumed to have emissivity ε_2 and ambient outside temperature of T_2. Convective losses from the surface will be a function of temperature difference $\Delta T = T_1 - T_2$, surface material, wind velocity and geometry. The net rate of heat loss from convection and radiation is given by

$$dQ/dt = \sigma A(\varepsilon_1 T_i^4 - \varepsilon_2 T_2^4) + hA(\Delta T)^{5/4} \tag{5}$$

where our convection equation is a simple model and h is a convection constant that depends on geometry, air flow and type of surface. Convection can be rewritten as

$$dQ/dt_{conv} = hA(\Delta T)^{5/4} = (h\Delta T^{1/4})A\Delta T = U_{conv}A\Delta T. \tag{6}$$

The parameter $U_{conv} = h\Delta T^{1/4}$ is relatively constant since $\Delta T^{1/4}$ varies slowly. Its inverse varies from $R_{conv-SI} = 0.04\text{--}0.2$ ($R_{conv-Eng} = 0.2\text{--}1$) with larger values outside in the wind and smaller values inside buildings.

U-Factor for Radiation

Net radiation flow from a surface at temperature T_1 located in outside ambient temperature T_2 is

$$dQ/dt = \sigma A \varepsilon(T_1^4 - T_2^4) = \Delta(A\varepsilon\sigma T^4) \approx (4\varepsilon\sigma T^3)A\Delta T = (U_{rad-SI})A\Delta T \tag{7}$$

with emissivity assumed equal, $\varepsilon_1 = \varepsilon_2 = \varepsilon$. Since temperature difference is typically much less than ambient absolute temperature ($\Delta T \approx 30$ K $\ll T_1 \approx T_2$), we can *linearize* the net radiation flow. The net radiation U-factor for room temperature $T_1 = 20°C$ (293 K) is

$$U_{rad-SI} = 4\varepsilon\sigma T_1^3 = (4\varepsilon)(5.7 \times 10^{-8})(293 \text{ K})^3 = \varepsilon(5.7). \tag{8}$$

A blackbody with $\varepsilon = 0.9$ gives $R_{rad-SI} = 0.2$ ($R_{rad-Eng} = 1$), which is similar to R_{conv} for conditions outside of buildings. A stove loses heat first by convection and radiation to the inside building surfaces, then by conduction through the walls, windows and infiltration, followed by convection and radiation away from the outside wall surfaces.

Annual Heat Loss

Annual heat loss is obtained by arranging the heat loss equation into a product of heat loss rate dQ/dt and time spent at that rate dt,

$$dQ = UA\Delta T \, dt. \tag{9}$$

Heating severity is proportional to the product $\Delta T \Delta t$. We can get a measure of the severity of the climate by essentially adding the hourly intervals of ΔT over a year. For heating, we retain just the terms with $T_{outside}$ colder than T_{base} (a defined base temperature) to get *degree-hours per year* (dh/yr):

$$dh/yr = \sum_{i=1}^{8760} (T_{base} - T_{outside})_i (1 \text{ hour}). \tag{10}$$

The base temperature T_{base} is usually defined as 65°F which takes into account 3°F *free temperature* from the internal heat of people and appliances. Division by 24 converts *degree hours* to *degree days*. The total annual heat loss is a summation over n paths,

$$Q_{total} = (dd/yr)(24 \text{ hr/day}) \sum_{j=1}^{n} U_j A_j. \tag{11}$$

The total heat loss must be increased by about 25% to account for infiltration losses. *Annual fuel consumption* is obtained by dividing total heat loss by furnace efficiency η (70–90%), which should be reduced to take into account heat duct losses of perhaps 20%. The concept of *heating degree days* is not useful for warmer climates because it ignores daytime storage of energy that is used in the early evening.

Infiltration Losses

About one-fourth of heat is lost from cold air infiltrating into a house through holes and small cracks. Heat transfer from infiltrating air is given by

$$dQ/dt_{infil} = (dm/dt)c\Delta T, \tag{12}$$

where dm/dt is the infiltration rate of air mass and c is the specific heat of air. The rate of exchanging air in buildings is described in terms of air changes per hour, or *ach*. A building with an infiltration rate $R_{ACH} = 1$ ach turns over 100% of its air in 1 hour. This gives

$$dQ/dt_{infil} = (V\rho)R_{ACH}c\Delta T, \tag{13}$$

where $V\rho$ is the mass of the air in the house (volume x density) . We "integrate" the above equation over a year using the degree-day statistic to get annual heat loss Q_{infil} for a 140-m^2 house in St. Louis:

$$Q_{infil} = (V\rho)R_{ACH}c(dd/yr)(24 \text{ hr/day}) \tag{14}$$

$$Q_{infil} = (140 \times 2.5 \text{ m}^3)(1 \text{ ach})(1.3 \text{ kg/m}^3)(1004 \text{ J/kg-}^\circ\text{C})(24 \text{ h/day})(2800 \text{ }^\circ\text{C-day/yr})$$

$$Q_{infil} = 3 \times 10^{10} \text{ J} = 30 \text{ MBtu/yr} = 5.5 \text{ bbl/yr.}$$

Infiltration costs the average homeowner whose furnace/duct efficiency $\eta = 2/3$ about $(5.5 \text{ bbl/yr})/(2/3) = 8$ bbl/yr. Infiltration costs the US about 800 Mbbl/yr = 2 Mbbl/day. If all houses were tightened by 50% to 0.5 ach, the nation would save roughly 1 Mbbl/day, but the level of radon and pollution would be increased. Air-to-air heat exchangers can used in superinsulated houses to retain 75% of the heat in exiting air.

Traditional Windows

Passive solar energy in buildings depends on glass that transmits sunlight readily, but strongly absorbs 10-μ–wavelength infrared radiation from objects at 300 K. The convection and radiation paths are in parallel and have same area, giving an effective resistance

$$R_{par} = \frac{R_{rad} R_{conv}}{R_{rad} + R_{conv}}. \tag{15}$$

Convection resistance is larger on inside surfaces than on outside surfaces. Also the radiation R value is not exactly the same for the two surfaces. *Ignoring these differences*, the total R-value for one pane of glass is the sum of two equal parallel resistances and the conductive resistance of a thin sheet of glass, or $R_{total} = 2R_{par} + R_{cond}$. Using approximate SI units, $R_{conv} = R_{rad} = 0.2$, $R_{par} = 0.1$ and $R_{cond} = 0.005$ for glass, we obtain $R_{total} = 2(0.1) + 0.005 = 0.205$, in fair agreement with measurements of $R_{SI} = 0.16$ ($R_{Eng} = 0.9$) for single-glazed windows.

The conduction R-value for a pane of glass can be ignored since it is much less than the convection and radiation R-values. The glass temperature is between inside and outside temperatures. Our approximation of equal R values for inside and outside surfaces give a glass temperature of half-way between inside and outside temperatures. A cold room at 60°F on a cold day of 0°F develops ice on the inside its single-glazed window since the median temperature is 30°F.

Double-glaze can be approximated by making the four parallel resistances to be the same, which is not true since still air between two sheets of glass makes convection less effective and outside convection is greater than inside convection. Ignoring the R-value of glass, we obtain $R_{total-SI} = 4 \times 0.1 = 0.4$ ($R_{total-Eng}$ 2.2), which is fairly close to the measured value of $R_{total-SI} = 0.3$ ($R_{total-Eng}$ 1.6). This theory doubles the R-value with the second pane of glass, which is too optimistic since the measured ratio is 1.7. Further savings are possible when low-emissivity (low-E) coatings are used to reflect (rather than absorb) IR back into the room. In addition pulled drapes reduce convection and radiation losses, particularly in the night when outside temperature is lowest.

Considerable energy is lost through windows. An average living unit has a floor area of 140 m^2 (1500 ft^2) and window area of 20 m^2 (15% of floor area, 225 ft^2) with an SI U-factor of 4 (English 0.7). The annual window energy loss for a 2800°C-day (5000°F-day) heating season in SI units is

$$(UA)(24)(dd/yr) = (4 \text{ W/m}^2\text{-}°C)(20 \text{ m}^2)(24 \text{ hr/day})(2800 \text{ °C-day}) = 1.9 \text{x} 10^{10} \text{ J/yr. (16)}$$

In practical units this is 18 MBtu/yr, or 3.2 bbl/yr of oil (equivalent). If furnace/duct efficiency is about 2/3, the total fuel used for windows is (3.2 bbl/yr)(3/2) ≅ 5 bbl/yr. For 100 million US residences the loss of energy through windows is 500 Mbbl/yr = 1.3 Mbbl/day, or 3% of total US energy use.

Low-E Windows

Convection is the dominant heat path for low-E coatings, constraining U-values to 30% of traditional glass. Low-E coatings are also useful in summer, as they reflect away IR emitted by warm surfaces outside the house and near-IR from the sun. Thus, low-E glass reduces heating in winter and air conditioning in summer. Window manufacturers started making low-E windows in 1991, capturing 35% of the US market by 1996.

Low-E windows use materials selected to transmit visible light and reflect, rather than absorb, infrared. Wide-band tin and indium oxide semiconductors can be doped to produce IR reflection. IR reflection can also be done with very thin layers (10–20 nm) of silver. Reflectivity of low-E windows is based on plasma physics. They exploit a concept known as the plasma frequency, which is the resonant frequency at which electrons will oscillate when displaced a tiny amount with respect to positive ions. In low-E windows, electromagnetic waves above the plasma frequency are transmitted, while EM waves below the plasma frequency are reflected. An example of this is the increased quality of radio transmission at night. Because darkness lacks ultraviolet, the nighttime ionosphere has fewer free electrons at 1000-km altitude, giving a lower plasma frequency of about 3 MHz. Since radio waves have a frequency under 3 MHz, they are reflected from the ionosphere, allowing radio signals to travel as far as 10,000 km. In low-E windows, the doping level in the semiconductors determines the free-electron density, analogous to how UV changes free-electron levels in the ionosphere. By increasing doping levels, one can increase the plasma frequency. The low-E materials SnO_2:F and In_2O_3:Sn reflect infrared at wavelengths longer than 1.5 μ, while transmitting visible light and near IR at wavelengths under 1.5 μ. To obtain the 1.5-μ transition, an electron density of 5 x 10^{26}/m^3 is chosen to give plasma frequency ($\lambda = c/f = 2\pi c/\omega$)

$$\omega_p = (ne^2/\varepsilon_o m)^{1/2} = 1.3 \text{ x } 10^{15} \text{ radians/sec,} \qquad (17)$$

where n is electron density, e is electron charge and m is electron mass, and ε_o is the permittivity of free space. There are two ways low-E windows can help. In the winter we want to reflect heat back into the room for wavelengths more than a micron or so. During the summer, we would like to reflect thermal IR at the same wavelengths. However, Sun produces considerable energy in the near IR at wavelengths shorter than one micron. There would be less airconditioning if we reflect more near IR away from windows. Present windows favor the winter situation.

Induced Infiltration

Houses typically are at partial vacuum, typically 4 Pascal below atmospheric pressure. The first cause is stack induced infiltration. The pressure at the top of a chimney or attic of height h is reduced with respect to the ground level by

$$\Delta p_{stack} = -\rho g h = -(1.3 \text{ kg/m}^3)(9.8 \text{ m/sec}^2)h = -13h \text{ Pascal}, \qquad (18)$$

with h in meters. If this reduction were the only effect, air would not rise, but remain in static equilibrium. But warmed air rises because its density is inversely proportional to its absolute temperature. Warm air rises because it is less dense, gravity pushes it upwards with a pressure difference proportional to the temperature difference.

$$\Delta p_{temp} \approx -13hT_o(1/T_o - 1/T_i) \approx -13h(T_i - T_o)/T_i, \qquad (19)$$

where T_o is outside temperature at the top of the stack, T_i is house inside temperature. At $T_i = 20°C = 293$ K and $\Delta T = T_o - T_i$,

$$\Delta p_{temp} = 0.044h\Delta T_i. \qquad (20)$$

For a two-story house with an attic or chimney at $h = 10$ m and $T_o = 0°C$ in winter, the thermal pressure difference is

$$\Delta p_{temp} = (0.044)(10 \text{ m})(-20 \text{ K}) = -9 \text{ Pascal}. \qquad (21)$$

The air infiltration flow rate (m³/sec) from Bernoulli's theorem is given by

$$F = area \text{ x } velocity = A(2\Delta p/\rho)^{1/2}. \qquad (22)$$

If stack loss is confined to a total area 0.1 m by 0.1 m, the air loss rate from a temperature-driven pressure difference of 9 Pascal is

$$F = (0.01 \text{ m}^2)(2 \text{ x } 9 \text{ Pa}/1.3 \text{ kg/m}^3)^{1/2} = 0.037 \text{ m}^3/\text{sec}. \qquad (23)$$

The energy loss rate is

$$P = \rho Fc\Delta T/\eta = (1.3 \text{ kg/m}^3)(0.037 \text{ m}^3/\text{sec})(1.006 \text{ kJ/kg-K})(20 \text{ K})/1 = 0.97 \text{ kW} \qquad (24)$$

for electrical heating with $\eta = 1$. The fuel rate is 1.2 kW for a furnace with $\eta = 0.8$. This leak costs 30 kWh/day or $3/day at 10¢/kWh. For natural gas, it is a loss of 10^5 Btu/day = 1 therm/day = $1/day.

Wind-induced infiltration: Wind blowing parallel to a surface reduces pressure *at* the surface due to the Bernoulli effect. A reduced pressure at the outside surface draws air from the inside through building leaks. The removed air is replaced by cold winter air or warm summer air. *Large area leaks* have turbulent flows proportional to the square root of the pressure difference between the inside and outside. *Minute area leaks* have viscous (laminar) air flow related to the square root of the pressure difference. Since buildings have both laminar and viscous flow, the flow rate has to be determined empirically. One way to do that is with a blower door, which can be mounted in the door of house, with all the other doors and windows closed. A blower door fan that produces air flow of 1–2 m^3/sec can create an overpressure of 100 Pascals within the building. A graph of flow rate F vs. pressure difference Δp is fit to $F = k\Delta p^n$, where n is between 0.5 for turbulent flow and 1.0 for laminar flow and k is a constant proportional to *effective* area. The curve is extrapolated to a 4 Pascal defined pressure difference to determine the *effective* loss area. Most US houses have leakage areas between 300 cm^2 and 1000 cm^2. Super–tight houses can be as low as 50 cm^2 and old leaky houses can be as high as 3000 cm^2.

Besides measuring leaks, blower doors also *discover* air leaks by over pressurizing houses, which forces air to escape through cracks. The trail of smoke coming from a miner's smoke stick locates leaks, which are then plugged with polystyrene foam. Leaks into an attic are discovered with smoke sticks by under pressurizing a house, that is, sucking air from the attic into the house. In this way blower doors are used to reduce infiltration losses, which at the extremes of weather, can cost $1500/year in Maine (heating) and Miami (cooling).

HOUSE DOCTORS

US buildings built before the oil embargo in a time of cheap energy often had no insulation in warm climates like California, or too little insulation in colder climes. A US house before the embargo using oil or gas heat consumed 73 million Btu/yr (77 GJ, 12.5 bbl of oil). In 1990 the average new US house consumed 40% less energy at 43 million Btu (45 GJ, 7.4 bbl). Air infiltration is responsible for 20% of the total loss, and hot air systems lose another 20% through duct leaks and furnace inefficiency.

In this section we consider the benefits a house gains from occasionally "seeing" a doctor, much as a human body needs occasional visits to a doctor. Princeton University's Center for Energy and Environmental Studies pioneered the concept of the house doctor in the 1970s. Prior to that time there was little scientific study on buildings because industry was very decentralized. The Socolow–Princeton group discovered that *attic bypass paths* allowed warm air to travel around ceiling insulation through channels in walls to cold attics. At Cal Poly San Luis Obispo, my student, Jim Woolaway, and I followed the Princeton and Berkeley (blower door) work by creating a house doctor laboratory for Cal Poly architecture students. The laboratory was named *The Arthur Rosenfeld House Doctor Laboratory* for the man who has been the main *leader* in improving energy use in US buildings. My university objected to my naming the *Rosenfeld* laboratory for a non-donor, but they never made me take the sign down! It is obvious that retrofitting an existing building is more complicated and expensive than constructing it correctly in the first place. However, some energy-saving, retrofit measures can be easily carried out. In the 1980s it was estimated that an investment of $1000 would save 25% of a home heating bill and $2000 would save 40%. We will describe the *blower-door* which measures infiltration and locates leaks.

Small blower doors are used to discover leaks in hot air duct systems. A Lawrence Berkeley National Laboratory group, led by Max Sherman, discovered that a "typical house with ducts located in the attic or crawlspace wastes approximately 20% to 40% of heating and cooling energy through duct leaks, and draws approximately 0.5 kilowatts more electricity during peak cooling periods. Sealing leaks could save close to 1 quad/year nationwide." The LBNL group devised a relatively simple repair to fill the leaks in the ducts. An aerosol sealant is sprayed into a closed, overpressurized duct system. The excess pressure pushes the sealant to the leaks, where it collects and plugs them. This is the same approach that is used to seal car radiators, in which a liquid sealant is added to the coolant, which is pushed into the leaks and plugs them.

Superinsulated houses reduce infiltration rates to as low as 0.1 air exchange per hour, but the resultant air may suffer from indoor pollution and radon. Air-to-air heat exchangers transfer the heat in warm exiting air to the cold and clean air entering from outdoors. The efficiency of an air-to-air heat exchanger is as

$$\eta = \Delta T_{\text{rise}} / \Delta T_{\text{in-outside}}, \tag{25}$$

where ΔT_{rise} is the temperature rise of the incoming air through the heat exchanger and $\Delta T_{\text{in-outside}}$ is the temperature difference between inside and outside air. Temperature efficiency varies between 50–85%, depending on design and weather conditions.

Other measurements and devices provided by house doctors include:

- thermoelectric meters to measure heat flow through walls
- infrared temperature scanners to find and quantify thermal leaks
- thermocouple temperatures over the 24-hour cycle
- kWh meters on appliances
- solar-flux gain and thermal loss meters on windows
- lumens/watt measurements on lighting.

Thermal Simulations

Engineers, architects and regulators rely on computer simulations to improve building designs and determine if a building satisfies state energy regulations. For example, results of simulations revealed a $9 million annual energy savings for a San Francisco federal building if it were to use natural ventilation without fans or air conditioning. This would be accomplished by taking advantage of interior heat that is absorbed during the day on exposed ceiling slabs, which then dissipate heat at night to warm the building. CALPAS (California Passive Solar), originally developed by Phil Niles of Cal Poly, was the first simulation computer program that simulated energy aspects of passive solar buildings. It was later enhanced by Bruce Wilcox of the Berkeley Solar Energy Group. Its progeny, *DOE-2* and *Energy Plus*, calculate energy loss in buildings to an accuracy of 10%, using the following features:

- solar gain from windows and skylights
- heat conduction through all surfaces (interior and exterior)
- infiltration of air
- heat gain from occupants, lights, and equipment.

An alternative type of calculation uses the analog of electrical *RC* networks to determine energy flow and energy storage in buildings. In the analog model *voltage* is the analog of *temperature*, electrical *current* is the analog of *heat flow*, *capacitance* is the analog of *thermal mass,* and electrical *resistance* is the analog of the product of *R-value and surface area* (but without i^2R losses). There is no thermal analog for electrical inductance.

Heat Island Mitigation

Urban centers of large cities have seen an increase in average higher summer temperatures prior to the 1950s. The trend has increased air conditioning loads and air pollution. The yearly high temperature in downtown Los Angeles at 4 PM rose 3.5°C (7°F) in the past 50 years. This is partially driven by the fact that black asphalt reflectivity is as low as 4%. The heat-island effect could be mitigated with a shift from green shingles to white shingles on house roofs, raising reflectivity from 18% to 43% and dropping roof temperature by 10°C (21°F). Higher reflection is possible with TiO_2 surfaces, which have a reflectivity of 83%, further reducing roof temperature by another 17°C (33°F). The Los Angles albedo could be raised by 0.08 by planting more trees and an increasing roof reflectivity over 25% of the city's area. These actions could lower LA downtown temperature in 20 years by 2.5°C (5°F), enough to save 1.5 GW_e peak power. In similar fashion, the US could reduce the national air conditioning peak load of 100 GW_e by 20% with increased reflectivity. See chapter in this book by Hashim.

BUILDING ENERGY STANDARDS

Many studies show a 25% energy savings on *new construction* could be realized for little or no *net initial cost*, since money could be saved by down-sizing air conditioners and heating plants. About 1.7 million new housing units are built each year, but this is only 2% of all housing stock and it will take 50 years to take full advantage of the energy savings. One-half of the states, which account for two-thirds of new homes, have mandatory energy building codes that are as restrictive as the standards recommended by the American Society of Heating, Refrigeration and Air Conditioning Engineers (ASHRAE). California led the nation with its 1975 energy laws, as it mandates the tightest of US standards. California building standards have to be shown to be cost effective for them to be adopted. A few cities require *existing houses* to fulfill a minimum energy standard before the houses can be resold. A much less restrictive approach is the use of labels that indicate a house's energy saving features and failures. New homes earn an *Energy Star* if they consume at least 30% less energy than energy code specifications for heating, cooling and water heating. This represents a savings of some $300–600/yr [0.3($1000–2000/yr)] from code level. See the appendix of this paper on LEED certification for buildings.

Heat/Cold Storage in Large Buildings

The first successful, proof of principle, passive solar, energy-independent building was built in Atascadero, California by Ken Haggard of Cal Poly University and Harold Hay. Analysis of the *Atascadero House* was carried out by Philip Niles, creator of the CALPAS passive solar simulation program.[2] Sweden has developed another approach to produce an almost heat-energy free building. Many Swedish office buildings need very little heat because they use excess daytime heat for nighttime warmth, storing energy in concrete floor and ceiling slabs. To dampen noise, concrete flooring is usually padded, an approach that prevents good thermal contact between the concrete and the inside air. However, Swedish Thermodeck® buildings transfer energy to the interior of the concrete through hollow-core tubes in floor slabs. The tubes are cast into the concrete slabs to reduce slab weight while minimizing mechanical deflections. (I-beams use the same approach to shift mass away from the center of the beam.) Although Stockholm, with 3580°C-day/yr (6444°F-day/yr), is colder than Chicago, Thermodeck buildings use only 4 kWh/ft^2 of electric resistance heating per year, a cost so low that it is not necessary to hook Thermodecks to the Stockholm district heating system. This system also works well in the summer, using night air to pre-cool buildings.

We begin with energy gains/losses to see if it is possible to operate Thermodeck buildings with essentially no added heat. A single-occupant office is 2.4 m wide by 4.2 m deep by 2.7 m high, for a 10-m^2 area and 27-m^3 volume. A cold day in Stockholm of –9°C (16°F) has an inside-to-outside temperature difference $\Delta T = 21°C - (-9°C) = 30°C$ (54°F). A person's body heat produces 100 W, and lights and machines in an office contribute an additional 300 W. Because Swedish offices must have windows, Thermodeck buildings have a large surface-to-volume ratio, increasing insulation needs. Each office has 1.5 m^2 of triple-glazed windows ($U_{SI} = 2$) and 5 m^2 of wall surface ($U_{SI} = 0.25$). Natural infiltration enters in a room at 5 m^3/hr during unoccupied hours. During the occupied hours infiltration is increased to 20 m^3/hr to ensure good air quality. Solar gain is a modest 30 W/room during winter occupied hours.

The heat-transfer rate from infiltration is

$$dQ/dt = \rho(dV/dt)c\Delta T, \qquad (26)$$

where air density $\rho = 1.3$ kg/m^3, air flow rate is dV/dt, air specific heat $c = 1000$ J/kg-°C and ΔT is the inside-to-outside temperature difference. The loss rate during the occupied the hours, when $dV/dt = 20$ m^3/hr $= 0.0056$ m^3/sec, is

$$dQ/dt = (1.3 \text{ kg/m}^3)(0.0056 \text{ m}^3/\text{sec})(1000 \text{ J/kg-°C})(-30 °C) = -218 \text{ W}. \quad (27)$$

98

The infiltration loss rate from a flow of 5 m^3/hr during unoccupied hours is 54 W. Window loss rate is 90 watts ($UA\Delta T = 0.25 \times 5 \times 30$). Wall loss rate is 38 watts (2 x 1.5 x 30).

During the day, 430 W of heat is gained in an office (person 100 W, equipment 300 W, solar 30 W), while at night the loss rate is 346 W (infiltration 218 W, wall 38 W, window 90). The net gain by day is 430 W – 346 W = 84 W. During the night there is no gain, while at night the loss rate is 182 W (infiltration 54 W, wall 38 W, window 90). The net loss by day is 430 W – 346 W = 84 W. Heat is transferred in the morning to raise the temperature above T_o from the evening hours. The temperature T of the room air is

$$T = T_o + (dQ/dt)t/C, \tag{28}$$

where C is the slab heat capacity/m^2 and dQ/dt is internal heat rate in W/m^2. The heat capacity of the 30-cm thick slabs is $C = 120$ Wh/m^2–°C, which includes an extra 20% to account for the heat capacity of walls and furnishings. Using these values, the time-dependence of the room temperature is

$$\text{occupied (8 W/m}^2) \quad T = T_o + 0.07t, \tag{29}$$

$$\text{unoccupied (–18 W/m}^2) \quad T = T_1 - 0.15t, \tag{30}$$

with t in hours. During a 10-hour workday room temperature remains fairly constant, rising by 1°C (0.07 x 10). Overnight room temperature drops by about 2°C (0.15 x 14), a drop that is lessened by adding a little heat to the room.

Thermal Storage to Reduce Peak Power

Daytime air conditioning could be significantly reduced by taking advantage of the night coolth and the fact that power is cheaper at night. During the deep air conditioning season, chillers are run at night, when electricity is cheaper, to pre-cool slabs. This approach also takes advantage of improved efficiencies from lower night temperatures. Implementing off-peak electricity usage may not give considerable energy savings in terms of kWh, but the tactic does save the expense of daytime peak power charges. Finally, this approach downsizes cooling systems by allowing them to be used on a 24-hour basis.

Internal heat gains are a dominant feature of large buildings, forcing the use air conditioning. This excess demand causes severe daytime summer peak power loads, as much as 2–3 times the nighttime load. The fraction of US homes with air conditioning reached 77% by 2001, increasing US peak demand by 2 GW$_e$/yr. The 5%/yr growth rate for new commercial buildings (replacement plus growth) adds 2.5 billion ft^2/yr of floor space, driving peak demand by 1.6 GW$_e$/yr. Residential and commercial air conditioning each used 80 GW$_e$, for a total peak load of 160 GW$_e$. This could be reduced with off-peak cooling with thermal storage.

The headquarters of the Alabama Power Company in Birmingham, Alabama, installed five large ice cells to contain 550 tonnes of ice for cooling for its 110,000-m^2

building. Ice storage on the basis of floor-area density basis is 5 kg per 1 m^2 floor space (550 tonne/110,000 m^2), with a stored energy density of

$$Q/m^2 = mL_{fusion} = (5 \text{ kg/m}^2)(3.4 \times 10^5 \text{ J/kg}) = 1.7 \text{ MJ/m}^2, \qquad (31)$$

where L_{fusion} is the latent heat to freeze ice. The electrical power/m^2 needed to make ice during 16 off-peak hours is (COP *coefficient of performance* of 2.5)

$$P/m^2 = (Q/m^2)/(COP)(\Delta t) = (1.7 \text{ MJ/m}^2)(1 \text{ kWh/3.6 MJ})/(2.5 \times 16 \text{ hr}) = 12 \text{ W}_e/m^2. \quad (32)$$

Total power required for ice storage is

$$P = (12 \text{ W}_e/m^2)(1.1 \times 10^5 \text{ m}^2) = 1.3 \text{ MW}_e. \qquad (33)$$

Without thermal storage it would take 2.8 MW_e to cool the building, more than twice the 1.3 MW_e used with ice storage. Coolth stored in ice supplies 2/3 of a day's cooling requirement with 1/3 coming from direct daytime cooling. Since ice storage covers 2/3 of daily summer heat gain, the total daily heat gain in summer is $(1.7 \text{ MJ/m}^2)/(2/3) =$ 2.6 MJ/m^2. The average summer heat input power/m^2 over 8 hours of daytime is

$$P/m^2 = (Q/m^2)/t = (2.6 \text{ MJ/m}^2)/(8 \text{ hr})(3600 \text{ sec/hr}) = 90 \text{ W/m}^2. \qquad (34)$$

SCALING LAW FOR BULDINGS

Energy use by large buildings is *load dominated* since large buildings have considerable internal heat from equipment, people and lighting. Big buildings have less surface area per unit volume, with internal energy supplying much of the surface losses. This makes large buildings fairly independent of climate while small buildings have energy bills proportional to the heating and cooling degree-days they experience. Energy use in houses is *skin dominated* since houses must replace heat losses through their envelopes. The physical difference between large and small buildings is easily seen through scaling law relations. Scaling laws for buildings determine the *free temperature* of buildings as a function of size, as well as for superinsulated houses.

Heat loss is proportional to a cubic building's surface area $6L^2$ and the temperature difference ΔT between inside and outside:

$$dQ/dt_{loss} = 6U_{eff}L^2\Delta T, \qquad (35)$$

where U_{eff} is the effective *thermal transmittance* ($U_{eff} = 1/R_{eff}$) for the building that *takes into account all energy leaks*. On the other hand, internal heat gain is proportional to building volume (floor area times a fixed ceiling height of about 3 m):

$$dQ/dt_{gain} = GL^3, \qquad (36)$$

where G is internal heat load per unit volume. A typical house has about 1 kW of free heat (3400 Btu/h), while office buildings typically have an internal gain flux on a floor-area basis of $f = 66$ W/m^2 (6 W/ft^2). The volume gain G is f/H where H is the height of one floor, obtained from

$$dQ/dT_{\text{gain}} = GL^3 = fnL^2 = fL^3/H, \tag{37}$$

where the number of floors $n = L/H$. Because of internal gains, the inside of a typical unheated house is 2°C (3–4°F) warmer than the outside temperature.

Let's consider the simple case of *walls without mass* and specific heat, ignoring the time dependence of warm-up and cool-down. For a building *without a furnace*, $\Delta T = \Delta T_{\text{free}}$. We can find the free temperature by requiring the gains and losses to be equal:

$$dQ/dt_{\text{gain}} = dQ/dt_{\text{loss}} = GL^3 = 6U_{\text{eff}}L^2\Delta T_{\text{free}}. \tag{38}$$

$$\Delta T_{\text{free}} = GL/6U_{\text{eff}}. \tag{39}$$

Thus, ΔT_{free} is proportional to the power density G and building length L. Large free temperatures are observed in big buildings (large L), buildings with very good insulation (small U_{eff}) and buildings with large internal heat loads (large G). We will apply the scaling model to large buildings (large L) and superinsulated houses with considerable insulation (small U_{eff}).

A temperature difference is needed to force heat power through the walls. Such is the case of elevated temperatures created by placing a blanket over a 100-W light bulb. The blanket's high thermal resistance blocks heat flow, raising the temperature of the interior of the blanket. The higher interior temperature pushes the 100 thermal watts through the blanket, but it may also start a fire. Free temperature ΔT_{free} in buildings can save considerable energy.

Thermostats do not call for extra heat until T_{outside} drops ΔT_{free} below $T_{\text{thermostat}}$. The outside temperature at which the furnace comes on (ignoring time delay from thermal inertia) is called the *balance point* of a building. The balance point temperature of a typical building is

$$T_{\text{balance}} = T_{\text{thermostat}} - \Delta T_{\text{free}} = 68°F - 3°F = 65°F = 18.3°C. \tag{40}$$

This temperature is universally chosen as the base temperature in the degree-day formula since the furnace turns on when temperature goes below T_{balance}. At the balance point, the internal heat gain without the furnace balances the heat losses. As building sizes increases, the increased dQ/dt_{gain} raises $\Delta T_{\text{free}} = GL/6U_{\text{eff}}$, lowering the value of T_{balance} at which heat is first needed.

At outside temperatures below the balance point, the net heat loss rate is

$$dQ/dt_{\text{loss}} - dQ/dt_{\text{gain}} = 6U_{\text{eff}}L^2(\Delta T - \Delta T_{\text{free}}) = 6U_{\text{eff}}L^2(\Delta T - GL/6U_{\text{eff}}) \tag{41}$$

where $\Delta T = T_{\text{thermostat}} - T_{\text{outside}}$. Note that length L appears in two ways: The *multiplicative* $U_{\text{eff}}L^2$ term, which is the *lossiness* of the building, increases with conductivity and size. The *subtractive* $GL/6U_{eff}$ term for free temperature reduces losses by increasing free temperature and lowering the balance point.[3]

Decreased *thermal transmittance* ($U_{eff} = 1/R_{eff}$) saves energy in two ways:

(1) Lowering the *multiplicative* U_{eff} saves energy proportionally since dQ/dt_{net} α $U_{eff}L^2$. A superinsulated house can have five times (or more) thermal resistance than a normal house, raising walls from R10 to R50. If this is done for all pathways, $U_{super} = U_{normal}/5$. The multiplicative role of U_{super}, reduces heating bills to 20% of their former value.

(2) The *subtractive* free temperature allows houses to be run at effectively lower temperatures. Let the internal energy of a *normal house* be 1 kW (3400 Btu/h) with lossiness $6U_{eff}L^2 = 1100$ Btu/h-°F. Free temperature is obtained from a heat balance:

$$1 \text{ kW} = 3400 \text{ Btu/hr} = 6U_{eff}L^2\Delta T_{free} = (1100 \text{ Btu/hr-°F})\Delta T_{free\text{-}normal}, \quad (42)$$

giving $\Delta T_{free\text{-}normal} = 3°F$ and $T_{balance\text{-}nomral} = 68°F - 3°F = 65°F$. A superinsulated house with 20% of its former lossiness ($6U_{eff}L^2 = 220$) has a ΔT_{free} that is five times larger than a normal house.

$$\Delta T_{free/super} = (U_{normal}/U_{super})\Delta T_{free/normal} = 5\Delta T_{free/normal} = 5\text{x}3°F = 15°F. \quad (43)$$

Considerable energy saving is seen by examining the *degree day distribution function* for a particular location. For days when $T_{outside}$ is greater than the balance point of the building, 100% of the energy is saved. On other days, a fractional energy savings is realized. For example, if the free temperature of a superinsulated house is 15°F, the balance point is $T_{balance} = T_{thermostat} - \Delta T_{free} = (68°F - 15°F) = 53°F$. The furnace is not engaged on days that are warmer than 53°F, which gives 100% savings. For a just-freezing day of 32°F, the furnace supplies 53°F – 32°F = 21°F of what a *typical house* needs, 65°F – 32°F = 33°F, which reduces normal heating bills to 21°F/33°F = 64%. But, we have not included the multiplicative factor of 20%, which gives the actual fuel bill reduction to 64%/5 = 13% of its normal value.

Large families with a greater number of occupants that produce and use more thermal power further enhance savings. Twice the thermal output doubles free temperature from 15°F to 30°F. Some super-super insulated houses in Saskatchuwan, Canada, with (10,000 °F–days/yr) use only $100 of natural gas to get through the winter. We now see that it is *theoretically possible* for a house with extremely small lossiness to function with the heat of "two cats fighting," but economics makes this only a pedagogical example.

PASIVE SOLAR BUILDINGS

Only thermally tight buildings can be successfully solar heated since solar flux is a low-density energy source with average power of 0.2 kW/m². Hence, it is good advice to "insulate before you insolate." By a quirk of nature, marvelous glass transmits

102

visible light from the sun, while absorbing infrared radiated from inside buildings. Coupling a glass filter with thermal mass gives us low-technology solar heat for buildings and water. Heat capacity of materials allows us to transfer heat from the hotter body to the colder body without moving parts. Passive solar is an *essentially* free energy source for buildings in warmer climates and it supplies a good boost in cold climates.

Solar energy results from a three-step fusion of four protons into a helium nucleus. The solar flux s_o is a result of the following parameters:

- mass of ^1H is 1.0078 AMU and ^4He is 4.0026 AMU.
- Earth-sun distance, the astronomical unit (AU), is 150 million km.
- Sun becomes a giant star at 10^{10} years when 10% of 2×10^{30} kg ^1H is consumed.

The mass of four protons is reduced 0.7% when converted to ^4He:

$$\Delta m/m = (4 \times 1.0078 - 4.0026)/(4 \times 1.0078) = 0.0071 = 0.7\%. \quad (44)$$

The available solar energy over the sun's lifetime is

$$\Delta M_{sun}c^2 = (0.007)(0.1 \times 2.0 \times 10^{30} \text{ kg})(3 \times 10^8 \text{ m/sec})^2 = 1.3 \times 10^{44} \text{ J}. \quad (45)$$

The sun's average power over its 10^{10} years is

$$\Delta E_{sun}/\Delta t = (1.3 \times 10^{44} \text{ J}/10^{10} \text{ yr})(1 \text{ yr}/3.2 \times 10^7 \text{ sec}) = 4.1 \times 10^{26} \text{ W}, \quad (46)$$

which gives a solar flux above Earth's atmosphere,

$$s_o = P_{sun}/4\pi(1 \text{ AU})^2 = (4.1 \times 10^{26} \text{ W})/(4\pi)(1.5 \times 10^{11} \text{ m})^2 = 1.5 \text{ kW/m}^2. \quad (47)$$

The actual solar flux is 10% smaller at

$$s_o = 1.367 \text{ kW/m}^2 = 434 \text{ Btu/ft}^2\text{-hr} = 0.13 \text{ kW/ft}^2 = 2.0 \text{ cal/min-cm}^2. \quad (48)$$

The solar flux at Earth's surface is reduced by three factors of 2: (1) daytime angles, (2) nighttime darkness and (3) reflection and absorption by atmosphere and clouds. The first factor of 2 comes from an average of $\cos\theta$ over a hemi-sphere, where θ is the sun's angle from the zenith position. The second 2 results from 12 hours of darkness for each average day. The combined factor of $2^2 = 4$ reduction is easily grasped. The area of Earth's disk intercepting sunlight is πR_E^2 (R_E = Earth radius), but rotation spreads sunlight over 24 hours onto Earth's $4\pi R_E^2$ spherical area, which is four times the disk area. The third factor of 2 is an average of atmospheric absorption and reflection from atmosphere and clouds. The three factors of 2 reduce above Earth solar flux from 1.37 kW/m^2 to 0.2 kW/m^2 for the lower–48 *average* solar power.

Solar flux at noon can be estimated from the *solar angle* θ, which is determined from a location's latitude, and day of the year. Knowledge of θ is used to determine atmospheric absorption/scattering, cosq for horizontal flux and sinq for vertical flux. Earth's spin axis is tipped 23° with respect to the *plane of the ecliptic.* The tip angle, combined with latitude angle θ_L, gives us the solar angle with respect to the zenith at solar noon (θ_{noon}) for four key days, the two solstice and two equinox days: $\theta_{noon} = \theta_L$ (spring and fall equinox), $\theta_{noon} = \theta_L - 23°$ (summer solstice) and $\theta_{noon} = \theta_L + 23°$ (winter solstice). San Diego at 33°N has θ_{noon} varying between 10° and 56°, while Seattle at 47°N has θ_{noon} varying between 24° and 70°. The value of θ_{noon} for other days is obtained by fitting a sine function to θ_{noon} values for equinox and solstice days. San Diego has 14 hours of sun in the summer and 10 hours in winter. Seattle's lower sun angle is countered by 16 hours of sun in the summer, but winter is both darker and shorter at 8 hours. Everywhere on equinox days the sun rises due east and 12 hours later it sets due west.

The solar flux above the atmosphere s_o is reduced to s_1 by absorption and scattering from air and clouds, an affect which increases with the zenith angle. Solar flux on a surface is further reduced to s_h by a factor of cosθ on horizontal surfaces and to s_v by a factor of sinθ on vertical surfaces facing the sun. A_o is an area that is orthogonal to solar rays. The horizontal surface that captures all the rays passing through A_o is a larger area A_h. The cosine of θ relates A_h to A_o by $\cos\theta = A_o/A_h$, giving $s_h = s_1\cos\theta$. For vertical surfaces facing the sun, such as windows for passive solar energy, solar flux s_1 is reduced to $s_v = s_1\sin\theta$. Solar energy gathered by solar collectors depends on many variables: latitude of collectors, time of day, season of the year, angle of collectors and materials and equipment used for the collectors. These results must be integrated over the daytime, since passive solar energy is concerned with integrated solar flux and not instantaneous flux. All of this can make calculations difficult, but our approximate methods obtain an accuracy of about 10%.

Atmospheric Transmission

We return to the reduction of s_o to s_1, as air and clouds reflect and absorb sunlight before it reaches the ground. When the sun is at solar angle θ from the zenith, its rays pass through more air than when it is at the zenith position. The surface mass density (kg/m^2) traversed by light is given in units of *n Earth air masses,* which increases from $n = 1$ at a vertical θ = 0° to infinity at a horizontal θ = 90° (flat Earth error). The ratio of path lengths in air gives

$$n = \sec\theta = 1/\cos\theta. \tag{49}$$

When the sun is low in the sky at θ = 60°, mass traversed is doubled, as in $n = 1/\cos60° = 1/0.5 = 2$. When the sun is very low at θ = 80°, the mass traversed is increased by $n = 1/\cos80° = 5.7$.

A doubling of air mass to $n = 2$ does not double absorption, as additional air absorbs less effectively. The solar flux absorbed, Δs, in a small amount of mass Δm is

$$\Delta s = -\lambda s \Delta m, \tag{50}$$

where λ is an absorption constant. This integrates to

$$s_1 = s_o e^{-\lambda m}, \tag{51}$$

where s_o is the initial solar flux. This is the orthogonal solar flux, with the area perpendicular to the solar rays. The path integral of mass density of air traversed by sunlight increases with θ,

$$m = nm_o = m_o \sec\theta \tag{52}$$

where m_o is the air mass traversed at $\theta = 0°$. This allows us to write a general expression for the solar flux at angle θ from the zenith position:

$$s_1 = s_o \exp(-\lambda m_o \sec\theta) \tag{53}$$

The value of λm_o is determined from the flux above the atmosphere ($s_o = 1367$ W/m^2) and the maximum flux at Earth's surface ($s_1 = 970$ W/m^2) when the *sun is in the zenith*, giving

$$s_1 = 970 \text{ W/m}^2 = 1367 \text{ W/m}^2 \exp(-\lambda m_o). \tag{54}$$

This gives $\lambda m_o = 0.34$, which gives solar flux at sea level as a function of θ,

$$s_1 = s_o e^{-0.34 \sec\theta} = s_o e^{-1/3\cos\theta}. \tag{55}$$

This can be corrected for site elevation, which depends on the exponential density of air as a function of elevation.

Angle of Collector

Horizontal collector flux is reduced from s_1 by $\cos\theta$,

$$s_h = (s_o \cos\theta) e^{-1/3\cos\theta}. \tag{56}$$

Passive solar energy in winter relies on vertical, south-facing glass, which has a flux at solar noon of

$$s_v = (s_o \sin\theta) e^{-1/3\cos\theta}. \tag{57}$$

Collectors are typically raised toward the south by an angle ϕ above the horizontal, giving a flux at solar noon of

$$s_{\text{noon}} = [s_o \cos(\theta - \phi)] e^{-1/3\cos\theta}.$$ (58)

Sun at $\theta = 0°$ and $60°$

When the sun is in the zenith position ($\theta = 0°$), solar flux is reduced to

$$s_{\text{zenith}} = s_o e^{-1/3\cos\theta} = s_o e^{-1/3\cos 0} = 0.72 s_o.$$ (59)

Thus, s_o is reduced by 28%, from 434 to 312 Btu/ft^2-hr (in units familiar to architects) and from 1.37 to 0.98 kW/m^2 (for electrical engineers). Sun rays at $\theta = 60°$ pass through 2 atmospheres, giving

$$s_1 = s_o e^{-1/3\cos 60} = s_o e^{-1/1.5} = 0.51(434 \text{ Btu/ft}^2\text{-hr}) = 220 \text{ Btu/ft}^2\text{-hr}.$$ (60)

The solar flux incident on horizontal collectors at solar noon is further reduced by $\cos 60° = 0.5$ to $s_H = 220/2 = 110$ Btu/ft^2-h. Note that flux on south-facing vertical windows at solar noon is not reduced nearly as much as on the horizontal, as $s_V = (0.51 s_o)(\sin 60°) = 220 \times 0.87 = 191$ Btu/ft^2-hr. The resilience of vertical flux in winter is a tremendous aid to passive solar heating. When snow is on the horizontal ground, the solarium with vertical glass can be warm. At noon on December 21 in Minneapolis (45° N, $\theta = 68°$), the horizontal flux is reduced to 66 Btu/ft^2-hr while south-facing windows receive a respectable 163 Btu/ft^2-hr on a cloudless day.

Integrated Solar Flux

Integrated solar flux over daytime hours is relevant for solar heated water or buildings. The thermal mass of water and cement respond slowly, as they are not responsive to instantaneous flux. A plot of solar flux versus time is approximately the first half-cycle of a sine function. Therefore, we approximate the *direct* solar flux as a sine function, ignoring the indirect, diffuse component, scattered from the entire sky:

$$s_{\text{direct}} = s_{\text{noon}} \sin(2\pi t/T).$$ (61)

The maximum amplitude s_{noon} is the flux on the south-facing collector at solar noon with sunrise at $t = 0$ and sunset at $t = T/2$. The value of T is set by location and time of year such that T/2 gives the length of the daylight period. Note that the equinox daylight is $T/2 = 12$ hours ($T = 24$ hours) at all latitudes. The summer day is $T/2 = 14$ hours at 30° latitude and 16 hours at 45° latitude. The winter day is $T/2 = 10$ hours at 30° and 8 hours at 45°. Note that vertical surfaces have a much higher flux in winter as compared to summer, due to the increased value of $\sin\theta$.

Horizontal integrated solar flux at equinox at 30°N with $T/2 = 12$ hours and $s_{\text{h-noon}} = 255$ Btu/ft^2-hr is

106

$$I = \int_0^{T/2} s_h \, dt = \int_0^{T/2} s_{\text{h-noon}} \sin(2\pi t/T) dt = s_{\text{h-noon}} T/\pi = 255 \times 24/\pi = 1950 \text{ Btu/ft}^2\text{-d.} \quad (62)$$

The equinox value is almost equally bracketed by the summer solstice value of 2800 Btu/ft^2-day and the winter solstice value of 970 Btu/ft^2-day. Horizontal collectors give very wide variations of integrated flux during a year, as the daily flux at 30° N latitude varies by a factor of 2.9 during the year while the collector at 45° varies by a factor of 8. Raised collectors at θ_L + 15° have a much smaller variation, with a factor of 1.4 at 30° and 2.6 at 45°. But this can be misleading if seasonal clouds are significant. The average daily–integrated flux is about one-fourth the sum of the daily–integral fluxes of two solstices and two equinoxes.

Since *active solar space heating* is needed only in the winter, the winter season is favored by raising the collector towards the south above the horizontal by 15° plus the local latitude. This angle is close to the extreme value of 23° plus the latitude (θ at solar noon on December 23). On the other hand, *solar hot water* is needed during the entire year so the collectors are raised at the latitude angle, which takes advantage of solar oscillation to either side of the equinox position. However, in the winter mode the collectors are usually needed at an angle of 15° plus latitude. This is the angle that is usually used for stationary collectors because (1) winter days are shorter, (2) the winter sun is weaker and (3) winter feed water is colder.

Passive Solar Window Gains and Losses

Glass transmits solar energy and reflects and retains infrared heat for basic passive solar energy as part of *glass plus mass*. The energy gains and losses through south-facing windows in winter are now calculated to evaluate the efficiency of passive solar energy. We begin again with a favorable location in a moderate climate, San Luis Obispo, California, which has few cloudy days in winter. We assume the following:

- double-glaze window, $U = 0.5$ Btu/ft^2-hr
- 55°F outside temperature in winter
- 90% transmission through
- winter flux, south-facing window, $s = (270 \text{ Btu/ft}^2\text{-hr})\sin(2\pi t/T)$ and $T/2 = 10$ hours.

The *daily heat loss* is

$$Q_{\text{loss}}/A = U\Delta T\Delta t = (0.5 \text{ Btu/ft}^2\text{-hr})(65°F–50°F)(24 \text{ hr}) = 180 \text{ Btu/ft}^2\text{-d.} \quad (63)$$

The *daily solar energy gain* is

$$Q_{\text{gain}}/A = 0.9 s_v T/\pi = (0.9)(270 \text{ Btu/ft}^2\text{-hr})(20 \text{ hr})/\pi = 1550 \text{ Btu/ft}^2\text{-d.} \quad (64)$$

The ratio $Q_{\text{gain}}/Q_{\text{loss}} = 1550/180 = 8$ is favorable and could be further improved by using drapes or R11 Venetian blinds at night, or by R4 windows. Passive solar heating can be used in more severe climates as a partial energy source that reduces energy demand.

THERMAL FLYWHEEL

Adobe buildings are similar to engine flywheels, which smooth power variations from internal combustion explosions. Massive adobe construction smoothes — or flattens — the temperature cycle of the Southwest by bringing excess day heat for cool nights, and excess night *coolth* to moderate warm days. Prior to the oil embargo, light, 2-by-4 construction without mass or insulation failed to smooth the temperature cycle.

To illustrate this idea, the thermal time constant of 9-inch diameter water pipes is determined to see if these pipes can adequately shift day heat into the night. To simplify the mathematics we idealize outside temperature as a square wave cycle with outside temperatures that jump 20°F above 70°F room temperature for 12-hour days, then 20°F below room temperature for 12-hour nights. How long will the water tubes retain warmth? Will the tubes slowly drop to a reasonable temperature of 70°F over a 12-hour period? Will the time constant τ be too short or will it be pleasant?

We ignore small temperature variations over storage volume, giving a stored energy

$$Q = Wc\Delta T, \tag{65}$$

where W is water weight, c is water specific heat and ΔT is temperature difference between water and room. For simplicity, the temperature differential $\Delta T = T_{tube} - T_{room}$ is denoted below as T, the temperature above room temperature. Heat loss from stored energy is the time derivative of Q, that is

$$dQ/dt = Wc\ dT/dt. \tag{66}$$

Heat loss comes mostly from radiation and convection:

$$dQ/dt \cong A(U_{conv} + U_{rad})\Delta T = AU_{total}\Delta T, \tag{67}$$

where A is tube surface area for a one-foot length and U_{total} is the sum of radiation and convection U-factors. Equating the surface loss rate to the loss rate of stored energy inside gives

$$dQ/dt = AU_{total}T = -Wc\ dT/dt, \tag{68}$$

which has an exponential solution, $T = T_o\ e^{-t/\tau}$ with a thermal relaxation time

$$\tau = Wc/AU_{total}. \tag{69}$$

Large heat capacity (Wc) gives a long relaxation time, while a large loss conductance (AU_{total}) gives a short relaxation time. The numerator weight x specific heat (Wc) is proportional to stored energy, and the denominator area x transmittance (AU_{total}) is proportional to energy loss rate of stored energy. The value of τ for 9-inch water tubes is

$$\tau = Wc/AU_{total} = (28 \text{ lb})(1 \text{ Btu/lb-}°\text{F})/(2.4 \text{ ft}^2)(1.0 \text{ Btu/ft}^2\text{-hr-}°\text{F}) = 12 \text{ hr}, \quad (70)$$

where weight is 28 pounds/ft, area is 2.4 ft^2/ft with $U_{total} = 1$. The 12-hour relaxation time is the correct choice to effect transfer of day heat to the night.

The energy stored per foot of tube, heated to 80°F, is

$$\Delta Q/\text{ft} = Wc\Delta T = (28 \text{ lb/ft})(1 \text{ Btu/lb-}°\text{F})(80°\text{F} - 70°\text{F}) = 280 \text{ Btu/ft}. \quad (71)$$

This energy/foot is shed overnight,

$$\Delta Q/\text{ft} = AU_{total}\Delta T\Delta t = (2.4 \text{ ft}^2)(1 \text{ Btu/ft}^2\text{-hr-}°\text{F})(80°\text{F}-70°\text{F})(10 \text{ hr}) = 240 \text{ Btu/ft}. \quad (72)$$

The heat loss through the house envelope during a 12-hour night, ΔQ/night, is

$$(600 \text{ Btu/hr-}°\text{F})\Delta T\Delta t = (600 \text{ Btu/hr-}°\text{F})(70°\text{F}-50°\text{F})(12 \text{ hr}) = 1.4 \times 10^5 \text{ Btu}. \quad (73)$$

The length of tube needed to replace the lost heat is

$$(1.4 \times 10^5 \text{ Btu/night})/(240 \text{ Btu/ft-night}) = 500 \text{ ft}. \quad (74)$$

A 1500-ft^2 house with eight rooms requires 60 ft of water tubes per room. By following the adage, "insulate before you insolate," tube length can be reduced by using more insulation and closing infiltration paths. The solar energy available to heat the water is about 50% of the integrated solar flux 2000 Btu/ft^2-day:

$$\Delta Q_{solar}/\text{ft} = \eta IA = (0.5)(2000 \text{ Btu/ft}^2\text{-day})(0.75 \text{ ft}^2/\text{ft}) = 750 \text{ Btu/ft}, \quad (75)$$

which is more than sufficient to raise the tube temperature to 80°F with 280 Btu/ft.

APPENDIX: LEED CERTIFICATION

The U.S. Green Building Council (USGBC) began in 1993 to establish programs and building standards to produce buildings that save energy and reduce impacts on the environment A large part of this effort has been to develop the criteria for energy-saving buildings under the framework of the Leadership in Energy and Environmental Design (LEED) program. This material in this appendix was obtained from USGBC web site at www.usgbc.org.

Forward from the Green Building Council

The built environment has a profound impact on our natural environment, economy, health and productivity. Breakthroughs in building science, technology and operations are now available to designers, builders, operators and owners who want to build green and maximize both economic and environmental performance. The U.S. Green Building Council (USGBC) is coordinating the establishment and evolution of a national consensus effort to provide the industry with tools necessary to design, build and operate buildings that deliver high performance inside and out. Council members work together to develop industry standards, design and construction practices and guidelines, operating practices and guidelines, policy positions and educational tools that support the adoption of sustainable design and building practices. Members also forge strategic alliances with key industry and research organizations, federal government agencies and state and local governments to transform the built environment. As the leading organization that represents the entire building industry on environmental building matters, the Council's unique perspective and collective power provides our members with enormous opportunity to effect change in the way buildings are designed, built, operated and maintained. The Council's greatest strength is the diversity of our membership. The USGBC is a balanced, consensus nonprofit representing the entire building industry, consisting of over 11,000 companies and organizations. Since its inception in 1993, the USGBC has played a vital role in providing a leadership forum and a unique, integrating force for the building industry

Committee–Based. The heart of this effective coalition is our committee structure in which volunteer members design strategies that are implemented by staff and expert consultants. Our committees provide a forum for members to resolve differences, build alliances and forge cooperative solutions for influencing change in all sectors of the building industry.

Member–Driven. The Council's membership is open and balanced and provides a comprehensive platform for carrying out important programs and activities. We target the issues identified by our members as the highest priority. We conduct an annual review of achievements that allows us to set policy, revise strategies and devise work plans based on member needs.

Consensus-Focused. We work together to promote green buildings and in doing so, we help foster greater economic vitality and environmental health at lower costs. The various industry segments bridge ideological gaps to develop balanced policies that benefit the entire industry.

The rating system is organized into five environmental categories: Sustainable Sites, Water Efficiency, Energy & Atmosphere, Materials & Resources, and Indoor Environmental Quality. An additional category, Innovation & Design Process, addresses sustainable building expertise as well as design measures not covered under the five environmental categories. The main issues covered are as follows: site selection, community connectivity, brownfield redevelopment, public transport, bicycle storage, low-emitting and fuel efficient vehicles, parking capacity, protect and restore habitat, maximize open space, stormwater quantity and quality, heat island, light pollution, water efficient landscapes, innovative wastewater, water–use reduction, optimize energy performance, on–site renewable energy, refrigerant management, measurement and verification, green power, collection of recyclables, rapidly renewable materials, certified wood, air delivery, ventilation, low–emitting materials, thermal comfort, and so forth.

The LEED for new construction ratings are awarded according to this scale:

Certified: 26-32 points
Silver: 33-38 points
Gold: 39-51 points
Platinum: 52-69 points

USGBC will recognize buildings that achieve one of these rating levels with a formal letter of certification and a mountable plaque.

A brief example of the many options to optimize energy performance: Whole Building Energy Simulation (1–10 Points). Demonstrate a percentage improvement in the proposed building performance rating compared to the baseline building performance rating per ASHRAE/IESNA Standard 90.1-2004 by a whole building project simulation using the Building Performance Rating Method in Appendix G of the Standard. The base–line energy performance is obtained from California Energy Code (Title XXIV) and other factors. The minimum energy cost savings percentage for each point threshold is as follows for new buildings (percent–saved and points awarded): (10.5%, 1 point), (14%, 2 points), (17.5%, 3 points), (21%, 4 points), (24.5%, 5 points), (28%, 6 points), (31.5%, 7 points), (35%, 8 points), (38.5%, 9 points), (42%, 10 points).

BIBLIOGRAPHY

E. Adams (ed.), *Alternate Construction*, John Wiley, New York (2000).
Amer. Coun. Energy Efficient Economy, *Energy Efficiency*, ACEEE, Washington, DC (1984–2008).
American Institute of Architects, *Energy Design Handbook*, AIA Press, Washington, DC (1993).
B. Anderson, *Solar Buildings and Architecture*, MIT Press, Cambridge, MA (1990).
D. Hafemeister, *Physics of Societal Issues*, Springer, New York (2007).
M. Krarti, *Energy Audit of Building Systems*, CRC Press, Boca Raton (2000).
E. Mazria, *The Passive Solar Energy Handbook*, Rodale, Emmaus, PA (1979).
A. Meinel and M. Meinel, *Applied Solar Energy*, Addison Wesley, Reading, MA (1979).
US Energy Info. Admin, *A Look at Residential Energy Consumption*, EIA, Washington, DC (1999).

---

[1] US home heating modes in (2000/1980/1960/1940) in % of use: natural gas (51/53/43/11), electricity (30/18/2/0), oil (9/18/32/10), bottled gas (7/6/5/0), coal (0.1/0.6/12/55), wood (2/3/4/23), solar (0.04/0/0/0) [US Census Bureau].

[2] P. W. Niles, "Thermal Evaluation of a house using a movable-insulation heating and cooling system," *Solar Energy* vol. 18, 413–419 (1976).

[3] In equation 41, it might seem that one could cancel the $6U_{eff}$ in front with the $6U_{eff}$ in the denominator of the second term. This is mathematically pleasant, but it can't be done since it is a physical two-step process negating the concept of free temperature.

Progress Towards Highly Efficient Windows for Zero–Energy Buildings

Stephen Selkowitz

Department Head
Building Technologies Department
Environmental Energy Technologies Division
Lawrence Berkeley National Laboratory
Berkeley, CA 94720

Abstract. Energy efficient windows could save 4 quads/year, with an additional 1 quad/year gain from daylighting in commercial buildings. This corresponds to 13% of energy used by US buildings and 5% of all energy used by the US. The technical potential is thus very large and the economic potential is slowly becoming a reality. This paper describes the progress in energy efficient windows that employ low-emissivity glazing, electrochromic switchable coatings and other novel materials. Dynamic systems are being developed that use sensors and controls to modulate daylighting and shading contributions in response to occupancy, comfort and energy needs. Improving the energy performance of windows involves physics in a variety of application: optics, heat transfer, materials science and applied engineering. Technical solutions must also be compatible with national policy, codes and standards, economics, business practice and investment, real and perceived risks, comfort, health, safety, productivity, amenities, and occupant preference and values. The challenge is to optimize energy performance by understanding and reinforcing the synergetic coupling between these many issues.

Fenestration Impacts on Building End–Use Energy Consumption

The buildings sector consumes 39% of the total primary energy use by the United States. This amounts to about 40 quads per year (1 quad = 10^{15} BTU/yr), or about 20 million barrels of oil equivalent per day (Mbble/yr). Energy demand of buildings consumes 71% of US electrical energy consumption and 54% of US natural gas consumption. The 39% of US energy use by buildings is approximately evenly divided between residential buildings and commercial buildings; residential buildings consume 21% of total US energy use, while commercial buildings consume 18%.

Windows and daylighting directly or indirectly impact about 50% of the building sector's energy consumption. For residences, windows influence the 32% of total energy consumed by heating and the 10% consumed by cooling. For commercial buildings, they directly impact the 16% of energy consumed by heating, and the 13% consumed by cooling, and can displace about one quarter of the 28% that is consumed by lighting by utilizing effective daylighting techniques. Figure 1 and Table 1 display

CP1044, *Physics of Sustainable Energy, Using Energy Efficiently and Producing It Renewably*
edited by D. Hafemeister, B. Levi, M. Levine, and P. Schwartz
© 2008 American Institute of Physics 978-0-7354-0572-1/08/$23.00

the overall end-use energy consumption data for buildings and the use attributable to windows.

Fenestration Impacts on
Building End Use Energy Consumption 4

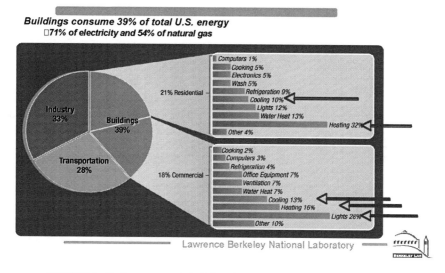

Buildings consume 39% of total U.S. energy
☐*71% of electricity and 54% of natural gas*

FIGURE 1. End–use energy in buildings by category. (Source: U.S. DOE)

	Residential	Commercial
Heating	1.65	0.96
Cooling	1.02	0.52
Daylighting		1.0
Total 2.67	plus	2.48 = 5.1 quads/yr

TABLE 1. End–use energy consumption (quads/year) due to windows in US buildings.

Energy required for windows in the current stock of commercial buildings requires 0.96 quads/year for heating and 0.52 quads for cooling, which costs owners about $20 billion/year. Figure 2 indicates how this consumption can be reduced with successively better window systems. If today's typical better designed window products were used on all existing commercial buildings, the consumption due to windows would be reduced by 15%. If low–emissivity windows were exclusively used, window consumption would be reduced by 50%. If electrochromic dynamic windows were used, consumption would be reduced by 70%. If highly insulated glazings were added to the dynamic windows, windows could be net gainers of energy by 0.15 quads. If integrated facades that control and modulate daylight are used, the gain could be increased to 1.1 quads. Overall these strategies suggest the potential to shift from an energy cost of 1.48 quads/yr to a net energy benefit of plus 1.1 quads, for a net gain of over 2.5 quads. These technical potentials will be difficult to achieve in the real world but it is important to understand that the technical potential exists to convert these windows to net energy suppliers to the building.

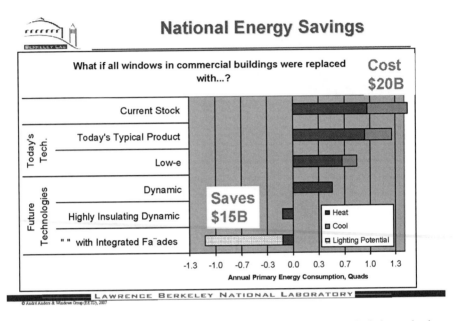

FIGURE 2. Energy savings (quads/year) in commercial buildings with advanced window technology.

Similar cumulative sector savings can be obtained in residential buildings. For example, consider the baseline as a hypothetical home in a northern climate with no windows, costing $1000/year for heating. The following heating season impacts were then determined with building energy simulation tools when the following windows were added: with single glazed with storm windows (typical of existing older houses), annual heating cost is $1310; and with double glazed windows with low-E coatings, the cost drops to $1120. If super-insulated triple pane windows are used the total heating cost is $960 per year, which is $340 less than the older windows and actually lower than the energy cost of a building without windows. The windows are able to "beat" the performance of a windowless house because the passive solar gain entering through the windows exceeds the thermal losses, thus making the window a "net energy supplier".

New Window Technologies

To better understand the performance of the new technologies it is useful to begin by defining the values of thermal resistance R and its inverse thermal transmittance ($U = 1/R$). The basic heat rate loss equation is as follows:

$$dQ/dt = UA\ \Delta T = (A/R)\ \Delta T,$$

where the overall heat loss rate dQ/dt in English (SI) units is BTU/feet²-hour (watts/meter²). Temperature difference ΔT is in Fahrenheit (Celsius). R value units are ft²·°F·h/Btu (K·m²/W) with English R values a factor of 5.675 larger than the SI values. U values are in units of Btu/ft² °F·h (W/K m²). An R1 window in English units is 1/5.675 smaller in SI at R0.2. For windows, convection and thermal radiation cannot be neglected. The overall window U-value contains all forms of energy transfer, namely thermal conductance, convection and long wave radiation, and by convention includes not only the intrinsic thermal properties of the window but the convective and radiative heat transfer associated with the indoor and outdoor interfaces with climate.

Advanced windows can become net–energy producers as shown in the simulated example above. In typical northern climates annual windows should approximately balance total winter energy gains and losses at U values below U0.2 English (R5 English) which will reduce total net winter energy consumption to zero. This is possible with technology available today and may become routinely available and cost effective in the next ten years. Typical U and R-values (English units) for a range of generic windows are as follows:

> Single glaze, U1, R1
> Double Glaze, U0.5, R2
> Low–E, U0.35, R3
> Highly Insulating Window, U0.17, R6
> "Superwindow", U0.10, R10

Markets change slowly because consumer demand moves slowly and it takes time for manufacturers to retool to make better products. U.S. Residential Window Markets have progressed substantially over the last 30 years and are expected to continue to change in response to changing market and environmental needs:

> o 1973: Typical Window:
> clear, single glazed,
> or double glazed or single with storm window in north,
> U (national average) = 0.85 (English units)

> o 2006: Typical Window:
> 95% of all windows sold are double glazed
> 60% have a low-E coating
> Net 30-65% energy savings vs. 1973 typical window
> U (national average) = 0.45 (English units)

> o 2020+: Future Windows for Zero–Energy Buildings:
> Zero net–energy (typical) in winter;
> U (typical) = 0.1-0.2 (English units)
> 80% cooling savings from dynamic solar control in summer

Vision for Windows:
energy losers, to neutrality, to net energy suppliers

From a purely scientific perspective we can define the requirements for a zero energy window, a window that imposes no net energy load on a building as follows:

- In heating climates it is necessary to reduce heat losses (U) so that useful transmitted solar energy balances and then <u>exceeds thermal loss</u>. To do this we need very low U but moderate solar gain.
- In cooling climates it is necessary to reduce cooling loads due to solar gain. A very low window solar heating gain factor (SHGC) is needed but the window must remain transparent for view. Static controls can never be optimal for all conditions so we must transition to dynamic controls, e.g. low transmission in summer, high transmission in summer.
- In mixed climates, it is necessary to have dynamic solar controls due to the hourly, daily and seasonal dynamics for optimization.
- In all climates, we should replace electric lighting with daylight provided by windows and skylights.
- Ultimately the window can become an electricity supplier if we use options that integrate photovoltaics with shading systems and glazing.

These functions can be achieved with the use of a wide range of technologies that are integrated into the glazing and/or a complete window system. (In this usage a "window" includes glazing elements, framing elements that hold the glazing and may open to permit air flow, and a wide range of interior and exterior shading and insulating systems that add thermal control and other functionality to window systems in buildings. A wide variety of coatings can be applied to glazings to save energy (Figure 3) but there is a wider "kit of parts" that can be applied at different physical scales from "Nano" to "Macro" to enhance performance.

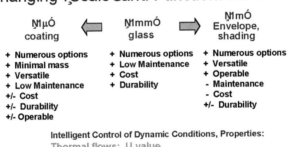

Glazing/Window Technology:
Changing Scale and Function: Kit of Parts

FIGURE 3. Coatings (micro sized), glass (mm sized) and envelopes (meter sized).

The best developed and most widely utilized coating technologies are low-emissivity coatings that reflect long-wave infrared energy. Low–E windows that reflect long-wave infrared already save considerable energy in heating climates (Figure 4). Later variations of the coatings add more layers that are tuned to shift the reflection edge to shorter wavelengths, reflecting not only the long-wave IR but the near IR as well, thereby reducing solar gain for better summer sun control.

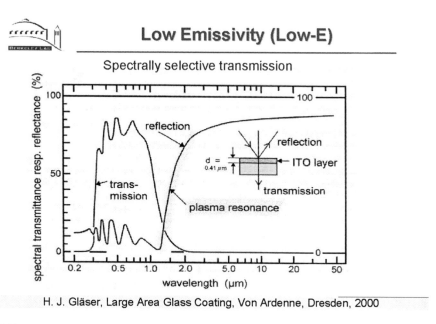

H. J. Gläser, Large Area Glass Coating, Von Ardenne, Dresden, 2000

FIGURE 4. Low–E windows: Infrared radiation below the plasma frequency of indium tin oxide is reflected, while visible and some near IR are transmitted.

Low-E coatings and other improvements have substantially reduced energy consumption of windows compared to prior standard practice. But the challenge of further improving window technology remains: Today's windows cost US consumers over $20 billion/year in energy costs, much of which could still be avoided. Low-emissivity coatings and gas fills are the most significant technology changes in the window industry in last 20 years but additional future innovation is needed to address the zero energy building challenge. In most cases solutions will come from clever technology advances that are coupled with insights into market factors that accelerate industry investment in these new technologies.

R&D and Market Issues

There are many factors which must be considered with developing better windows in a cost-competitive market: Coating design, window thermal performance optimization, manufacturing technology, durability, integration into a complete window, rating and labeling performance, field test to verify performance vs. climate, and building applications. In the near term, the zero-energy window heating objective requires a U less than U0.15 (0.8 W/m^2-K). In the longer term, the target is for a U less than U0.1 (0.5 W/m^2-K). These performance levels can be reached by adapting existing low-E/gas fill technology or by exploring new approaches. Longer term options include a low-E window that is evacuated with the two glass plates separated by small, low conductance spacers. In this design an edge seal that has a long life without leakage, able to withstand operating stresses and low cost is the critical factor. In this configuration heat transfer occurs largely by long wave radiation, and by conduction through the spacers and the glass edges. The physics is well understood-the engineering and cost challenges are significant. At least one manufacturer has a vacuum window on the market although its cost is high and performance is not much better than standard low-E windows. Another approach uses microporous, optically clear aerogels whose intrinsic thermal conductance is very low, but whose solar transmission is high. Lab prototypes have been fabricated but commercial windows are not yet available despite many years of R&D.

New Technology Options

In order to provide cost effective, high performance products in the near term variants on existing technologies show great promise. Significant further reductions in heat loss can be achieved with triple glazed units with two low-emissivity coatings, low conductance gas fills, "warm edge" low conductance spacers and insulated frame systems. Triple glazed, low-E windows exist on the market today (and are widely used in Europe) but they are heavy and costly, and more prone to gas leakage with four sets of edge seals. To address cost and durability issues a design with a center "floating" thin glass or rigid plastic layer was employed with multiple coatings and gas fill. These non-structural center layers look promising and are being explored by several manufacturers. From a market perspective they build on the existing industrial infrastructure that manufactures low-E coatings and warm edge technology, and thus could have market impacts in a relatively short time frame if adopted by manufacturers.

While heat loss is the largest energy impact of windows the demographics of the last three decades have resulted in a boom in construction in the sun-belt where the key energy impacts are cooling. Even in the northern half of the country most new housing is air-conditioned, increasing the energy impacts of typical glazing solutions.

Glass properties can be altered to control solar energy inputs: It is relatively easy to reduce solar gain with tinted or reflective glass. But most markets value view and transparency. We should keep in mind that very low solar transmittance may increase winter heating costs as discussed in the sections above. The challenge is thus twofold: to manage solar gain from winter to summer, and even when it is desirable to reduce solar gain, we still want to transmit daylight.

Total solar heat gain is determined largely by the solar radiation passing through the window (visible light and near IR transmittance). In addition to direct solar transmittance, one must add the solar radiation absorbed within the glazing system and redirected to the indoor space by all modes of heat transfer ("inward flowing fraction").

The ideal energy-saving window combines low U-value with controllable or switchable solar gain. The goal is to admit sunlight and daylight when desired or advantageous, but to modulate it or reject it when it will increase energy use. Looking at the optical properties of glass alone, the objective would be to make the window transmittance switchable to optimize net solar gain. Figure 5 shows a hypothetical switchable coating that might be either transmissive or reflective in the solar spectrum out to 3-microns and always reflective in the long-wave range beyond 4-microns. Such a glazing would always reflect long wave IR but would modulate its solar gain and daylight transmittance over a wide range. The challenge is to find materials systems that allow this wide dynamic range, as well as meeting market criteria for appearance, durability and affordability.

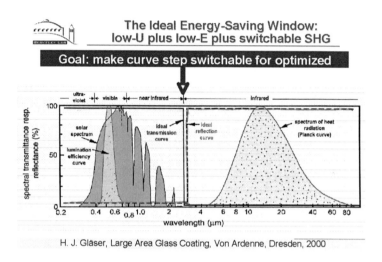

H. J. Gläser, Large Area Glass Coating, Von Ardenne, Dresden, 2000

FIGURE 5. Goal is to control transmittance across solar spectrum.

Emerging Options

The optimal solar optical properties for a window depend on the geographic location, climate, orientation, season, building type as well as occupant use patterns. The best solution is a window whose properties are variable and controllable, as discussed above.

Switchable glazings can have two or more states for optical transmission, a clear state and a "dark" state to control glare and solar transmission. The triggering is integrated into the glazing and can be actively or passively actuated. An early example was photochromic sunglasses. The glass or plastic darkens or clears automatically in response to incident sunlight intensity. Thermochromic glazings respond to temperature and become dark as they heat up. Smart coatings for dynamic control of windows must balance cooling and daylighting, but also view and glare. Due to these often contradictory needs the optimal performance will come from switchable coatings that can be actively controlled via electrical means. These provide flexible and optimized control for all climate, building and occupant conditions. Normally the control comes from two transparent electrical conductors that sandwich an optically active layer or stack of coatings. This active control comes at a cost- the additional transparent electrical conductors and the associated wiring and controls infrastructure. Active control can be achieved using several different materials approaches, using liquid crystals, suspended particle displays, or electrochromic coatings.

Liquid crystal and Suspended particle devices both change transmission by altering particles that align in response to an electric field between transparent thin films. Transmission can be switched faster than with other chromic techniques. This can be integrated in smart window solutions. They are available but are still costly today and energy efficiency and lifetime have not yet been proven.

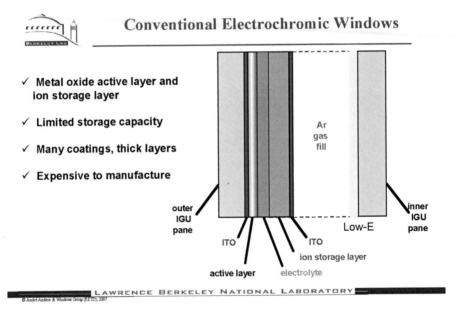

FIGURE 6. Conventional electrochromic windows using a multilayer absorptive coating design on one glass surface in the insulated glass unit (IGU).

Electrochromic devices are the most promising approaches with several products now on the market. Ions are reversibly inserted or extracted from an active electrochromic layer; upon insertion a reversible color change occurs. Most of these coatings are absorptive in the low transmission state, with a dynamic visible transmittance range of approximately 4% to 60%. Because they are absorptive the inner-most ITO layer is a low-E surface which helps reject the absorbed energy. Continuing R&D is aimed at reducing manufacturing costs, and improving switching speed. Many materials exhibit electrochromic effects but only a few are suitable for use as switchable windows. New organic materials may help reduce cost and new reflective devices may also improve dynamic range and performance.

Although progress to market has been slow we expect that further material science breakthroughs in this field of active coatings will produce new efficiency options in the years ahead. In addition to this materials work, a number of other supporting R&D activities are needed, as illustrated in Figure 7, to ensure that promising coatings find their way to market and energy savings applications in building

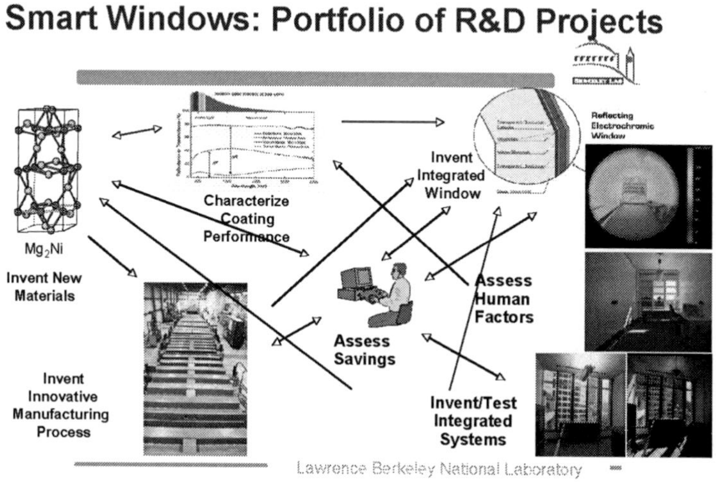

Smart Windows: Portfolio of R&D Projects

FIGURE 7. Smart Windows: Portfolio of R&D Projects

Window/Daylighting for New York Times Headquarters

The functionality of "smart glazings" can be captured today using automated shades and blinds. Architects and building owners increasingly are interested in using daylighting concepts in office buildings to offset electric lighting needs but must control solar gain and glare while admitting the daylight. Over the last 4 years our group at LNBL partnered with the New York Times and their design team to develop an automated shading and photocell-controlled daylight dimming system for their new all glass, 52 story headquarters in Manhattan. When approached by the owner we considered the lighting and daylighting issues including color, glare, privacy, peak load, energy savings, occupant impact, integrated control of electronically–controlled windows, lighting system, solar heat gains, and visual comfort. From our previous experience we knew that energy savings and demand management can be controlled with active facades with daylighting controls. By using sensors to control lighting and motorized shades which reduce solar inputs, electric lighting has been reduced on our past projects by up to 75% and air-conditioning has been reduced by 25%.

The New York Times wanted a highly glazed façade that would give workers good views and would allow the city to see "news" at work. They developed a façade with good spectrally selective glass with low solar heat gain but a high daylight transmittance, and further used exterior fixed shading to reduce solar gains and diffuse some of the daylight entering the space. But these solutions alone were not enough to guarantee good energy performance and comfort. In order to address issues of glare, cooling, and visibility it was necessary to develop an integrated, automated shading and dimmable lighting system, which was reliable and robust. We hoped to develop solutions that would have widespread use beyond this single building. The challenge was to develop a workable, affordable integrated hardware/software solution that could be guaranteed to work in practice.

Intelligent dynamic facades can provide excellent cooling load control and admit daylight without glare. Daylighting can meet illumination requirements and save energy, while providing for a pleasant work environment to enhance comfort and performance but it is essential that glare be controlled. The visual tasks may vary such that reflected glare on computer screens from bright windows can be a critical design issue. The absolute brightness of window view can cause visual discomfort. There are several potential glare sources in such a space, from the direct sunlight, the brightness of the sky and reflected sunlight from adjacent buildings.

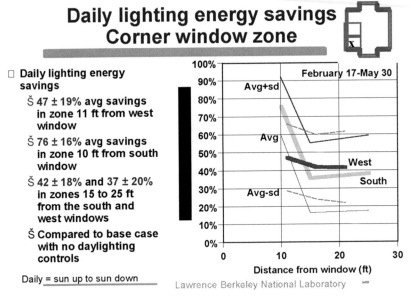

FIGURE 8. Savings of daily lighting energy due to daylighting, as a function of distance from the window, measured in the New York Times daylighting mockup.

The overall façade design had a layered approach. An external sun shading layer with fixed ceramic rods provided some solar shading and light diffusion. The glazing system used a low-E glass, with spectrally selective thermal control, which provided good insulating value and excellent solar gain control while admitting adequate daylight. The interior shading system was dynamic with a motorized shade system to control solar intensity and to control glare. Because of the complexity of the system the Times built a 5000 square foot mockup to develop and test these control integration concepts. Figure 8 shows measured data from the mockup, showing that lighting can be reduced by 30–90% at a distance of 10 feet form the window and 20 to 60% at 20 feet from the window. Based on the mockup testing detailed performance specifications were developed for bidding. The winning suppliers then installed the new systems in the mockup for further evaluation and test. Extensive computer simulation was employed to ensure that solutions that worked in the mockup would be functional throughout the building- on all floors and orientations. Commissioning tools were developed so that the owner could determine that these operable systems were working as planned once installed. Overall the project looks very successful, both for the New York Times as an owner and because it has advanced the application of emerging technology to wider markets. This single large project enabled the two major technology suppliers to develop new systems with new functionality based on our research results that are now offered commercially to all building owners.

CONCLUSIONS

We have defined a path towards future high performance windows. Advances in materials science and their underlying physics is important to develop new coatings and innovative devices. The materials and components must be integrated into dynamic and intelligent systems that control the envelope, lighting, and HVAC to achieve the energy savings and other market benefits needed. Greater functionality for energy and comfort can be obtained with equal or lower net system cost and better reliability. This can displace more energy consuming HVAC and lighting system with smarter life-cycle integration. These are essential elements of a zero energy/carbon neutral future.

There are many benefits of high performance window systems. In addition to energy savings, comfort for the occupant with with improved satisfaction and task performance is paramount. These designs add value for the building owner and reduce operating costs. And they substantially reduce energy use and associated greenhouse gas emissions for the planet.

REFERENCES

Additional technical details on window energy efficiency and material presented in this paper can be found at: http://windows.lbl.gov and at:
New York Times project: http://windows.lbl.gov/comm_perf/newyorktimes.htm
Electrochromics project: http://windows.lbl.gov/comm_perf/electrochromic

Appliances: Designs and Standards for Sustainability

James E. McMahon

Energy Analysis Department
Environmental Energy Technologies Division
Lawrence Berkeley National Laboratory
Berkeley, CA 94720

Abstract. Buildings consume 40% of US energy and produce 39% of US carbon dioxide. These numbers can be dramatically reduced with improved appliance efficiency. For example, energy use by the average new refrigerator dropped about 70% from 1974-2002, thanks to improved materials, technologies and designs. In this chapter, I review progress in gas furnaces, air conditioning and lighting, as well as the trends in refrigerators and freezers. The goal of zero net-energy buildings appears possible. In the future, buildings might consume perhaps 70% less energy than today due to efficient building components, appliances, equipment and lighting; systems integration; better controls; and behavioral changes. The remaining 30% energy needs could be supplied by low- or no-carbon energy sources.

Energy Use In Buildings

The buildings sector consists of 116 million US residences with 169 billion square feet of floorspace and 77 billion square feet of commercial space. Industrial buildings are not included here. In the US, buildings consume about 39 quads per year, or 40% of US total primary energy consumption, as illustrated in Figures 1 and 2. The cost of this energy is $370 billion/year, 2.5% of the gross domestic product (GDP). Most of this energy is supplied by either electricity or natural gas, so that buildings account for 71% of US electricity and 54% of US natural gas usage. For residences, the major uses of energy are heating (32%), water heating (13%), lighting (12%) and cooling (10%). For commercial buildings, the major uses are lighting (28%), heating (16%), and cooling (13%). Annual buildings costs for construction, remodeling and energy total $1.5 trillion (10% of GDP), including $370 billion for energy. On a per-capita basis, this is over $4000/year including over $2000/year for new construction, $1000/year for remodeling-renovation and $1000/year for energy.

CP1044, *Physics of Sustainable Energy, Using Energy Efficiently and Producing It Renewably*
edited by D. Hafemeister, B. Levi, M. Levine, and P. Schwartz
© 2008 American Institute of Physics 978-0-7354-0572-1/08/$23.00

Total End-Use Energy Consumption
1949 – 2004

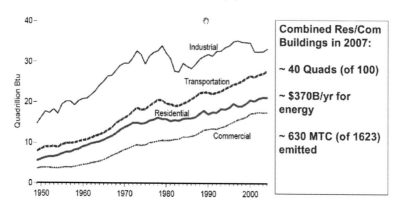

Combined Res/Com Buildings in 2007:

~ 40 Quads (of 100)

~ $370B/yr for energy

~ 630 MTC (of 1623) emitted

FIGURE 1. Trends in end–use energy consumption for industry, transportation and buildings (1949–2004).

Buildings' Energy Consumption by End Use

Buildings consume 39% of total U.S. primary energy
 • 71% of electricity and 54% of natural gas

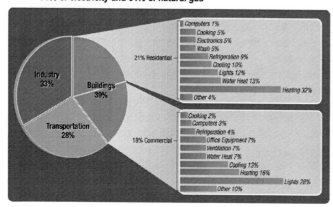

FIGURE 2. Energy consumption by buildings, broken down by appliance use.

U.S. CO$_2$ Emissions

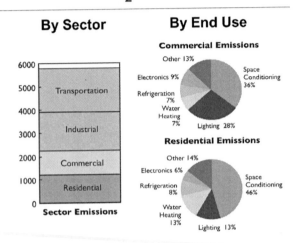

FIGURE 3. US carbon dioxide emissions by sector (millions of tons per year) and by end use within commercial and residential buildings.

As shown in Figure 3, the buildings sector in 2004 produced 630 million tonnes of carbon per year (2.31 billion tonnes/year of CO$_2$), which is 39% of US carbon emissions of 1623 million tonnes of carbon (5.95 billion tonnes/year of CO$_2$). US buildings alone are responsible for more CO$_2$ than the total energy use by any country in the world, except China and the US. Note that the largest contribution to CO$_2$ emissions from this sector comes from space conditioning (heating and cooling).

Had the US economy as a whole not made significant changes starting in the early 1970s, energy use would have been 170 quads/year by 2005, as illustrated in Figure 4. Before 1973, energy use per unit of GDP, a measure of an economy's energy intensity, was dropping by 0.4% per year. After 1973, the drop in energy intensity was much greater, 2.1%/yr. Because of this reduced energy intensity, US energy consumption in 2005 was 100 quads/yr, or 70 quads/year lower than it would have been. The avoided use of 70 quads/yr can be thought of as a very large wedge, as described in the chapter by Rob Socolow in this book. The additional 25 quads/year to grow from 75 quads/yr in 1973 to 100 quads/year in 2005 was met with new domestic and foreign supplies. The 70 quads/year saved is the energy equivalent of removing one billion cars from the highways. Without this reduction, the total cost of energy would have been about $1.7 trillion/year (in 2000 dollars), compared to the actual cost of $1.0 trillion/year.

Energy efficiency R&D has yielded net benefits to the US economy. The National Research Council estimated these benefits in a 2001 study, *Energy Research at DOE; Was It Worth It?* Using their method, we estimate that the economic benefits from energy efficiency standards were the largest of all the DOE energy efficiency programs. The efficiency standards saved a net of $48.5 billion in lifetime savings.

R&D provided new technologies, including the electronic fluorescent ballast, which saved $15 billion, followed by advanced window coatings which saved $8 billion. The DOE–2 Buildings Design Tool saved a significant amount, but the exact amount is hard to quantify. It's certain, however, that without computer simulation tools like DOE–2, the energy gains in new buildings would not have taken place. The research done to develop the technologies was paid for many times over. To achieve widespread implementation, it was necessary to have both research and policy.

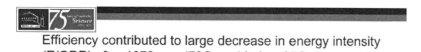

Efficiency contributed to large decrease in energy intensity (E/GDP) after 1973   (70Q avoided vs 25Q new supply)

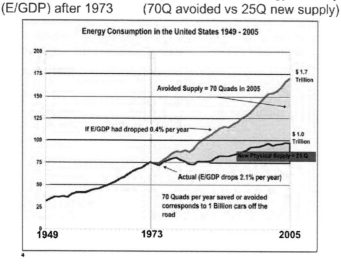

FIGURE 4. End–use energy efficiency contributed to avoiding a demand of 70 quads/year by 2005, while 25 quads of new supply were obtained from foreign and domestic sources.
Source: Art Rosenfeld

Minimum Energy Performance Standards

Although commonly referred to as "appliance standards," energy efficiency standards are regulations applied to about 50 product types that consume energy, mostly in buildings. These include normal home appliances (like refrigerators and clothes washers), electric motors, lighting equipment, furnaces, air-conditioners, heat pumps, water heaters and so forth. The declines in per-unit energy consumption for new refrigerators, central air conditioners and gas furnaces are displayed in Figure 5. The bottom curve is for refrigerators, which grew 10% in size, but had a 75% reduction in energy use from 1972-2001, in part due to mandatory standards. The open arrows are the effective dates for state standards, while the solid arrows are the effective dates for the national standards, which were called for in updates to the

Energy Production and Conservation Act (EPCA) of 1975. The middle curve for central air conditioning systems shows a drop in unit energy consumption to 60% of the 1972 consumption.[1] The standard that took effect in 2006 dropped the AC energy consumption still further, to 50%. The top curve for gas furnaces shows a drop of 25% in the period 1972-2001.

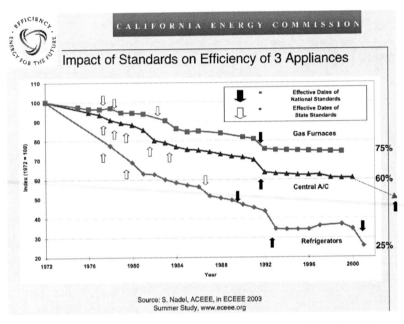

FIGURE 5. Reduction in energy use for refrigerators, central air conditioning and gas furnaces. The open arrows and solid arrows are the effective dates for state and national standards, respectively.

We examine the case of energy standards for refrigerators in more detail. Because residential electricity prices had been declining from 1960 to 1973, refrigerators were not primarily designed and constructed for energy efficiency. The average 1974 refrigerator consumed 1825 kWh/year (Figures 6 and 7). This level was adopted as the starting point standard by the California Energy Commission in 1977 to give manufacturers time to improve their products. California tightened the standard to 1500 kWh/year in 1979 and to 1000 kWh/year in 1987. US standards reduced unit consumption to under 700 kWh/year in 1993 and about 500 kWh/year in 2001. Refrigerator energy use declined on average by 4%/year between 1974 and 2002. Note that the volume of refrigerators increased by 10% from 1975, yet energy use dropped by 74% (Figure 7). Note also that the cost of a refrigerator in 1983 dollars dropped by a factor of three, while energy costs to operate a refrigerator dropped by a factor of three or four. This is the poster child for increased energy efficiency, but there are other success stories.

---

[1] Based on laboratory tests of new equipment, without accounting for changes in building construction or occupant behavior.

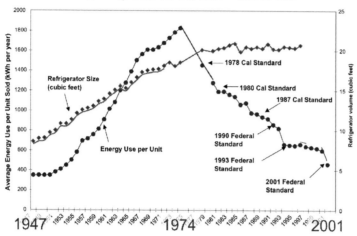

US New Refrigerator kWh/year Declined 74%
Annual Drop from 1974 to 2001 = 4% Per Year

FIGURE 6. New refrigerator energy use declined by 4% per year (average) from 1974 to 2002 while the average volume grew.

1975 Refrigerators: Before the oil embargo, refrigerators were designed with thin walls having little insulation to permit greater internal volumes and smaller initial costs. In the 1930s, refrigerator heat exchangers were placed in domes on top for good heat exchange, but they were subsequently considered unsightly. By 1975, the heat exchanger was placed on the back of the refrigerator, typically located next to a wall, reducing heat transfer. The motors and compressors were placed below the box, where they leaked heat upwards into the box, requiring further refrigeration. The 1975 refrigerator had 57% heat gain through the walls, 23% through interior heaters and fans, and 12% through the gasket area. Only 6% of heat gained was from food cooling, and door opening caused only 2%. Automatic-defrost heaters consumed 110 kWh/year and freezer case heaters, which prevent icing, consumed 60 kWh/year (compared to roughly 1800 kWh/yr for the entire refrigerator). The motor–compressor units of these refrigerators were poor, some with a coefficient of performance (COP) of 0.9. Laboratory tests to measure the energy efficiency of refrigerators were being developed.

Today's Refrigerators: New refrigerators that consume under 500 kWh/year have improved in quality while making the following changes:

- Higher quality insulation. Polyurethane foam insulation blown into the walls by robots replaced manually-installed fiber-glass insulation, saving energy and labor costs
- Thicker insulation: 5 cm of foam insulation at an R-value of 3.3 in SI units or 2.5 cm gas-filled panels at an R-value of 2.5.[2]
- More efficient motors and compressors
- Greater condenser area, which lowers temperatures from 122°F to 101°F
- Greater evaporator area, which lowers temperature from 18°F to 9°F
- Elimination of case heaters (or addition of a switch)
- Double gaskets on doors.

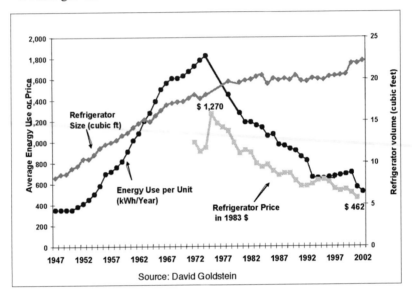

FIGURE 7. US refrigerator prices (in 1983 dollars), sizes and energy–intensity (1947–2001).

Life Cycle Cost

One way to determine whether energy efficiency improvements in general (and mandatory standards in particular) are cost effective is to determine the minimum life cycle cost, which includes the initial cost, the cumulative cost of energy and the maintenance costs. The future costs and benefits must be discounted and converted into values comparable to the present time, where they are totaled. Great care must be taken to choose a meaningful discount rate. Too high a discount rate ignores the value of future benefits; too low a discount rate ignores the returns that investments can obtain in other economic areas. Figure 8 displays the life–cycle costs (LCC) for

---

[2] H. Geller, "Progress in the Energy Efficiency of Residential Appliances and Space Conditioning Equipment" in *Energy Sources: Conservation and Renewables, AIP Conference Proceedings* 135 (1985), edited by D. Hafemeister, B. Levi and H. Kelly.

refrigerators as calculated in 1989 and again in 1995 as part of the U.S. Department of Energy analysis of projected impacts from regulations that were expected to take effect in 1993 and 2001, respectively. The 1989 LCC curve starts at $1420 for an annual energy use around 950 kWh/yr, corresponding to mandated efficiency requirements for 1990, and drops to a minimum at $1220 near 650 kWh/yr and rises to $1300 near 500 kWh/year, as increasingly costly measures are employed to make the refrigerator more efficient. The 1989 LCC curve gives a projected 27% reduction in energy consumption at the standard level of 690 kWh/year mandated for 1993, compared to designs expected to be sold in 1990. The 1995 LCC curve gives a further 30% reduction in energy consumption corresponding to the standard of 475 kWh/year for 2001 compared to 1993. The LCC curve calculated in 1995 begins at $1270 because of the progress in refrigerator design made by that date. It drops to $1120 for annual energy use of about 450 kWh and rises to $1130 for annual energy use of 425 kWh/year. The standard that took effect in 2001 based on the 1995 analysis at 475 kWh/year exceeded the *maximum technologically feasible* level estimated in 1989.

Energy-Efficiency is a Renewable Resource

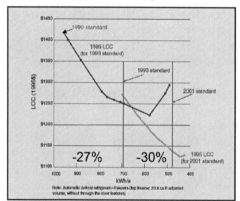

Updated 2001 standards exceeded the maximum technologically feasible level of a few years earlier.

The maximum technology kWh/a in refrigerators changed 14% in 6 years (2.5%/a) from 495 kWh/a (1989) to 425 kWh/a (1995)

Average standards, % change, effective date:
690 kWh/a, -27% (from 1990), 1993
475 kWh/a, -30% (from 1993), 2001

FIGURE 8. Energy Efficiency is a Renewable Resource. Updated 2001 standards exceeded the maximum technologically feasible level of a few years earlier. The maximum technology kWh/a in refrigerators dropped 14% in 6 years (2.5%/a) from 495 kWh/a (1989) to 425 kWh/a (1995). The average standards dropped 27% to 690 kWh/a from 1990 to 1993 and they dropped another 30% to 475 kWh/a inform 1993 to 2001, The life-cycle cost curves for refrigerators calculated are based on technology available in 1989 and 1995. Each of the points on the curves represents a specific action, such as a better compressor or more insulation.

The *maximum–technology* level for refrigerators was reduced by 14% in 6 years from 495 kWh/year (estimated by manufacturers in 1989 planning for production in 1993) to 425 kWh/year in 1995 (planning for production by 2001). The appliance manufacturers made these changes in the context of a series of government regulations.

The phrase *low–hanging fruit* has often been applied to this example of reducing the energy consumption of refrigerators. In fact, there have been a series of sequential harvests of energy efficiency for household refrigerators. In that sense, efficiency is a renewable resource. In the dynamic world of competitive – and regulated – manufacturing, additional advances were made over time, taking advantage of new materials, designs or production processes. The average rate of improvement for refrigerators was 4%/year from 1974 to 2002. In my view, the opportunities exist for a continued rate of reduced energy consumption for most appliances. A 3.5% reduction per year, if continued for 20 years of research, amounts to halving the energy needed by a given appliance. A recent federal law has required DOE to examine refrigerators again to decide whether additional changes that are technologically feasible, economically justified and save significant energy should be mandated.

The US has over 200 million refrigerators and freezers in use. The amount of energy saved by lowering consumption from 1826 kWh/year to 475 kWh/year is 270 billion kWh/yr. This is worth $27 billion/year at an electricity cost of 10 cents/kWh. A one-GWe power plant at an 80% capacity factor produces 7 billion kWh/yr. These improvements avoided the construction of forty large power plants of one GWe, obtained by dividing energy saved per year by energy per power–plant year (270 billion kWh/year divided by 7 billion kWh/GWe-year).

Availability of Cost–Effective Energy Efficiency

Several studies have estimated the amount of energy reductions available by improved energy efficiency on a cost-effective basis. In 2000, the Interlaboratory Working Group report estimated that a 20% reduction from business as usual was possible by 2020–2025 (Interlaboratory Working Group, 2000). The American Solar Energy Society estimated in 2007 that a 23% reduction was possible by 2025 by means of energy efficiency (American Solar Energy Society, 2007). This second study was done seven years later and also estimated a 20-25% reduction, but starting from a more efficient baseline. The observation that even after significant efficiency improvements, the same magnitude of potential is available again suggests that energy efficiency is a renewable source. In 2005, it was discovered that a new potential for saving electricity was available through water conservation, because using less water means avoiding the energy used to move (pump) and treat the water and resulting wastewater. The cost of saving energy via water conservation was preliminarily estimated to be 50% lower than the cost of saving electricity in existing utility programs in California. This is an example of looking for – and finding – new opportunities. Lighting upgrades using on-the-shelf technologies can save 19% of national lighting energy, or 130 billion kWh/year, as illustrated in Table 1 and Fig. 9. This corresponds to the saving of 17 one–GWe power plants at a capacity factor of

80%. The lighting upgrades to affect these gains include electronic ballasts in fluorescent lamps, compact fluorescent bulbs, ceramic metal halide lamps and high–intensity discharge lamps.

TABLE 1. Lighting upgrades can save 120 billion kWh/year in offices, homes, stores and roadways. The 19% savings is equivalent to removing 16 million cars from the road in terms of carbon reductions. [Personal communication: Francis Rubinstein, LBNL]

Potential 19% National Lighting Savings

Sector	Lighting Upgrade Measure	Estimated Energy/Cost Savings	Carbon Reduction (MMTCe)	Equivalent Cars Removed
Offices	Replace T-12 magnetic with controlled T-8 electronic	35 BkWh $2.6 billion	7	4.7 million
Homes	Replace Incandescent Bulbs with Energy-Efficient Lamps	55 BkWh $4 billion	11	7.3 million
Stores	Replace PAR/R-lamps with Ceramic Metal Halide	10 BkWh $750 million	2	1.3 million
Roadways	Replace Mercury lamps with modern HIDs	20 BkWh $1.5 billion	4	2.7 million
Total	**All Measures**	**120 BkWh $10 billion**	**24**	**16 million**

FIGURE 9. Modernizing US lighting can save 19% of lighting energy: (left) lighting demand in 2007 was 633 billion kWh/year and (right) it could be 513 billion kWh/year in 2017.

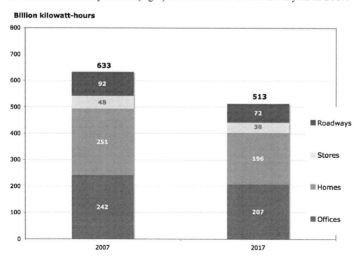

Billion kilowatt-hours

133

Cost of Conserved Energy

There are two basic ways to calculate the economic return on investments to save energy. The first is to calculate the payback period, the amount of time to repay the investment. This method does not account for future energy costs or the equipment lifetime. The second, and preferable approach, is to determine the cost of conserved energy, which can be compared to the current cost of energy supply. If the total cost of saving a kilowatt-hour is less than the local cost of supplying a kilowatt-hour, the measure is cost-effective and may be worth doing. For an investment of about $150, refrigerators in the 1970s saved 1300 kWh/yr, giving a simple payback period (no discounted benefits) of

$$T = \text{cost/annual saving} = \$150/(1300 \text{ kWh/yr})(\$0.085/\text{kWh}) = 1.4 \text{ years}$$

for electricity at $0.085/kWh. The cost of conserved energy (CCE) is defined as

$$CCE = \text{incremental equipment cost/lifetime energy savings.}$$

An investment of $150 for a refrigerator that lasts at least 10 years can save 1300 kWh/yr over older models. Calculating the simple CCE (no discounted benefits and other costs) shows that this is an excellent monetary investment with

$$CCE = \$150/(10 \text{ year} *1300 \text{ kWh/yr}) = 1.2¢/\text{kWh},$$

which is much less than residential electricity at 8.5¢/kWh (2004). At today's electricity prices, this investment looks even better.

Figure 10 displays estimates of the cost of conserved energy for various conservation actions (the bars): HVAC (heating, ventilation and air conditioning), ventilation, lighting, refrigeration and electronics. All of these actions have a CCE that is less than the marginal price of electricity (6–8.5 cents/kWh). Figure 11 is a similar display of the CCE for home appliances for the year 2010. These are all cost-effective investments since the CCE's (1 to 7 cents/kWh) are below the price of residential electricity (8.5 cents/kWh when this work was done).

Figure 12 displays calculations (McKinsey Global Institute, 2007) which compare the cost of carbon reduction (euros per tonne of CO_2) to the amount of atmospheric CO_2 avoided (% of the present rate of emissions). The bars to the left are below the zero-cost line, showing that both carbon and money are saved. The bars further to the right are above the zero-cost line, implying that carbon is saved, but at a cost. If the areas above and below the line are equal, there is no net cost for the total reduction in CO_2 (the sum of the two areas).

Cost of Conserved Energy (CCE) is Lower than Electricity Price for Many Energy Efficiency Increases
(Commercial, 2010)

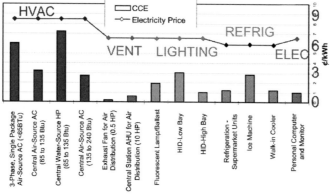

Source: National Commission on Energy Policy, 2004

FIGURE 10. Cost of conserved energy for HVAC, ventilation, lighting, refrigeration, and electronic equipment. The cost of electricity is based on marginal costs, varying between 6 and 8.5 cents/kWh.

Cost of Conserved Energy (CCE) is Lower Than Electricity Price for Many Energy Efficiency Increases (Residential, 2010)

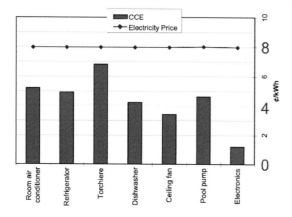

Source: National Commission on Energy Policy, 2004

FIGURE 11. Cost of conserved energy for residential appliances.

EE reduces carbon and saves money

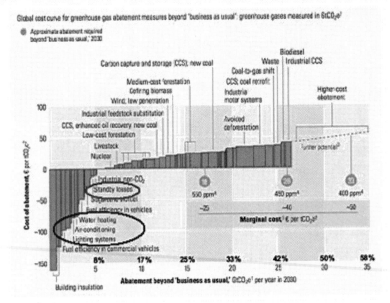

FIGURE 12. The vertical axis is the cost of carbon avoidance in euros per tonne of CO_2 equivalent. The horizontal axis is the percent reduction in global emissions of CO_2 or the equivalent reductions in other greenhouse gases. The circled items are examples of cost-effective energy efficiency improvements. [McKinsey Global Institute, 2007]

The transition to a society in which target reductions in greenhouse gas (GHG) emissions due to energy use in buildings are met is sketched in Figure 13. Energy efficiency progress is taking place, in parallel with progress in sustainable generation of electricity. These two areas, enhanced end-use efficiency and renewable energy sources, will work together to reduce GHG emissions. The vertical bars depict energy demand, which is being reduced over time with improved technologies. The rising curve is the production of sustainable energy produced with little or no net carbon emissions (labeled "on-site energy supply). The triangle that diminishes from left to right represents the amount of purchased energy from conventional energy suppliers, which is being reduced to zero. By 2020, we expect to have zero net-energy new houses that meet aggressive GHG reduction targets.[3]

---

[3] "Zero net-energy" houses produce at least as much energy onsite as they purchase from commercial suppliers.

Efficiency and carbon-neutral supply are complements

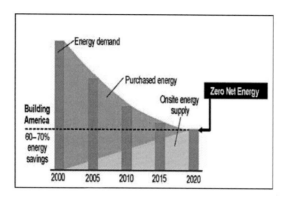

FIGURE 13. The illustrative vertical bars depict energy demand, which is being reduced with improved energy efficiency. The rising triangle is the production of sustainable energy produced without carbon emissions. The triangle that diminishes from left to right is the amount of purchased energy, which is being reduced to zero with reduced demand and renewable energy production.[4]

The Future of Energy Efficiency

Current best practices in new buildings can reduce carbon emissions by at least 70% in homes and 60% in office buildings. This result is from the following government projections and plans:

- DOE's Building America plan has a goal of achieving 70% energy–consumption reduction by 2020 compared to code requirements.

- Leadership in Energy and Environmental Design (LEED) certifies energy reductions by 60% for new commercial buildings compared to code requirements.

- California plans to achieve zero net–energy for new residential buildings in 2020 and for new commercial buildings in 2030.

These ambitious goals assume that we can save 60-70% energy with better technology, controls, and systems integration, which can then be followed with no—or

---

[4] R. Anderson, C. Christensen, and S. Horowitz, "Program Design Analysis using BEopt Building //Energy Optimization Software: Defining a Technology Pathway Leading to New Homes with //Zero Peak Cooling Demand" (Preprint, Conference Paper NREL/CP-550-39827), May 2006.

low—GHG energy for the remaining 30–40% energy. A whole-building approach with photovoltaics can save more than can be saved with improvements in building equipment alone. A National Renewable Energy Laboratory study in 2006 by R. Anderson, C. Christensen and S. Horowitz (Gallagher, 2006) shows that it is possible today to build a relatively large house of 2592 square feet in Sacramento at an incremental cost about 5% above code that can achieve the following:

- Zero peak cooling demand,
- Reduction in annual heating energy by 70%,
- Reduction in annual cooling energy by 60%,
- Reduction in total source energy use by 60%.

Additional savings are possible from the following systems affecting energy use in buildings:

•Efficient *data centers* reduce requirements for electricity and cooling.

•*Digital networks* offer opportunities to maximize comfort and utility while minimizing energy.

•*Combined heat and power* can improve efficiency and reduce carbon emissions.

•Neighborhood systems (e.g., *district heating*) help manage thermal energy.

•*Micro-grids* provide local power, can differentiate power qualities, and can improve reliability.

•*Demand response* incorporates price signals, communicated between utilities and consumers to address the current demand situation.

In discussing buildings, we need to keep in mind that size does matter. We need to pay attention to absolute energy consumption (and GHG emissions), not just efficiency.

Table 2 above indicates that large savings in greenhouse gases, electricity and fuels are possible on a global end-use basis. There are many policy paths that can turn this projection into a reality at the state and national level. Policy responses to climate change include energy efficiency as a major element.

At the federal level, changes in energy efficiency can be affected in a number of ways:

•Labels
--Energy Guide
--Energy Star
•Mandatory Energy Performance Standards
--US
--Others (China, Australia, EU, Canada, Mexico)
•Greenhouse gas cap-and-trade program or taxes or fees
•Tax credits
--To manufacturers
--To consumers

TABLE 2. Global potential for reductions in greenhouse gases, electricity and carbon dioxide that are possible in residential and commercial buildings.

Global Potential of Energy Efficiency Standards and Labeling Programs
- Across all countries, potential to reduce GHG emissions

		Residential	Commercial
Percent change in GHG emissions	From electricity	25%	11%
	From fuels	8%	3%
Cumulative	CO_2 Gt To 2030	1.5 (11)	0.4 (3.6)

Equivalent to 25% of IPCC "zero cost" potential in 2020, 33% in 2030.
The rest can be achieved with building codes, utility programs, incentives, behavior.

M. McNeil, V. Letschert, S. de la Rue du Can, LBNL· personal communication, February 27, 2008
Work in progress for Collaborative Labeling and Appliance Standards Programs (CLASP)

R&D investment in energy efficiency in the US buildings sector has been low relative to the importance of buildings' energy use, but private "Clean Tech" investing is increasing. It was $5.18 billion in 2007, up from $3.6 billion in 2006. The investments focused on

- energy efficiency
- water efficiency
- renewable energy

Annualized returns on clean energy investments in 2007 were high, as evidenced by the fact that Clean Technology Investments (CTIUS) had returns of 43%, much higher than NASDAQ at 10.6% and S&P 500 at 5.5%. Innovation in these areas is expected to increase. There is increasing financial and political pressure for energy efficiency. Greenhouse gas markets were a $30 billion business in 2006, and $80 billion in 2007.

CONCLUSIONS

Energy efficiency has proven itself for thirty years. Many technologies have been shown to be technically feasible and economically justified, and have been widely adopted. Public and private investment to research additional opportunities and to implement known solutions is increasing. Clean technology venture capital investments are up. Energy efficiency is the fastest, most cost-effective option to reduce carbon emissions. Energy efficiency is a renewable resource with the potential to provide most – but not all - of the solution for reducing GHG emissions. Complimentary approaches, such as low-GHG energy supplies and behavioral changes, will be necessary to leverage the energy efficiency gains in order to achieve the necessary aggressive GHG reduction targets.

REFERENCES

•American Solar Energy Society, "Tackling Climate Change in the U.S." (2007).
•K. Gallagher, J Holdren, and A. Sagar, "Energy-Technology Innovation" in *Ann. Rev. Environ. Resources* 31, 193-237 (2006).
•Interlaboratory Working Group on Energy-Efficient and Clean-Energy Technologies, Department of Energy, "Scenarios for a Clean Energy Future" (2000).
•McKinsey Global Institute, "Curbing Global Energy Demand Growth: The Energy Productivity Opportunity" (2007).
•National Research Council, "Energy Research at DOE: Was It Worth It?," National Academy Press, Washington, DC (2001).

Energy Efficient White LEDs for Sustainable Solid–State Lighting

Steve DenBaars

Solid State Lighting and Energy Center
Materials Department
Electrical and Computer Engineering Department
University of California
Santa Barbara, CA 93106

Abstract. The R&D–level white–LED single–lamp luminous efficacy has reached levels as high as 152 lumens/watt at low currents, which greatly exceeds the bare–bulb incandescent lamp efficacy of 17 lumens/watt. Commercial–based white LED lamp fixtures are typically much lower, in the region of 64 lm/W, due to several issues associated with scaling up to higher currents, heat, and optical losses. These issues will be solved by employing higher light extraction, better optical designs, new chip designs, and better heat sinking. The long lifetime of LEDs (100,000 hours) and their higher efficiencies could lead to considerable maintenance and energy–saving benefits for consumers.

INTRODUCTION

Lighting consumes over 20% of all electricity produced in the US. That corresponds to over 25 Quads of primary fuel (with an associated 410 million tons of carbon emissions), at an annual cost of ~$300 billion. Currently, conventional lighting sources include incandescent lamps and fluorescent lamps, which are rather inefficient at converting electricity to light, *i.e.*, 2-4% (7-15 lumens/watt (lm/W)) for incandescent, and 15-20% (50-80 lm/W) for fluorescent. Solid-State Lighting (SSL) using white LED lamps is gaining immense popularity because of its high efficiency 40% (150 lm/W) and the ability to produce bright directional light. LEDs are now essential for backlighting personal cell phones, camera flash, and LCD displays and televisions. LEDs are also valuable in directing public safety with brighter traffic signals, streetlights, and automobile headlights. It is used in architectural lighting, equipment indicators, and other large displays. This chapter will discuss the practicality of LED lighting and how it can significantly increase energy efficiency for homes and businesses. White-light LEDs are 8 times more efficient than incandescent light bulbs in converting energy to light. This higher efficiency can result in substantial amounts of energy savings, which can be a significant contribution in solving the world's current energy crisis. It also helps to reduce greenhouse gases

CP1044, *Physics of Sustainable Energy, Using Energy Efficiently and Producing It Renewably*
edited by D. Hafemeister, B. Levi, M. Levine, and P. Schwartz
© 2008 American Institute of Physics 978-0-7354-0572-1/08/$23.00

produced by the lower energy requirements needed for lighting. For instance, the long LED lifetime of 100,000 hours will lower costs of lamp replacement.

The only barrier that currently slows the entrance of white LEDs into the market is their higher cost. Our research shows us that costs are dropping, and we expect to see white LEDs make significant inroads into the marketplace in the next five years.

To understand how LEDs differ from an average incandescent light bulb, the overall structure of the light emitting diode must be examined critically. LEDs produce light by combining positive and negative charges inside an active layer in the LED chip This active layer is sandwiched between a p–type material, with excess holes, and an n–type material, with excess electrons. A voltage (supplied either by a battery or by a small DC power supply) biases the p–n junction in a forward direction. Holes are then drawn from the p–type material, while electrons are drawn from the n–type material into the active region. (See Figure 1). When both holes and electrons radiatively recombine, light is created. Typically, the device center consists of a multiple quantum well structure, with light-guiding layers and cladding layers on either side to confine the light and the carriers near the active region.

For producing blue LEDs, the active material is gallium nitride. Indium gallium nitride is used with a 3.2–volt battery (or a power supply) applied between the top and bottom of a stack of gallium nitride. As illustrated in Figure 2, members of the nitride family emit all colors. In this figure, photon energy varies between 2 eV in the blue–visible to 6 eV in the deep ultraviolet (650 to 200 nm wavelength).

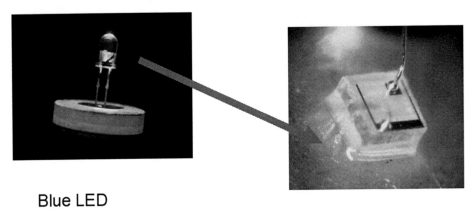

Blue LED

FIGURE 1. Blue LED is composed of a single crystal chip of GaN. When a voltage is applied across the crystal, holes from the p-type layer radiatively combine in the active region with electrons from the n-type layer, producing light.

There exist two primary methods of constructing white light:

- One method is to combine green, blue and red LED's in a ratio that gives white light. However, this approach currently hampered by the relatively weak performance of green LEDs compared to ratio of the red and the blue lights.

- A second approach is to use blue LED's to fluoresce a set of properly–chosen phosphors, which emits yellow light and thus combined creates white light. This approach is viable, producing white light in lamps with a greater efficiency in lumens per watt than compact fluorescents. The only drawback to this superior technology is the cost of the bulb. In some cases, it 50 times more expensive than of that of an incandescent bulb.

Advantages of LED light technology:

- **LEDs considerably have a longer life than ordinary incandescent light bulbs.**
  One LED has a lifetime of 100,000 hours (equivalent to 11 years). This lifespan is significantly longer than an incandescent lamp's lifetime of 1000 hours (1.4 months) and compact fluorescents at 10,000 hours (1.1 years). Longer life reduces unnecessary bulb replacements, waste, and labor. LED technology is important for large companies and governments in executing all aspects of efficiency.

- **LEDs have a robust structure and size, allowing it to sustain functionality.**
  Its unique structure gives the LED a strong advantage in maintaining its primary function of emitting bright light. Since there are no moving parts, no glass, and no filaments, the structure is more durable than other bodies of light. In addition, LEDs are fairly small in size. An average package is 5 mm in diameter.

- **LEDs are extremely energy efficient.**
  Compared to fluorescent and incandescent light bulbs, LEDs radically use 50 to 90% less energy. Thus, LEDs have the potential of significantly decreasing power supply and in turn saving enormous amounts of the nation's energy costs.

- **LEDs are non-toxic.**
  Unlike compact fluorescents, LED lights do not contain mercury and are therefore non-hazardous to the environment and public safety.

- **LEDs are versatile and have fulfilled various society needs.**
  LEDs are available in a variety of colors and can be pulsed.

- **LEDs keep a cooler temperature than other lights.**
  LEDs emit less heat radiation than incandescent or high–intensity discharge lamps.

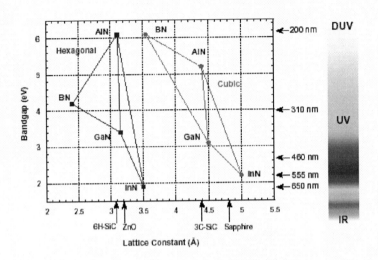

FIGURE 2. Bandwidth versus lattice constant for a variety of nitride compounds and for two different crystalline structures. A mixture of two compounds, such as GaN and InN results in a material whose bandgap and lattice constant fall along the line joining the two compounds. For example, material with a ratio of .1In/.9Ga will emit light with a wavelength of 390 nm. The family spans wavelengths ranging from the green–visible to the deep–ultraviolet.

ENERGY–EFFICIENT WHITE LEDS

The United States alone consumes about 500 billion kWh per year of lighting (which is 22% of the US grid). Because incandescent bulbs are notoriously inefficient, they have been banned in Australia, and will undergo further scrutiny in this country. Currently, conventional lighting sources include incandescent lamps and fluorescent lamps, which are rather inefficient at converting electricity to light, *i.e.*, 2-4% (7-15 lumens/watt (lm/W)) for incandescent, and 15-20% (50-80 lm/W) for fluorescent. The luminous efficacy (proportional to efficiency) of an incandescent bulb depends on the operating conditions and wavelength. A 100–watt incandescent bulb creates some 1700 lumens to give 17 lumens per watt. Since the 100–watt bulb is hotter than a 60–watt bulb, it radiates a larger fraction of energy in the visible region. Therefore, it is no surprise that the 60–watt bulb produces only 12 lumens per watt. In all, the efficiency of an incandescent bulb is alarmingly low.

The evolution of luminosity for various light sources is shown in Figure 3. The incandescent bulb, invented by Thomas Edison in 1879, had improved its efficiency until 1925. Since then, the bulb has remained constant. Fluorescent tubes were invented in 1937 and had improved to an estimate of 100 lumens per watt by 2000. Sodium and metal halide lamps have continued to evolve, but are not suitable for homes and businesses due to their color. Semiconductor lamps were invented in 1965

144

and have recently experienced a spurt of improvement. Current commercially available GaN–based visible optoelectronic devices[1] grown along the polar [0001] *c*-direction utilize active regions composed of InGaN/GaN multi-quantum wells (MQWs). These blue emitting devices are combined with green-yellow and red phosphors to produce white light. Another approach involves the combination of red, green and blue semiconductor LEDs to span the visible spectrum to create white light with both a high theoretical efficiency and full color flexibility.[2]   White LEDs are typically in the range of 50-80 lm/W for mass produced lamps at high currents (350mA). At R&D levels the University of California at Santa Barbara has developed a GaN white-light LED that yields 152 lumens per watt at low currents (20mA).

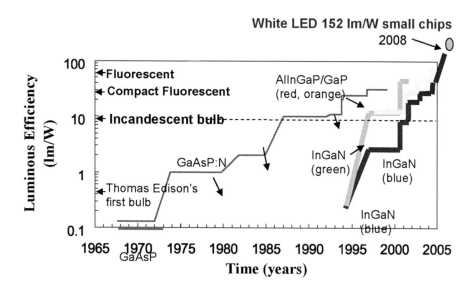

FIGURE 3. Evolution of luminosity for various LED light sources (1965–2007).

The efficacy of a full lighting system is less than the efficacy of the bare bulb. This is because the fixture absorbs some of the light. This effect can be less for some LED lamps since they can be projected in a desired direction. The lighting system efficacy for six different luminaire types is given in Table 1. The bare bulb luminosity is reduced by the fixture efficiency to obtain the useable lumens per watt. Note that fluorescent tubes have a higher luminosity (80 lm/W) than the compact fluorescent (45 lm/W). And the tubes have a higher fixture efficiency (77%) than the compact lamp (33%). This gives a dramatic difference in the usable luminosity with the tubes at 62 lm/W vs. compacts at 15 lm/W. At this point LED lighting fixtures have about the same luminosity of 64 lm/W as the fluorescent tube, but the LED is expected to surpass this value

TABLE 1. Usable lumens per watt for 6 luminaire types. The results are obtained after optical losses, heating effects, and converter losses. Because of the high directionality of LEDs good useable light with high luminous efficacy is produced.

Lighting System Efficacy

Luminaire Type		Lumens Per Watt	Fixture Efficiency	Usable Lumens Per Watt
Halogen Incandescent		17	45%	8
Compact Fluorescent		45	33%	15
150 W Cobra Head Type II Streetlight (HPS)		91	50%	46
400W HID w/Glass Housing (MH)		70	54%	38
XLamp LED Lighting Fixture		71	*90%*	*64*
T8 Fluorescent Tube		80	77%	62

Courtesy Cree Inc

White LEDs should be entering the home and office lighting market. The best white LED lamps in research samples need only seven watts to produce the same light as a 60–watt incandescent bulb. The corresponding compact fluorescent requires about fifteen watts. If a 150 lm/W white LED system were developed and employed in the United States, it would alleviate the need of one hundred and thirty-three new power stations. It would also eliminate the harmful emissions of 258 million metric tons of carbon per year.

Heat can be a problem for LED lamps. A system that operates at 25 °C can produce 100 lm/W, but if it overheats to 60 °C, luminosity can drop to 70 lm/W. There is considerable variation in the LED lighting system thus far produced (Table 2). LED system efficacy from three different companies varied by a factor of four. Such labeling might become more standardized with the institution of Energy Star Ratings for Solid-State Lighting, scheduled for September 2008.

TABLE 2. LED lighting system efficacy vs. system efficacy.

Company	lm/W	fixture efficiency	system efficacy
LED A	80 lm/W	80%	64 lm/W
LED B	45 lm/W	50%	22.5 lm/W
LED C	50 lm/W	30%	15 lm/W
CFL	60 lm/W	50%	30 lm/W

NEW APPLICATION AREAS

LEDs can also aid plant growth by providing the light frequencies that enhance photosynthesis and growth. It is known that chlorophyll has a second distinct absorption peak in the vicinity of 450 nm (blue light region) in addition to the first absorption peak near 660 nm (red light region). Blue light is also indispensable for morphologically healthy growth for plants, while red light contributes to photosynthesis.

Solar energy obtained from photovoltaics can be used to operate LEDs in rural, off-grid situations. The Light-Up-the-World Foundation has deployed these PV/LED systems in the field.[3] Kerosene lighting and firewood are used by 1/3 of the world in rural locations. These fires have caused countless fires, and fires produce light inefficiently at 0.03 lumens/watt. The average villager spends 10 to 25% of his annual income on kerosene. LED lighting costs are effective on an annual basis, with a payback period of only 6 months. Therefore LEDs have paved the way to positive social results and assisting humanitarian issues.

CONCLUSION

Growing research and interest on white LEDs has overwhelmingly advanced within the past couple of years. It began in the mid-1990s, with demonstrations of high-brightness green and blue LEDs by UCSB's Shuji Nakamura, who was involved with Nichia Lab, Japan at the time. The R&D–level white–LED single–lamp efficacy is now at 150 lumens/watt, which greatly exceeds the bare–bulb compact fluorescent efficacy of 50 lumens/watt. Commercial–based white LED lamp fixtures are much lower than 150 ln/W, in the region of 64 lm/W. This is due to several factors that need further research and development. The areas for further research are the fixture efficiency, heat sinking to avoid higher temperatures, and scaling production to obtain the benefits of mass production.

REFERENCES

[1] S. Nakamura and G. Fasol, *The Blue Laser Diode*, Springer, Heidelberg (1997).

[2] J. M. Phillips, M. E. Coltrin, M. H. Crawford, A. J. Fischer, M. R. Krames, R. Mueller-Mach, G. O. Mueller, Y. Ohno, L. E. S. Rohwer, J. A. Simmons, and J. Y. Tsao, "Research challenges to ultra-efficient inorganic solid-state lighting", *Laser & Photonics Review* 1, 307 (2007).

[3] See *Light-Up-the-World Foundation*, www.lutw.org.

Heating, Ventilating, and Air-Conditioning: Recent Advances in Diagnostics and Controls to Improve Air-Handling System Performance

C.P. Wray[a], M.H. Sherman[a], I.S. Walker[a], D.J. Dickerhoff[a]
and C.C. Federspiel[b]

[a]Lawrence Berkeley National Laboratory
1 Cyclotron Road MS 90R3074, Berkeley, California 94720
[b]Federspiel Controls
701 El Cerrito Plaza, El Cerrito, California 94530

Abstract. The performance of air-handling systems in buildings needs to be improved. Many of the deficiencies result from myths and lore and a lack of understanding about the non-linear physical principles embedded in the associated technologies. By incorporating these principles, a few important efforts related to diagnostics and controls have already begun to solve some of the problems. This paper illustrates three novel solutions: one rapidly assesses duct leakage, the second configures ad hoc duct-static-pressure reset strategies, and the third identifies useful intermittent ventilation strategies. By highlighting these efforts, this paper seeks to stimulate new research and technology developments that could further improve air-handling systems.

INTRODUCTION

Although the energy efficiency of many heating, ventilating, and air-conditioning (HVAC) system components in buildings has substantially improved over the past 30 years (e.g., furnaces, chillers, air-handler drives), there is still a need to make other equally critical components more efficient (e.g., the air-distribution system, which delivers heating, cooling, and ventilation air to occupied spaces).

There are multiple problems that lead to poorly-performing air-handling systems. Examples include poor installation quality (e.g., duct joints are poorly sealed, especially downstream of terminal boxes and in exhaust systems); inadequate controls (e.g., unnecessarily closed terminal box dampers restrict flow and result in large pressure differences across the fan); unnecessarily restrictive codes and standards (e.g., requiring constant mechanical ventilation airflows, which can actually reduce indoor air quality in cases such as during periods of high outdoor ozone concentrations); and the lack of reliable diagnostic tools and procedures for system commissioning (e.g., industry-standard duct leakage test procedures cannot be used for ducts downstream of terminal boxes).

CP1044, *Physics of Sustainable Energy, Using Energy Efficiently and Producing It Renewably*
edited by D. Hafemeister, B. Levi, M. Levine, and P. Schwartz
© 2008 American Institute of Physics 978-0-7354-0572-1/08/$23.00

Together, myths and lore about system performance and a lack of understanding about the non-linear physical principles embedded in air-handling system technologies contribute to many of these problems. For example, a common myth is that air leaking from variable-air-volume (VAV) supply ducts located in the ceiling return plenums of a large commercial building does not matter, because the ducts are inside the building. In fact, however, the ducts are outside the conditioned space, the leakage short-circuits the air distribution system, supply fan airflow increases to compensate for the undelivered thermal energy, and fan power increases considerably.

By incorporating the physical principles that govern air-handling system performance, a few important efforts related to diagnostics and controls have already begun to solve some of these problems. They provide a solid foundation for further improvements to air-handling system performance in new and existing buildings. Three examples are:

1. A simple technique for rapidly measuring duct leakage in houses[1,2]. Most existing test methods use duct pressurization to precisely measure the aggregated size of leaks (but not the flow through them at operating conditions). Because duct system pressures are not uniform and knowing the location and size of hundreds of individual leaks is practically impossible, there are large uncertainties associated with this measurement technique. The new test method, called DeltaQ, instead measures the change in flow through duct leaks as the pressure across them is changed. The pressure difference changes are created by pressurizing and depressurizing the whole house (including the ducts) using a blower door. This new method provides duct leakage estimates during system operation and uses equipment and techniques already familiar to building technicians.

2. A new, widely-applicable diagnostic-based duct-static-pressure control strategy for VAV systems[3,4]. A correlation between duct pressure and flow is determined from data obtained by measuring these parameters at multiple duct pressures over the fan's operating range, while the terminal box dampers attempt to control flow in response to an approximately constant thermostat setpoint. The goal is to define the regions where all terminal boxes are in control (dampers modulating) and where the boxes are starved (one or more dampers are wide open). The control region correlation can be implemented within an existing direct-digital-control (DDC) system or, when pneumatic controls are present, in a control system overlay. The system can then operate with the terminal box dampers as wide open as possible while still maintaining thermal control.

3. A new dimensionless model of ventilation efficacy[5]. This model allows one to assess whether an intermittent ventilation strategy will provide the same or better indoor air quality compared to the common continuous strategy specified by codes and standards. In particular, using the model, one can determine which intermittent ventilation strategies are effective for ventilation-

related load-shifting and for providing temporary protection against poor outdoor air quality.

By providing highlights of these three efforts, with a particular emphasis on the non-linear physical principles embedded in these technological solutions, this paper seeks to stimulate new research and technology developments that could further improve air-handling system performance.

DELTAQ DUCT LEAKAGE TEST

Field tests in thousands of houses, several hundred small commercial buildings, and a few dozen large commercial buildings suggest that duct leakage is widespread and large. It is often 25 to 35% of system airflow in small commercial buildings, and can be as large as 10 to 25% in houses and large commercial buildings. Consequently, assessing duct leakage is a key factor in determining energy losses from forced air heating and cooling systems. Specifically, for energy calculations, it is the duct leakage airflow to outside of the conditioned space *at operating conditions* that is required. Using the common industry method of duct pressurization and assuming pressures at leaks is not a reliable indicator of determining whole-system leakage flow; instead, leakage flows themselves need to be measured[6].

Just like a building envelope leakage test, the DeltaQ test measures the pressure difference across the envelope while simultaneously measuring the airflow through the blower that is used to change the envelope pressure difference. In the DeltaQ test, pressurization and depressurization tests are performed twice: once with the air handler off and again with the air handler on. This procedure is used because the magnitudes (and for some leaks, the direction) of duct leak airflows are different when the air-handler is off or on. At each envelope pressure difference (ΔP) in the air-handler off and on tests, there are a pair of blower door flow data; the difference between each pair gives the DeltaQ (ΔQ).

Figure 1 shows an example pair of test points for a pressurization test. Return and supply ducts located outside the building envelope connect the air-handler to the conditioned space. The red (upper) text denotes the flows and pressures in the air-handler off condition; the green (lower) text denotes the air-handler on condition. In this case, ΔQ is -80 cfm.

Figure 2 shows a set of blower door flows together with the corresponding envelope pressures for a series of pressurization and depressurization tests. The data for each of the four parts of the test were acquired using a "ramping" technique[2], which uses the blower door to gradually increase the envelope pressure difference from zero to a peak value over a period of about 90 seconds and then gradually decrease the pressure difference back to zero over the following 90 seconds.

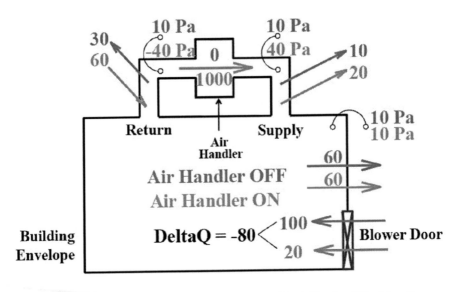

FIGURE 1. Example of DeltaQ Airflows and Pressures with Air-Handler Off and then On.

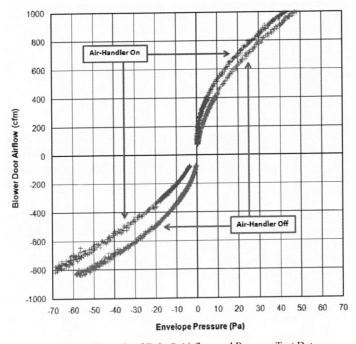

FIGURE 2. Example of DeltaQ Airflow and Pressure Test Data.

152

Converting the test data to duct leakage flows requires the use of the flow-balance-based DeltaQ model[1,7,8] (Eq. 1) and the use of regression routines to determine the model parameters that best fit the measured data (ΔQ and ΔP).

$$\Delta Q(\Delta P) = Q_s \left[\left(1 + \frac{\Delta P}{\Delta P_s} \right)^{n_s} - \left(\frac{\Delta P}{\Delta P_s} \right)^{n_s} \right] - Q_r \left[\left(1 - \frac{\Delta P}{\Delta P_r} \right)^{n_r} + \left(\frac{\Delta P}{\Delta P_r} \right)^{n_r} \right] \quad (1)$$

In particular, the unknowns in Eq. 1 are:

Q_s = the supply duct leakage flow,
Q_r = the return duct leakage flow,
ΔP_s = the characteristic pressure difference between the supply duct and house,
ΔP_r = the characteristic pressure difference between the return duct and house,
n_s = the supply duct leakage pressure exponent, and
n_r = the return duct leakage pressure exponent.

The characteristic pressures, ΔP_s and ΔP_r, sometimes can be measured, typically at the air-handler plenums or some fraction thereof, but in most situations the pressure at the leak site is unknown and it is advantageous to let these pressures be determined by best fits to the ΔQ and ΔP measurements. Analysis of many DeltaQ tests has shown that fitting to the measured data is more robust if the duct leakage pressure exponents are fixed[9,10]. In particular, experiments to characterize the pressure exponents for a wide range of duct configurations have shown that a value of 0.6 is suitable for most duct systems. However, if it is known that most of the leakage is orifice-like or is a disconnected duct, then a pressure exponent of 0.5 is preferred.

Because many data pairs are analyzed for the ramping technique, Walker and Dickerhoff[2] have found that standard least squares routines can take significant time to achieve a solution (several hours in some cases). To reduce the time requirements, they developed a new "pressure scanning" technique that takes 10 seconds or less. This technique applies Eq. 1 to fixed characteristic supply and return pressure combinations. For each supply and return pressure pair, the least squares error is calculated by comparing the estimated ΔQ to the measured ΔQ. The supply and return pressure combination that generates the smallest error is the solution to the DeltaQ equation, together with their corresponding leakage airflows.

Figure 3 shows ΔQ as a function of the envelope pressures. This figure also shows the calculated the fitted ΔQ curve based on Eq. 1. In this case, the ΔQ curve fits the data well and the corresponding leakage estimate is 130 cfm for the supply ducts and 17 cfm for the return ducts. For a system that has a 1000 cfm flow through the air-handler, these results indicate that the supply ducts are leaky (13% leakage) and need to be sealed; in contrast, the return ducts are tight (about 2% leakage).

Field tests have shown that the DeltaQ repeatability depends on both the envelope leakage and weather conditions. In general, a leakier building envelope and windier weather can lead to greater uncertainty. For these tests, the building envelope seems to dominate and a reasonable rule of thumb is that the repeatability is about 1% of the envelope airflow at 50 Pa. Future efforts should focus on field testing this diagnostic in more houses to confirm these preliminary findings, and to extend its applicability to small commercial buildings, which are similar in many respects to houses.

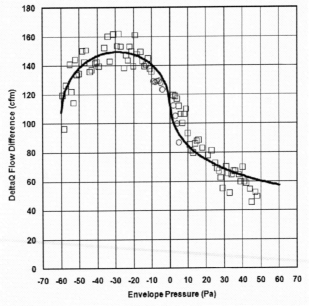

FIGURE 3. Example of Calculated DeltaQ Data and Fitted DeltaQ Line.

DUCT STATIC PRESSURE DIAGNOSIS AND CONTROL

VAV air-handling systems regulate the flow through the fan to maintain a fixed static pressure in the main supply duct. This strategy ensures that terminal boxes have enough pressure to operate properly, but it is inefficient because the duct static pressure will be higher than necessary most of the time. Considerable energy savings can be achieved if the duct pressure is reduced at part load. For example, Lorenzetti and Norford[11] showed that fan energy consumption in VAV systems could be reduced by 19 to 42% with static pressure reset (SPR). Federspiel[3] found that reducing the pressure in response to reduced supply airflow could reduce fan power consumption by 26% and cooling power by 17%.

The SPR concept is not new, but it is still not widely used for several reasons:

1. SPR traditionally has required a networked DDC system with digital controls at terminal boxes, and is susceptible to box control failures. Even today, many systems have pneumatic controls rather than digital controls.

2. Some SPR strategies require terminal box damper position sensors. It is not common, however, for terminal boxes to have these sensors. When present, these sensors may be insufficiently accurate to support an SPR strategy.

3. SPR adds control complexity. DDC systems may not have sufficient bandwidth and programming capabilities to support the additional demands of SPR.

4. Many SPR strategies adjust (reset) the static pressure using a feedback loop that regulates the most-open terminal damper to a nearly open position (e.g., 90% open). SPR strategies that use feedback are difficult to tune.

Engineers have overcome these problems by inventing ad hoc SPR strategies that reset static pressure based on some measurable quantity that is related to the load (e.g., flow, outdoor temperature, time, or a combination of these). Ad hoc reset strategies have the advantage of not requiring digital terminal controls, can be implemented through control overlays, and cannot destabilize the static pressure loop because they do not involve feedback. However, they must still be configured (typically carried out through an iterative trial and error process).

Federspiel[3,4] has recently developed a diagnostic-based method to configure ad hoc SPR strategies so that they can achieve improved performance. The method involves defining two modes of operation. When the duct static pressure is high enough that all of the terminal boxes are meeting the load, the system is operating in the *controlled* mode. When one or more terminal box dampers are 100% open and not meeting the load, the system is in the *starved* mode. The following 2 to 4 hour long test procedure is used to collect data while the system is forced to operate in both modes with fixed thermostat settings:

1.1. Start at a sufficiently high pressure (e.g., the system design pressure) and wait for the terminal box dampers to reach equilibrium (e.g., 15 minutes). Record the flow through the supply fan and the duct static pressure. The velocity pressure near the fan inlet can be used as a surrogate for the flow through the supply fan.

1.2. Reduce the duct static pressure by a small amount (e.g., 25 Pa), wait for the dampers to reach equilibrium again, and then record the time, supply flow, and static pressure. Repeat Step 2 until the supply flow is less than a pre-determined limit (e.g., 70% of the starting flow).

1.3. Increase the pressure by a small amount (e.g., 25 Pa), wait for the dampers to reach equilibrium again, and then record the time, supply flow, and static pressure. Repeat Step 3 until the pressure equals the starting pressure.

An example dataset is shown in Figure 4. The lowest supply duct pressure that keeps all the terminal boxes in control is the critical pressure. This pressure is at the intersection of the controlled and starved data. A dual-mode analytical model can be applied to the test data to determine the critical pressure.

Eq. 2 represents the controlled model for supply flow:

$$Q_c = Q_0 + C_p P^n + C_t \tau \qquad (2)$$

155

The first term (Q_o) on the right-hand side of Eq. 2 is a constant that represents the cumulative flow rate through the terminal box dampers in the controlled mode that is independent of pressure and time. Using the cumulative flow is an important advantage because it makes the resulting SPR strategy more robust in response to individual terminal box failures. The second term accounts for duct leakage upstream of terminal boxes. C_p is a leakage coefficient, P is the duct static pressure, and n is the leak pressure exponent (typically 0.5). The third term accounts for the loads and therefore the supply flow changing over the course of the diagnostic test in response to changing zone temperatures. C_t is the rate of change of the supply flow due to changing load and τ is time.

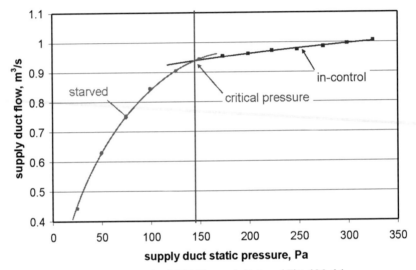

FIGURE 4. Example of SPR Diagnostic Data and Fitted Models.

When the duct static pressure drops below the critical pressure, the flow through the dampers becomes a function of the pressure and Eq. 2 no longer applies. In this starved mode, the supply flow can be modeled using Eq. 3:

$$Q_S = \left(C_0 P^n + C_1 P^{1+n} + C_2 P^{2+n} \right) \left(1 + \frac{C_t \tau}{Q_0} \right) + C_p P^n \tag{3}$$

Eq. 3 has three additional parameters: the polynomial coefficients C_0, C_1, and C_2, which are used to model the flow coefficient as a quadratic function of pressure. The term $C_t \tau / Q_o$ compensates for the fact that only a fraction of the terminal flows (those of unstarved terminals) may be changing with time in response to changing loads.

To determine the various coefficients for each model and then the critical pressure, the following least squares analysis procedure can be applied to the diagnostic data:

    2.1.   Assign the first N high-pressure points at the beginning of the test (from Step 1.2) and the M low-pressure points at the end of the test (from Step

1.3) to the controlled model. Estimate the coefficients in Eq. 2 using least squares. Determine if the time-dependent term can be dropped from the model using a t-test with a decision probability of 0.02.

2.2. Assign the remaining data points to the starved model (Eq. 3). Estimate the three polynomial coefficients of the starved model using the coefficients determined from the controlled model (Step 2.1).

2.3. Compute the variance of the combined residuals.

2.4. Repeat steps 2.1 through 2.3 for all allowable values of N and M, and choose the values of N and M that produce the lowest variance.

2.5. Determine the pressure at which the flow predicted by the starved model (Eq. 3) equals the flow predicted by the controlled model (Eq. 2).

For the data shown in Figure 4, the critical pressure determined using this procedure is 145 Pa. Substantial fan energy savings are possible by implementing SPR for this system, compared to simply operating at the design pressure of about 325 Pa. Specific savings will depend on the pre-control design flow and static pressure, the temporal load distribution, and the efficiency of the fan, motor, and drive.

Future efforts should focus on simulations and field experiments to assess the range of savings that can be achieved using this technique in comparison to the savings associated with conventionally configured ad hoc reset strategies and common zone-level DDC-based strategies.

EFFICACY OF INTERMITTENT VENTILATION

Ventilation is principally used to maintain acceptable indoor air quality (IAQ) by controlling indoor contaminant concentrations. A key step in designing a ventilation system is determining the correct amount of ventilation and the optimal system with which to provide it. There is no shortage of guidance on how much ventilation to use and codes and standards typically require constant ventilation. There are, however, a variety of reasons why the ventilation should not always be constant. For example:

- There may be periods of the day when the outdoor air quality is poor and one wishes to reduce the amount of outdoor air entering the building;
- Economizer operation can excessively ventilate a space from the point of view of indoor air quality; energy savings can be achieved by using lower ventilation rates at other times that take into account the excess ventilation;
- Demand charges or utility peak loads may make it advantageous to reduce ventilation for certain periods of the day; and
- Some HVAC equipment, such as residential or small commercial systems that link ventilation to heating and cooling system operation, may make cyclic operation more attractive than steady-state operation.

If ventilation rate and contaminant concentration were linearly related, the average concentration would be proportional to the arithmetic average ventilation and straightforward methods could be used to determine the effectiveness of intermittent ventilation. Unfortunately, ventilation and concentration are dynamically and inversely related through the mass continuity equation, which leads to a typically non-linear relationship between ventilation and concentration.

Sherman and Wilson[12] and Yuill[13,14] solved the continuity equation for the general case and defined temporal ventilation effectiveness (i.e., efficacy), ε, as a measure of how good a time-varying ventilation pattern is at providing acceptable IAQ. In particular, the efficacy is the ratio of the required constant ventilation rate to the actual ventilation rate; it links the equivalent (or desired) steady-state ventilation air-change rate (A_{eq}), the actual rates of over-ventilation and under-ventilation air-change rates (A_{high} and A_{low}), and the fraction of time that the space is under-ventilated (f_{low}):

$$\varepsilon = \frac{A_{eq}}{f_{low}A_{low} + (1 - f_{low})A_{high}} \qquad (4)$$

If one has an independent measure of the efficacy, it can be used with Eq. 4 to determine the range of acceptable design parameters. When such a measure is unavailable, a dimensionless closed-form expression can be used to determine the efficacy:

$$\varepsilon = \frac{1 - f_{low}^2 N \cdot \coth(N / \varepsilon)}{1 - f_{low}^2} \qquad (5)$$

where the nominal turn-over (N) in Eq. 5 is defined as:

$$N \equiv \frac{(A_{eq} - A_{low}) \cdot T_{cycle}}{2} \qquad (6)$$

T_{cycle} in Eq. 6 is the ventilation strategy cycle time (i.e., the combined period of a repeating high and low ventilation pair), which typically has units of hours. Figure 5 displays the relationship between the dimensionless parameters in Eq. 5.

The nominal turn-over is dimensionless, but we can use Eq. 6 to define a critical time, which is twice the equilibrium turn-over time (or 2 divided by the net equilibrium air change rate):

$$\tau_{critical} \equiv 2 / (A_{eq} - A_{low}) \qquad (7)$$

Variations in ventilation that happen in cycle times that are short compared to this critical time are effectively averaged out, but cycles that are longer than the critical time may be quite inefficient.

When the nominal turn-over is much less than unity (N<<1), the cycle time is short compared to the critical time and the efficacy is always high (close to unity). Almost any pattern of ventilation will work temporally efficiently. The efficacy approaches unity approximately linearly as can be seen by taking that limit of Eq. 5:

$$\varepsilon \approx 1 - f_{low}^2 N^2 \qquad (8)$$

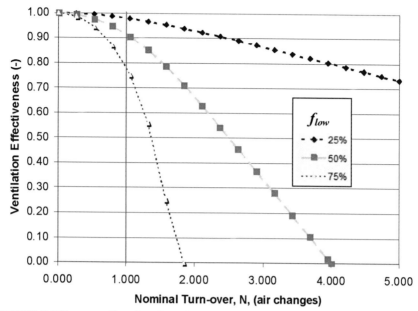

FIGURE 5. Efficacy as a Function of Nominal Turn-Over and Under-Ventilation Time Fractions.

For higher nominal turn-over (N>>1), the cycle time is long compared to the critical time and the effectiveness depends strongly on the fraction of time that the space is under-ventilated. Eq. 5 can be approximated in this regime as follows:

$$\varepsilon \simeq \frac{1 - f_{low}^2 N}{1 - f_{low}^2} \qquad (9)$$

In the latter case, especially for the large under-ventilation fractions (e.g., 75%), the effectiveness drops to zero at some point and certain combinations are simply not possible. That is, one cannot always find an on-cycle air change rate that can be used intermittently to provide ventilation equivalent to the continuous ventilation case.

When there is significant under ventilation, neither of these approximate expressions works very well in the critical regime near N=1.

As an example application of these principles, consider the case where one wants to be able to shut down the ventilation system for a period of time during the day to coast through some period of high cost or high outdoor pollution (called notch ventilation). Figure 6 plots the fractional increase in ventilation equipment capacity that is necessary to accommodate a notch of various sizes. The calculations were carried out using Eq. 5 and 6 for two air change rates (0.5 ach and 1.0 ach) for the case where there was no infiltration during the off period and the case where there was 0.2 ach of infiltration during the off period.

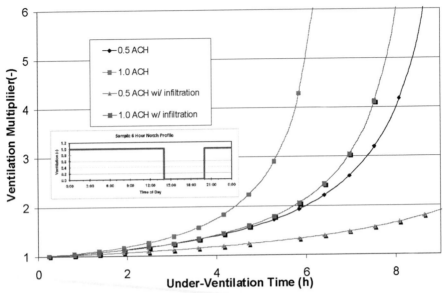

FIGURE 6. Fractional Increase in Ventilation Capacity Needed for an
Under-Ventilation (Notch) Strategy, with and without 0.2 ach of Infiltration.

If the objective is to achieve a substantial period with the ventilation system off
(e.g., at least 6 hours per day) and not to radically increase the size of the ventilation
system (e.g., ventilation multiplier of 2), then the ventilation requirement of the space
should not be more than about 0.5 ach. For low-density spaces (e.g., offices and large
homes), one can achieve this goal; for high density spaces (e.g., conference rooms and
theaters), one may not.

Figure 6 also shows that the presence of infiltration during the notch when the
ventilation system is off can be quite helpful. For example, if the ventilation
requirement is 0.5 ach and there is 0.2 ach of infiltration when the ventilation system
is off, one can have a notch that is almost 8 hours per day for only about a 50%
increase in ventilation system capacity.

Future efforts should focus on non-steady-state and/or multilevel modeling and
experimental verification, as well as specific application development.

SUMMARY

We have illustrated three novel diagnostic and control technologies that allow one
to improve the performance of air-handling systems: one assesses duct leakage, the
second configures an ad hoc duct-static-pressure reset strategy, and the third
determines what intermittent ventilation strategy is equivalent to a constant ventilation
strategy.

The DeltaQ duct leakage test described in this paper has been developed over the
past several years as an alternative to duct pressurization testing. With the addition of
the new ramping technique for obtaining measured data, resolution has been increased

at low envelope pressures and the test is quicker and easier to perform. Future efforts should focus on field testing this diagnostic in more houses to confirm preliminary findings that repeatability is affected more by envelope leakage than by wind-induced pressure fluctuations, and to extend its applicability to small commercial buildings.

The standard way to reset supply duct static pressure duct pressure is to control the most-open terminal damper to a nearly open position. Most systems can't measure terminal damper positions or do not have DDC capabilities, so the pressure is either not reset or suboptimal ad hoc reset strategies are used. The short, simple functional test and data processing technique described in this paper can, however, be used to configure these systems to improve their performance. Future efforts should focus on simulations and field experiments to assess the savings that can be achieved using this technique in comparison to the savings associated with conventional ad hoc reset strategies and common zone-level DDC-based strategies.

Ventilation standards and guidelines rarely address intermittent ventilation, so it is not always clear when one is allowed to average out variable ventilation rates and over what period of time one can do this. The model described in this paper can be used directly to address these issues. For example, there is a critical time for averaging the effects of intermittent ventilation. Any variations that happen faster than that critical time are essentially averaged out; things that happen slower than that critical time do not get averaged out and can lead to poor performance, which means increased costs to provide acceptable ventilation. Future efforts should focus on non-steady-state and/or multilevel modeling and experimental verification, as well as specific application development.

ACKNOWLEDGMENTS

This work summarized in this paper was supported by the Assistant Secretary for Energy Efficiency and Renewable Energy, Office of the Building Technologies Program, U.S. Department of Energy under Contract No. DE-AC02-05CH11231; by the California Institute for Energy Efficiency (CIEE), which is a research unit of the University of California; and by the Energy Innovations Small Grants (EISG) Program, which is part of the California Energy Commission's Public Interest Energy Research Program.

REFERENCES

1. I.S. Walker, D.J. Dickerhoff, and M.H. Sherman, "The DeltaQ Method of Testing the Air Leakage of Ducts", *ACEEE Summer Study Proceedings*, Washington, D.C.: American Council for an Energy Efficient Economy, LBNL 49749, 2002.
2. I.S. Walker and D.J. Dickerhoff, "Field and Laboratory Evaluation of a New Ramping Technique for Duct Leakage Testing", Berkeley, CA: Lawrence Berkeley National Laboratory Report LBNL-62262, 2007.
3. C.C. Federspiel, "Energy-Efficient Air-Handling Controls," Sacramento, CA: California Energy Commission Publication P500-03-052F, 2003.
4. C.C. Federspiel, "Detecting Critical Supply Duct Pressure", *ASHRAE Transactions*, Vol. 111, Part 1, 2005.
5. M.H. Sherman, "Efficacy of Intermittent Ventilation for Providing Acceptable Indoor Air Quality". Berkeley, CA: Lawrence Berkeley National Laboratory Report LBNL-56292, 2004.
6. C.P. Wray, R.C. Diamond, and M.H. Sherman, "Rationale for Measuring Duct Leakage Flows in Large Commercial Buildings", *Proceedings – 26th AIVC Conference*, Brussels, Belgium, LBNL-58252, September 2005.

7. D.J. Dickerhoff, I.S. Walker, and M.H. Sherman, "Validating and Improving the DeltaQ Duct Leakage Test", *ASHRAE Transactions*, Vol. 110, Part. 2, LBNL 53959, 2004.

8. I.S. Walker, M.H. Sherman, and D.J. Dickerhoff, "Reducing Uncertainty for the DeltaQ Duct Leakage Test", *Proceedings - ASHRAE/DOE/BTECC Thermal Performance of the Exterior Envelopes of Buildings IX*, LBNL-53549, 2004.

9. I. Walker, M. Sherman, J. Siegel, D. Wang, C. Buchanan, and M. Modera, "Leakage Diagnostics, Sealant Longevity, Sizing and Technology Transfer in Residential Thermal Distribution Systems, Part II", CIEE Residential Thermal Distribution Systems Phase VI Final Report, Berkeley, CA: Lawrence Berkeley National Laboratory Report LBNL 42691, December 1998.

10. J. Siegel, J. McWilliams, and I. Walker, "Comparison Between Predicted Duct Effectiveness from Proposed ASHRAE Standard 152P and Measured Field Data for Residential Forced Air Cooling Systems", Berkeley, CA: Lawrence Berkeley National Laboratory Report LBNL-50008, 2002.

11. D.M. Lorenzetti and L. K. Norford, "Pressure Setpoint Control of Adjustable Speed Fans", *Journal of Solar Energy Engineering*, 116, 1994.

12. M.H. Sherman and D.J. Wilson, "Relating Actual and Effective Ventilation in Determining Indoor Air Quality", *Building and Environment* 21 (3/4), 1986.

13. G.K. Yuill, "The Variation of the Effective Natural Ventilation Rate with Weather Conditions", *Proceedings - Renewable Energy Conference '86*. Solar Energy Society of Canada Inc., 1986.

14. G.K. Yuill, "The Development of a Method of Determining Air Change Rates in Detached Dwellings for Assessing Indoor Air Quality", *ASHRAE Transactions*, Vol. 97, Part 2, 1991.

Technologies and Policies to Improve Energy Efficiency in Industry

Lynn Price

Energy Analysis Department
Environmental Energy Technologies Division
Lawrence Berkeley National Laboratory
Berkeley, CA 94720

Abstract. The industrial sector consumes nearly 40% of annual global primary energy use and is responsible for a similar share of global energy-related carbon dioxide (CO_2) emissions. Many studies and actual experience indicate that there is considerable potential to reduce the amount of energy used to manufacture most commodities, concurrently reducing CO_2 emissions. With the support of strong policies and programs, energy-efficient technologies and measures can be implemented that will reduce global CO_2 emissions. A number of countries, including the Netherlands, the UK, and China, have experience implementing aggressive programs to improve energy efficiency and reduce related CO_2 emissions from industry. Even so, there is no silver bullet and all options must be pursued if greenhouse gas emissions are to be constrained to the level required to avoid significant negative impacts from global climate change.

INTRODUCTION TO INDUSTRIAL ENERGY EFFICIENCY

The industrial sector consumes nearly 40% of annual global primary energy use and is responsible for a similar share of global energy-related carbon dioxide (CO_2) emissions. Even so, energy efficiency in industry is a neglected topic by scholarly groups, most likely because industrial energy efficiency is a very broad and complicated topic. Also, there is an underlying assumption that industry has both the financial incentive and technical capability to use energy efficiently, and therefore industrial energy efficiency doesn't require much further study. However, studies and experience suggest otherwise; there is still a large gap between actual and best practice in terms of the implementation of cost-effective energy efficiency measures in industry. This chapter summarizes the status of global industrial energy use and related CO_2 emissions. This is followed with a review of technical solutions for a number of energy-intensive industries, such as steel and cement. Lastly, policy options and progress are discussed, with examples from selected countries.

CP1044, *Physics of Sustainable Energy, Using Energy Efficiently and Producing It Renewably*
edited by D. Hafemeister, B. Levi, M. Levine, and P. Schwartz
© 2008 American Institute of Physics 978-0-7354-0572-1/08/$23.00

Between 1971 and 2004, industry's share of global primary energy (which includes the energy consumed to generate and distribute secondary energy such as electricity and petroleum products) dropped from 40% to 37%. During the same period, transportation energy use increased from 18% to 22%, residential building energy use dropped slightly from 30% to 29%, commercial building energy use rose from 9% to 11%, and agricultural energy use remained constant at 3% (see Figure 1). Figure 2 shows CO_2 emissions from the same five end-use sectors (de la Rue du Can and Price, 2008). Industrial energy-related CO_2 emissions were 9.9 Gt CO_2 in 2004, of which direct emissions were 5.1 Gt CO_2 and the remainder were indirect emissions from electricity generation, transmission, and distribution along with other indirect emissions. Industrial energy use and energy-related CO_2 emissions have grown rapidly in developing countries, where they increased from 18% of global emissions in 1971 to 53% in 2004 (Bernstein et al., 2007).

Global Primary Energy Use by Sector, 1971-2004

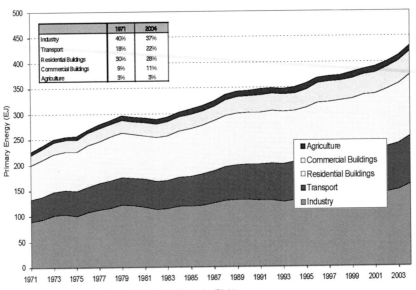

Source: de la Rue du Can and Price, 2008; Price et al., 2006, based on IEA data.
Primary energy includes energy used to produce electricity and heat. Biomass energy included.

FIGURE 1. Global primary energy use by the industrial, transportation, residential buildings, commercial buildings, and agricultural end-use sectors (1971–2004).

The share of energy consumption by end-use sectors is plotted for the world, the U.S., China, and California in Figure 3 (de la Rue du Can and Price, 2008; Murtishaw et al., 2005; Price et al., 2006; US EIA, 2007; NBS, 2005). Due to the high level of demand for energy for both construction of its own infrastructure and manufacturing products for global markets, industrial energy use in China is over 60% of total primary energy consumption, compared to the world (39%), the US (33%) and California (15%).

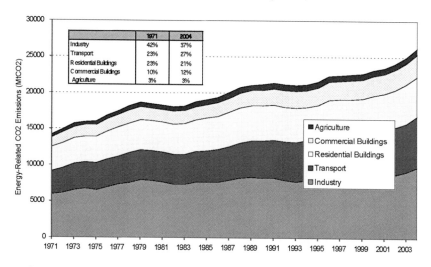

Energy-Related Carbon Dioxide Emissions by Sector, 1971-2004

	1971	2004
Industry	42%	37%
Transport	23%	27%
Residential Buildings	23%	21%
Commercial Buildings	10%	12%
Agriculture	3%	3%

Source: de la Rue du Can and Price,2008; Price et al., 2006, based on IEA data.

FIGURE 2. Energy-related carbon dioxide emissions of the industrial, transportation, residential buildings, commercial buildings, and agricultural end-use sectors (1971–2004).

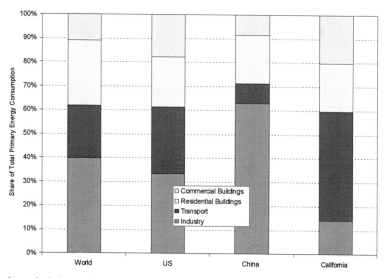

Note: industry includes agriculture
Sources: de la Rue du Can and Price, 2008; Murtishaw et al., 2005; Price et al., 2006; US EIA, 2007; NBS, 2005

FIGURE 3. Energy consumption by sector for the world, the US, China and California.

Additional detail on the energy used for specific industrial sub-sectors is provided in Figure 4. While the production of chemicals, petrochemicals, and primary metals such as steel dominate in the U.S. and China – as well as worldwide – California's industrial energy use is consumed for the production of non-metallic minerals like cement, as well as food and beverages, and electric and electronic equipment such as semiconductors that make up a large share of the "other" sector shown in the figure (IEA, 2007; Murtishaw et al., 2005).

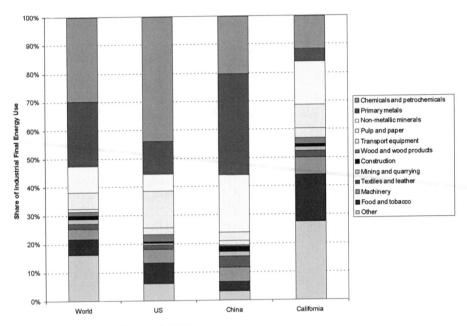

Sources: IEA, 2007; Murtishaw et al., 2005

FIGURE 4. Industrial energy consumption for 12 sub–sectors for the world, the US, China and California.

Potential to Save Energy

The conventional wisdom is that industry is already relatively energy efficient. Many studies and actual experience, however, indicates that there continues to be significant potential to reduce the amount of energy used to manufacture most commodities. The savings potential estimated by the International Energy Agency for five industrial subsectors is considerable: 13% to 16% for chemicals and petrochemicals, 9% to 40% for iron and steel, 11% to 40% for cement, 15% to 18% for pulp and paper, and 6% to 8% for aluminum (IEA, 2007; IPCC, 2007). In addition to sector-specific energy efficiency opportunities, there are also potential savings from improvements that are common to many industries such as motor and steam systems, increased use of combined heat and power, process integration, increased recycling, and energy recovery.

A recent study of the potential for improving the energy efficiency of industry in California identified 56 electricity and 36 natural gas energy–efficiency technologies and measurse for California's manufacturing sector. These measures are estimated to have an economic potential of saving 4.4 million metric tons of CO_2 (MtCO2) through 2016 (2 Mt CO_2 from electricity savings and 2.4 Mt CO_2 from natural gas savings), which represents savings of 15% of electricity and 13% of natural gas from their present baseline use (KEMA, 2006).

CO_2 emissions from the industrial sector in the U.S. can be reduced by between 10% and 29% below a business-as-usual baseline using policies to improve industrial energy efficiency through increased implementation of efficient practices and technologies such as preventative maintenance, pollution prevention and waste recycling (e.g. steel, aluminum, cement, and paper), process control and management, steam distribution system upgrades, improved energy recovery, cogeneration (CHP), and drive system improvements. A large share of the efficiency improvements can be achieved by retiring old process equipment and replacing it with the state-of-the-art equipment, especially for many capital-intensive industries (IWG, 2000; Worrell and Price, 2001).

Both the U.S. Department of Energy (DOE) and the U.S. Environmental Protection Agency (EPA) have a number of programs designed to provide information to various industrial sectors regarding energy efficiency improvement opportunities. The U.S. EPA has published a number of guidebooks which identified 90 energy-saving technologies and measures for the petroleum refining industry (Worrell and Galitsky, 2005), 102 for pharmaceuticals manufacturing (Galitsky et al., 2005), 150 for food processing (Masanet et al., 2007), 40 for cement manufacturing (Worrell and Galitsky, 2004), 114 for glassmaking (Worrell et al., 2007), 45 for breweries (Galitsky et al., 2003), and 93 for vehicle assembly (Galitsky and Worrell, 2003). DOE's Industrial Technologies Program provides many software tools for assessing energy efficiency of motors, pumps, compressed air systems, process heating and steam systems, as well as Sourcebooks that provide information on these industrial systems and a Quick Plant Energy Profiler software tool that helps industrial plant personnel understand how energy is being used at their plant and how they may save energy and money. Fact sheets or brochures contain information on energy efficiency methods, technologies, processes, systems and programs, or provide results from demonstration projects or annual reports. The DOE also provides case studies that describe energy-efficiency demonstration projects in operating industrial facilities in the aluminium, chemicals, forest products, glass, metal casting, mining, petroleum, steel, cement, textiles, and other sectors and tip sheets, technical fact sheets and handbooks, and market assessments for industrial systems. A recent DOE report identified about 90 new technologies for aluminum, chemicals, forest products, glass, metal casting, plastics, mining, petroleum refining, steel (U.S. DOE, 2007).

The Intergovernmental Panel on Climate Change (IPCC) recently summarized the available options for reducing greenhouse gas emissions in the industrial sector in the following categories: energy efficiency, fuel switching, power recovery, renewable energy sources, feedstock change, product change, material efficiency, non–CO_2

greenhouse gases and CO_2 sequestration (see Table 1). The IPCC estimates that the potential to reduce emissions in 2030 is 2.0 to 5.1 gigatonnes (Gt) CO_2-equivalent/year at a cost of $100 per ton of CO_2-equivalent compared to a relatively low emissions business-as-usual scenario. Much of this potential is available at lower costs and is found in the steel, cement, and pulp and paper industries (Bernstein et al., 2007).

TABLE 1. Selected Examples of Industrial Greenhouse Gas Emission Mitigation Measures.
[Source: IPCC, 2007].

Sector	Energy efficiency	Fuel switching	Power recovery	Renewables	Feedstock change	Product change	Material efficiency	Non-CO_2 GHG	CO_2 sequestration
Sector wide	Benchmarking; Energy management systems; Efficient motor systems, boilers, furnaces, lighting and HVAC; Process integration	Coal to natural gas and oil	Cogeneration	Biomass, Biogas, PV, Wind turbines, Hydropower	Recycled inputs				Oxy-fuel combustion, CO_2 separation from flue gas
Iron & Steel	Smelt reduction, Near net shape casting, Scrap preheating, Dry coke quenching	Natural gas, oil or plastic injection into the BF	Top-gas pressure recovery, Byproduct gas combined cycle	Charcoal	Scrap	High strength steel	Recycling, High strength steel, Reduction process losses	n.a.	Hydrogen reduction, Oxygen use in blast furnaces
Non-Ferrous Metals	*Inert anodes*, Efficient cell designs				Scrap		Recycling, thinner film and coating	PFC/SF₄ controls	
Chemicals	Membrane separations, Reactive distillation	Natural gas	Pre-coupled gas turbine, Pressure recovery turbine, H₂ recovery		Recycled plastics, biofeedstock	Linear low density polyethylene, high-performance Plastics	Recycling, Thinner film and coating, Reduced process losses	N₂O, PFCs, CFCs and HFCs control	Application to ammonia, ethylene oxide processes
Petroleum Refining	Membrane separation, Refinery gas	Natural gas	Pressure recovery turbine, hydrogen recovery	Biofuels	Bio-feedstock		Increased efficiency transport sector	Control technology for N₂O/CH₄	From hydrogen production
Cement	Precalciner kiln, Roller mill, *fluidized bed kiln*	Waste fuels, Biogas, Biomass	Drying with gas turbine, power recovery	Biomass fuels, Biogas	Slags, pozzolanes	Blended cement *Geo-polymers*		n.a.	O₂ combustion in kiln
Glass	Cullet preheating Oxyfuel furnace	Natural gas	*Air bottoming cycle*	n.a.	Increased cullet use	High-strength thin containers	Re-usable containers	n.a.	O₂ combustion
Pulp and Paper	Efficient pulping, Efficient drying, Shoe press, Condebelt drying	Biomass, Landfill gas	*Black liquor gasification combined cycle*	Biomass fuels (bark, black liquor)	Recycling, Non-wood fibres	Fibre orientation, Thinner paper	Reduction cutting and process losses	n.a.	O₂ combustion in lime kiln
Food	Efficient drying, Membranes	Biogas, Natural gas	Anaerobic digestion, Gasification	Biomass, Biogas, Solar drying			Reduction process losses, Closed water use		

Industrial Energy Efficiency Policies and Programs[1]

Barriers to the implementation of industrial sector energy efficiency and greenhouse gas emission mitigation options include slow capital stock turnover, a lack of willingness to invest in efficiency and mitigation options, lack of information and high transaction costs, profitability barriers, lack of skilled personnel to install the measures, and other market barriers. Policies and programs designed to address these barriers and encourage adoption of energy efficiency and emissions mitigation options

---

[1] Some of the material in this section has been previously published in McKane et al., 2007.

include regulations and standards, energy and/or CO_2 taxes, emissions trading, agreements and target-setting, emissions reporting, benchmarking, audits or assessments, and information dissemination and demonstration.

Target-setting agreements, also known as voluntary or negotiated agreements, have been used by a number of governments as a mechanism for promoting energy efficiency within the industrial sector. A recent survey of such target-setting agreement programs identified over 20 energy efficiency or GHG emissions reduction voluntary agreement programs in 18 countries, including countries in Europe, the U.S., Canada, Australia, New Zealand, Japan, South Korea, and Chinese Taipei (Taiwan) (Price, 2005).

International best practice related to target-setting agreement programs involves establishment of a coordinated set of policies that provide strong economic incentives as well as technical and financial support to participating industries. Effective target-setting agreement programs are based on signed, legally-binding agreements with realistic long-term (typically 5-10 year) targets, require facility- or company-level implementation plans for reaching the targets, require annual monitoring and reporting of progress toward the targets, include a real threat of increased government regulation or energy/GHG taxes if targets are not achieved, and provides effective supporting programs to assist industry in reaching the goals outlined in the agreements.

The essential steps for reaching a voluntary agreement are the assessment of the energy-efficiency potential of the industrial facility as well as target-setting through a negotiated process. Participation by industries is motivated through the use of both incentives and disincentives. Supporting programs and policies, such as facility audits, assessments, benchmarking, monitoring, information dissemination, and financial incentives all play an important role in assisting the participants in understanding and managing their energy use and GHG emissions in order to meet the target goals. Some of the more successful voluntary agreement programs include trading as well as the use of a mechanism to reduce environmental regulations or taxes for participants.

Netherlands – Long-Term Agreements and Energy Benchmarking Covenants

In The Netherlands, voluntary agreements – called the Long-Term Agreements (LTAs) -- between the government and industrial sectors consuming more than 1 petajoule (PJ) per year were established in support of achieving an overall national energy-efficiency improvement target of a 20% reduction in energy efficiency between 1989 and 2000. Each industry association signed an agreement with the Dutch Ministry of Economic Affairs committing that industry to achieve specific energy efficiency improvements by 2000. In total, 29 agreements were signed involving about 1000 industrial companies and representing about 90% of industrial primary energy consumption in The Netherlands. The average target was a 20% increase in energy efficiency over 1989 levels by 2000. The LTA program ended in 2000 with an average improvement in energy efficiency of 22.3% over the program period (Nuijen, 1998; Kerssemeeckers, 2002; Ministry of Economic Affairs, 2001).

Evaluations of the LTAs found that the agreements helped industries to focus attention on energy efficiency and find low-cost options within commonly used investment criteria (Korevaar et al., 1997). Various support measures were implemented within the system of voluntary agreements (Rietbergen, et al., 1998). It is difficult to attribute the energy savings to a specific policy instrument; rather, it is the result of a comprehensive effort to increase implementation and development of energy-efficient practices and technologies in industry by removing or reducing barriers. This emphasizes the importance of offering a package of measures that includes financial, technical, and informational assistance instead of a set of individual measures. A recent evaluation calculated that the cost of the LTAs was about $10-$20 per tonne of CO_2 reduced, depending upon whether full costs of all subsidies are included (Blok et al., 2004).

Following the LTAs, the Dutch government established a second LTA program – referred to as the Long-Term Agreements 2 (LTA2) program – for smaller businesses and industry. The LTA2 program, which runs from 2001 to 2012, differs from the first LTAs in that the LTAs were a voluntary agreement between Ministries and sectors, while the LTA2s are an agreement between individual businesses, sectors, and competent authorities. The energy-efficiency target for a business or sector is set based on the results of an independent research assessment. A 2005 evaluation of the program indicated that 34 sectors were participating, representing a total of 906 companies. The industrial companies participating in this program achieved an energy efficiency improvement of 19.1% compared to 1998 (the reference year) (SenterNovem, 2006). The energy efficiency improvements made by these companies during the 2001-2004 period were equivalent to an emissions reduction of 2.8 Mt CO_2 (SenterNovem, 2005).

United Kingdom Climate Change Agreements

In 2000, the United Kingdom Climate Change Program was established to meet the country's Kyoto Protocol commitment of a 12.5% reduction in GHG emissions by 2008-2012 relative to 1990 and a domestic goal of a 20% CO_2 emissions reduction relative to 1990 by 2010 (DEFRA, 2006). The Climate Change Levy -- an energy tax applied to industry, commerce, agriculture, and the public sector – is a key element of this program. The revenues from the levy are returned to the taxed sectors and used to fund programs that provide financial incentives for adoption of energy efficiency and renewable energy (DEFRA, 2004). Through participation in the Climate Change Agreements (CCAs), energy-intensive industrial sectors established energy efficiency improvement targets and companies that meet their agreed-upon target are given an 80% discount from the Climate Change Levy. There are 44 sector agreements representing about 5,000 companies and 10,000 facilities. Companies that exceed their targets will have excess carbon allowances which they are allowed to trade with companies that do not meet their targets through the UK Emissions Trading Scheme (DEFRA, 2005a).

Table 2 shows that during the first target period (2001-2002) total realized reductions were nearly three times higher than the target for that period (Future Energy Solutions, 2004). Industries underestimated what they could achieve via energy efficiency. When negotiating the targets, most companies believed that they were already energy-efficient, but when they actually managed energy because of the CCA targets, companies saved more than they thought that they could, especially through improved energy management (Pender, 2004). Industry realized total reductions that were more than double the target set by the government during the second target period and that were nearly double the target during the third target period (DEFRA, 2005b; Future Energy Solutions, 2005; DEFRA, 2007). Industry is saving over $832 million/year on the avoided energy costs as a result of meeting the CCA targets, in addition to the savings on the Climate Change Levy itself.

TABLE 2. Results of the UK Climate Change Agreements: Periods 1-3. Source: DEFRA, 2007. Note that adjustments to the target have been made due to significant changes in the steel sector; see referenced material for details

Absolute Savings from Baseline	Actual ($MtCO_2$/year)	Target ($MtCO_2$/year)	Actual minus Target ($MtCO_2$/year
Target Period 1 (2001-2002)	16.4	6.0	10.4
Target Period 2 (2003-2004)	14.4	5.5	8.9
Target Period 3 (2005-2006)	16.4	9.1	7.3

China – Top-1000 Energy-Consuming Enterprises Program

Between 1980 and 2000, China's energy efficiency policies resulted in a decoupling of the traditionally linked relationship between energy use and gross domestic product (GDP) growth, realizing a four-fold increase in GDP with only a doubling of energy use. However, during China's transition to a market-based economy in the 1990s, many of the country's energy efficiency programs were dismantled and between 2002 and 2005 China's energy use increased significantly, growing faster than GDP. Continuation of this trend in increased energy consumption relative to GDP growth – given China's stated goal of again quadrupling GDP between 2000 and 2020 – will lead to significant demand for energy, most of which is coal-based. The resulting local, national, and global environmental impacts could be substantial.

In 2005, realizing the significance of this situation, the Chinese government announced an ambitious goal of reducing energy consumption per unit of GDP by 20% between 2005 and 2010. One of the key initiatives for realizing this goal is the Top-1000 Energy-Consuming Enterprises program. The comprehensive energy consumption of these 1000 enterprises accounted for 33% of national and 47% of industrial energy usage in 2004 (see Figure 5). Under the Top-1000 program, 2010 energy consumption targets were determined for each enterprise. The goal of the Top-1000 program is for the participating enterprises to save 100 million metric tons of coal equivalent from the expected 2010 energy consumption of these 1000 enterprises.

Reported savings in 2007 indicate that the program is on track to reach this goal, which – if achieved – will save between 250 and 300 Mt CO_2 in 2010, contributing somewhere between 10% and 25% of the savings required to support China's efforts to meet a 20% reduction in energy use per unit of GDP by 2010 (Price et al., 2008).

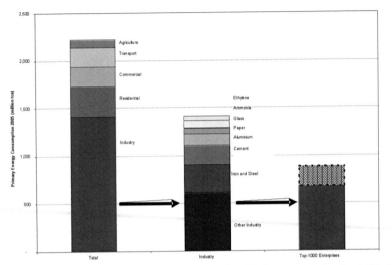

FIGURE 5. Energy Consumption of China, China's Industrial Sector, and the Top-1000 Energy-Consuming Enterprises, 2005. Note: Top-1000 program energy consumption is typically reported in final energy units (dark blue box). The shaded area provides the Mtce equivalent of electricity generation, transmission, and distribution losses so that the Top-1000 program can be compared in primary energy terms with the other two bars. Industry sub-sector breakdown based on LBNL LEAP model, not Chinese statistics.

CONCLUSIONS

While there are no "silver bullets" for improving energy efficiency and reducing greenhouse gas emissions in the industrial sector, it is clear that there are hundreds of emission reduction technologies and measures for industry. The key issue is how to realize significant implementation of these technologies and measures. Industry excels at producing specific commodities, not at saving energy or reducing greenhouse gas emissions. Many policies and programs exist to motivate and assist industries in saving energy and reducing emissions. Some of the most successful programs involve setting clear and ambitious targets and providing government support for industries to reach their goals. Only through continued implementation of energy-efficiency and greenhouse gas emissions mitigation technologies and measures – often spurred by government and industry programs – will the industrial sector be able to contribute its share of the significant level of emissions reductions required to avoid significant negative impacts from global climate change.

REFERENCES

L. Bernstein, J. Roy, K. Delhotal, J. Harnisch, R. Matsuhashi, L. Price, K. Tanaka, E. Worrell, F. Yamba, F. Zhou, S. de la Rue du Can, D. Gielen, S. Joosen, M. Konar, A. Matysek, R. Miner, T. Okazaki, J. Sanders and C. Sheinbaum Parado, *Industry*, Ch. 7, Intergovernmental Panel on Climate Change, *Climate Change 2007: Mitigation of Climate Change*, Cambridge Univ. Press (2007).

K. Blok, H.L.F. de Groot, E.E.M. Luiten and M.G. Rietbergen, *The Effectiveness of Policy Instruments for Energy-Efficiency Improvements in Firms: The Dutch Experience*, Dordrecht, The Netherlands, Kluwer Academic Publishers (2004).

S. de la Rue du Can and L. Price, "Sectoral Trends in Global Energy Use and Greenhouse Gas Emissions," *Energy Policy* 36(4), 1386-1403 (2008).

Department of Environment, Food, and Rural Affairs (DEFRA), *Climate Change Agreements: The Climate Change Levy* (2004), http://www.defra.gov.uk/environment/ccl/intro.htm.

Department of Environment, Food, and Rural Affairs (DEFRA), *UK Emissions Trading Scheme* (2005a), http://www.defra.gov.uk/environment/climatechange/trading/uk/index.htm.

Department of Environment, Food, and Rural Affairs (DEFRA), *News Release: Industry Beats CO_2 Reduction Targets* (21 July 2005b), http://www.defra.gov.uk/news/2005/050721b.htm.

Department of Environment, Food, and Rural Affairs (DEFRA), *Climate Change: The UK Programme* (2006), http://www.defra.gov.uk/environment/climatechange/uk/ukccp/pdf/ukccp06-all.pdf.

Future Energy Solutions, AEA Technology, 2005, *Climate Change Agreements – Results of the Second Target Period Assessment*, Version 1 (2005), http://www.defra.gov.uk/environment/climatechange/uk/business/ccl/pdf/cca-jul05.pdf.

C. Galitsky and E. Worrell, Energy Efficiency Improvement and Cost Saving Opportunities for the Vehicle Assembly Industry: An ENERGY STAR Guide for Energy and Plant Managers, Berkeley, CA, Lawrence Berkeley National Laboratory (LBNL-50939) (2003).

Future Energy Solutions, AEA Technology, *Climate Change Agreements – Results of the First Target Period Assessment*. Version 1.2 (2004), http://www.defra.gov.uk/environment/climatechange/uk/business/ccl/pdf/cca-aug04.pdf.

Future Energy Solutions, AEA Technology, 2005. *Climate Change Agreements – Results of the Second Target Period Assessment*. Version 1 (2005), http://www.defra.gov.uk/environment/climatechange/uk/business/ccl/pdf/cca-jul05.pdf

C. Galitsky, E. Worrell, N. Martin and B. Lehman, *Energy Efficiency Improvement and Cost Saving Opportunities for Breweries*. Berkeley, CA: Lawrence Berkeley National Laboratory (LBNL-50934) (2003), http://www.energystar.gov/ia/business/industry/LBNL-50934.pdf.

C. Galitsky, S. Chang, E. Worrell and E. Masanet, Energy Efficiency Improvement and Cost Saving Opportunities for the Pharmaceutical Industry: An ENERGY STAR Guide for Energy and Plant Managers. Berkeley, CA: Lawrence Berkeley National Laboratory (LBNL-57260) (2005), http://www.energystar.gov/ia/business/industry/LBNL-57260.pdf

Interlaboratory Working Group on Energy-Efficiency and Clean Energy Technologies, *Scenarios for a Clean Energy Future*. Oak Ridge, TN, Oak Ridge National Laboratory and Berkeley, CA, Lawrence Berkeley National Laboratory (2000).

Intergovernmental Panel on Climate Change, *Climate Change 2007: Mitigation of Climate Change*. Cambridge University Press (2007).

International Energy Agency, *Tracking Industrial Energy Efficiency and CO_2 Emissions*, Paris, IEA (2007).

KEMA, with assistance from Lawrence Berkeley National Laboratory and Quantum Consulting, *California Industrial Existing Construction Energy Efficiency Potential Study*. Oakland, CA, KEMA (2006).

M. Kerssemeeckers, *The Dutch Long-Term Voluntary Agreements on Energy Efficiency Improvement in Industry*, Utrecht, The Netherlands: Ecofys (2002).

E. Korevaar, J. Farla, K. Blok and K. Schulte Fischedick, *A Preliminary Analysis of the Dutch Voluntary Agreements on Energy Efficiency Improvement*, The Energy Efficiency Challenge, Proc. 1997 ECEEE Summer Study, Splinderuv Mlyn, Czech Republic (9-14 June 1997).

E. Masanet, E. Worrell, W. Graus and C. Galitsky, Energy Efficiency Improvement and Cost Saving Opportunities for the Fruit and Vegetable Processing Industry: An ENERGY STAR Guide for Energy and Plant Managers, Berkeley, CA, Lawrence Berkeley National Laboratory (LBNL-59289) (2007), http://ies.lbl.gov/iespubs/LBNL-59289.pdf.

A. McKane, L. Price and S. de la Rue du Can, *Policies for Promoting Industrial Energy Efficiency in Developing Countries and Transition Economies*, Vienna, United Nations Industrial Development Organization (2007).

Ministry of Economic Affairs, *Long-Term Agreements on Energy Efficiency: Results of LTA1 to Year-End 2000*, The Hague, Ministry of Economic Affairs (2001).

S. Murtishaw, L. Price, S. de la Rue du Can, E. Masanet, E. Worrell and J. Sathaye, *Development of Energy Balances for the State of California*. Sacramento, CA, California Energy Commission (500-2005-068) (2005).

National Bureau of Statistics, *China Statistical Yearbook 2004*, Beijing, NBS (2005).

W. Nuijen, "Long Term Agreements on Energy Efficiency in Industry," in Martin et al., (eds.) *Industrial Energy Efficiency Policies: Understanding Success and Failure*, Proceedings of a Workshop Organized by the International Network for Energy Demand Analysis in the Industrial Sector, Utrecht, The Netherlands (11-12 June 1998) (LBNL-42368) (1998), http://ies.lbl.gov/iespubs/42368.pdf.

M. Pender, *UK Climate Change Agreements*. Presentation at the Workshop on Industrial Tax and Fiscal Policies to Promote Energy Efficiency (24 May 2005) (2004), http://ies.lbl.gov/mariepender.

L. Price, "Voluntary Agreements for Energy Efficiency or Greenhouse Gas Emissions Reduction in Industry: An Assessment of Programs Around the World," *Proceedings of the 2005 ACEEE Summer Study on Energy Efficiency in Industry*. Washington, DC, American Council for An Energy-Efficient Economy (2005), http://ies.lbl.gov/iespubs/58138.pdf.

L. Price, S. de la Rue du Can, J. Sinton, E. Worrell, N. Zhou, J. Sathaye and M. Levine, *Sectoral Trends in Global Energy Use and Greenhouse Gas Emissions*, Berkeley, CA, Lawrence Berkeley National Laboratory (LBNL-56144) (2006).

L. Price and X. Wang, "Constraining Energy Consumption of China's Largest Industrial Enterprises Through the Top-1000 Energy-Consuming Enterprise Program," Proceedings of the 2007 ACEEE Summer Study on Energy Efficiency in Industry. Washington, DC, American Council for An Energy-Efficient Economy (LBNL-62874) (2007).

L. Price, X. Wang and Y. Jiang, China's Top-1000 Energy-Consuming Enterprises Program:Reducing Energy Consumption of the 1000 Largest Industrial Enterprises in China. Berkeley, CA, Lawrence Berkeley National Laboratory (2008).

M. Rietbergen, J. Farla, and K. Blok, "Quantitative Evaluation of Voluntary Agreements on Energy Efficiency," in Martin et al. (eds.), *Industrial Energy Efficiency Policies: Understanding Success and Failure*. Proceedings of a Workshop Organized by the International Network for Energy Demand Analysis in the Industrial Sector. Utrecht, The Netherlands (11-12 June 11-12 1998) (LBNL-42368) (1998), http://ies.lbl.gov/iespubs/42368.pdf.

SenterNovem, *Long Term Agreements on Energy Eff. in The Netherlands: Results for 2004* (2005), http://www.senternovem.nl/mmfiles/3MJAF05.03%20LTA%20Results%20for%202004_tcm24-175780.pdf.

SenterNovem, *Long Term Agreements on Energy Efficiency in The Netherlands: Results for 2005* (2006), http://www.senternovem.nl/mmfiles/2MJAF0638_LTA_Results_for_2005_UK_tcm24-209539.pdf.

U.S. Department of Energy, Industrial Technologies Program, *Energy Technology Solutions: Public-Private Partnerships Transforming Industry*. Washington DC: U.S. DOE (2007), http://www1.eere.energy.gov/industry/bestpractices/pdfs/itp_successes.pdf.

E. Worrell and L. Price, "Policy Scenarios for Energy Efficiency Improvement in Industry," *Energy Policy* 29, 1223-1241 (2001) (LBNL-49138).

E. Worrell and C. Galitsky, *Energy Efficiency Improvement Opportunities for Cement Making: An ENERGY STAR Guide for Energy and Plant Managers*, Berkeley, CA: Lawrence Berkeley National

Laboratory (LBNL-54036) (2004), http://www.energystar.gov/ia/business/industry/LBNL-54036.pdf.

E. Worrell and C. Galitsky, *Energy Efficiency Improvement and Cost Saving Opportunities for Petroleum Refineries: An ENERGY STAR Guide for Energy and Plant Managers,* Berkeley, CA, Lawrence Berkeley National Laboratory (LBNL-56183) (2005).

E. Worrell, C. Galitsky, E. Masanet and W. Graus, Energy Efficiency Improvement and Cost Saving Opportunities for the Glass Industry: An ENERGY STAR Guide for Energy and Plant Managers. Berkeley, CA, LBNL (LBNL-57335) (2007), http://ies.lbl.gov/iespubs/LBNL-57335.pdf.

U.S. Energy Information Administration, *Annual Energy Outlook 2007* (2007), http://www.eia.doe.gov/oiaf/archive/aeo07/pdf/appendixes.pdf.

Safe and Affordable Drinking Water for Developing Countries

Ashok Gadgil

Environmental Energy Technologies Division (Mailstop 90R3058)
Lawrence Berkeley National Laboratory
Berkeley, CA 94720

Abstract. Safe drinking water remains inaccessible for about 1.2 billion people in the world, and the hourly toll from biological contamination of drinking water is 200 deaths mostly among children under five years of age.[1] This chapter summarizes the need for safe drinking water, the scale of the global problem, and various methods tried to address it. Then it gives the history and current status of an innovation ("UV Waterworks™") developed to address this major public health challenge. It reviews water disinfection technologies applicable to achieve the desired quality of drinking water in developing countries, and specifically, the limitations overcome by one particular invention: UV Waterworks. It then briefly describes the business model and financing option than is accelerating its implementation for affordable access to safe drinking water to the unserved populations in these countries. Thus this chapter describes not only the innovation in design of a UV water disinfection system, but also innovation in the delivery model for safe drinking water, with potential for long term growth and sustainability.

SUMMARY

Safe drinking water remains inaccessible for about 1.2 billion people in the world, and the hourly toll from biological contamination of drinking water is 200 deaths mostly among children under five years of age.[1] Furthermore, there is increasing official recognition that the 1.2 billion number is a serious underestimate.[2] This chapter contains a summary of the importance of drinking water quality, the scale of the global problem, and various technical approaches to make water safe for drinking. Then this paper describes, and places in context, a UV-water disinfector developed for poor communities in the developing countries. It disinfects about 1 tonne of water per hour, using about 50 watts of electricity at a cost of 5 cents per tonne of water, enough for daily drinking water use of 2000 people (at 10 liters per person per day). As a result, UV Waterworks offers the first practical means of providing many communities in developing countries with readily accessible disinfected drinking water.

CP1044, *Physics of Sustainable Energy, Using Energy Efficiently and Producing It Renewably*
edited by D. Hafemeister, B. Levi, M. Levine, and P. Schwartz
© 2008 American Institute of Physics 978-0-7354-0572-1/08/$23.00

Specific capabilities of the UV Waterworks device include:

- Works with un-pressurized water
- Deactivates all pathogenic waterborne bacteria and viruses, and also cryptosporidium cysts
- Does not need a skilled operator
- Maintenance is only every three months
- Disinfection is rapid (water passes through the unit in 12 seconds)

Limitations of the device include:

- Does not treat non-biological water pollutants (e.g., chemicals, minerals)
- Does not remove turbidity (if the water is turbid, pre-filtration is needed)
- Does not produce potable water from sewage or wastewater
- Does not provide residual protection – water can be re-contaminated through improper handling
- Disinfection is not suitable for long term water storage (so water should be used within a couple of days of treatment)
- Does not work without electricity. Where grid electricity is unavailable, PV panels have been used without any problems, but raise the cost of disinfection from 20 cents to 1 dollar per person per year (these cost include full annualized costs including electricity, consumables, and amortization.)

Before we discuss the innovation, it is useful to place it in context with brief reviews of drinking water quality, scale of the problem, and recommended standard treatment options for obtaining safe drinking water.

DRINKING WATER QUALITY

WHO recently reported that diarrhea and dysentery claimed the lives of approximately 2 million people in 2005, the vast majority being children under the age of five in developing countries.[1] At any given time, about half the population in the developing world is suffering from one or more of the six main diseases associated with inadequate services for water and sanitation. Because poor people (commonly with meager savings, and without access to good health care, and without benefits such as sick leave) are most likely to rely on poor-quality water, the ill health caused by such water further perpetuates poverty. Unsafe water and lack of sanitation trigger a vicious cycle that hampers economic and social development. The United Nations estimated that in 2005 water-related illnesses cost nearly 450 million days of school and as much as 5 percent of GDP for some developing countries.[3]

Improved longevity, reduced infant mortality, health, productivity, and material well-being are generally recognized as essential measures of development. Developing country populations generally have poor ranking on these indices compared to those of industrial countries. Development literature suggests that availability of plentiful and safe water for domestic use and adequate sanitation to dispose of waste are fundamental to the development process, with benefits, such as labor productivity, spread across all sectors. The history of public health measures in the industrial countries documents the dramatic rise in life-spans that followed the public supply of safe drinking water, country after country.

Different developing countries define differently what constitutes safe drinking water. Recognizing that each country will have its own priorities regarding public investments in development infrastructure, the WHO does not directly recommend enforceable national water quality standards, instead recommending guidelines for drinking water quality. Guidelines are released as a set of three volumes on a rolling basis. The most recent edition (2004) of the first volume of the guidelines contains the actual recommended maximum acceptable values for water contaminants, and the second volume (most recent edition is 1996), Health Criteria, and Other Supporting Information, discusses the detailed reasoning behind the recommendations, for each substance covered in the first volume.

Potable or drinking water is defined as having acceptable quality in terms of its physical, chemical, and bacteriological parameters so that it can be safely used for drinking and cooking. A daily per capita consumption of 2 liters is the generally accepted value for a person weighing 60 kg, however it is recognized that there could be substantial variation in this value depending on the climate, the individual's level of physical exertion, and local cultural norms.[4] A maximum of 8 liters is the expected annual-average base daily consumption for drinking and cooking.[5] Depending on circumstances, a much larger amount of water could come in contact with skin or cooking utensils daily and must also have minimal contamination.

SCALE OF THE GLOBAL PROBLEM

Definitions of "access" to safe drinking water differ substantially since different countries determined them individually. Walking distance or duration (time) from the household to the potable water source is the principal criterion used for defining access (or lack of it), particularly for rural populations. Several of these definitions are very liberal and consider access to mean a one-way walk of up to a kilometer to a water source. Furthermore, the definition of "safe" drinking water itself is left to the individual country. Some countries consider access to exist if an "improved" water source is available within a kilometer of the dwelling. Definitions of "improved" sources include hand-dug wells, and protected springs, with no further water treatment. Even with these definitions about 20 percent of the developing world's population (1.2 billion people) lacked access to safe drinking water according to the WHO recommendations set in 1994, and as noted earlier, this number is a serious underestimate. The problem afflicts both urban and rural areas, both of which may lack effective large-scale water supply infrastructures. As of 2002, the World Health Organization reported that 31 percent of the rural population in the developing world did not have access to safe and convenient sources of water and 32 percent of the world's urban population lived in slums that lacked running water, electricity or even permanent structures.[6]

It should not come as a surprise, then, that UNICEF estimates the effort spent annually in developing countries for fetching water to be 10 million person-years. The toll on human health, productivity and well-being is almost unfathomable. The negative impact for vulnerable populations through lack of access to safe water, which is typically one among several obstacles to development they face, merits a significant investment of resources and expertise to address this issue.

Analysis of World Bank's own experts (e.g., reference 7) shows that in the developing countries the politically powerful and easy-to-supply population segments that are generally the first (and sometimes the only ones) that receive water services, leaving behind the more distant and politically weak ones. The investments, and the political will to make them, will be harder to come by for the hardest to reach of the unserved population segments, thus slowing the rate of progress as countries approach their goal for water coverage for all. Lastly, appropriate water supply and management technologies for city slums and distant rural villages are likely to be quite different, and need to be conceptualized differently. We can expect significant changes to local water resources in many places from global climate change and resulting weather variability in the coming decades. This may change agricultural runoff, sedimentation, erosion, and ground water recharge rates.

Already in parts of the developing world rates of ground water extraction, especially from intensive agricultural irrigation, far exceed the rates of ground water recharge. The resulting drop in the ground water table and water scarcity has implications for health and hygiene and drinking water access. In coastal areas, excessive pumping of ground water causes intrusion of seawater into fresh water aquifers, making well waters too saline for drinking.

In southern Bangladesh and parts of the Indian state of West Bengal, close to a hundred million people obtaining drinking water from shallow tube wells are now exposed to high levels of arsenic. This arsenic has leached into the water from local geology. The geochemical mechanism of this leaching is not fully understood. However, this has led to widespread arsenic poisoning among the exposed people. WHO experts project that, under the default scenario, arsenic poisoning will account for about 10 percent of adult deaths in these populations in the coming decades. This crisis has been correctly called the largest mass poisoning in the history of mankind.[8]

STANDARD WATER TREATMENT TECHNOLOGIES

The developing countries – particularly their rural areas – are a graveyard of technologies developed by well-meaning engineers and scientists in the industrial countries! This fact is recounted and described in many relevant textbooks and reviews on appropriate or intermediate technologies for development. To provide relevant background for the invention of UV Waterworks, it is useful to quickly review what is known about the standard methods of drinking water treatment, and see why we, as human society on the planet, continue to lose 200 children per hour to what seems to be entirely preventable matter of microbial pathogens in drinking water.

Protection and Storage

The most important way to obtain safe drinking water for a community is to protect its source from fecal contamination and to sufficiently isolate it from dumping of household garbage, industrial waste, mining and quarrying activities, and runoff of agricultural fertilizers, herbicides, and pesticides. Literature recognizes that for the major killers—the fecally transmitted pathogens in drinking water supplies—no single disinfection practice is failsafe. Disinfection-resistant pathogen strains exist, and more may emerge in the future. WHO experts therefore recommend multiple barriers to the potential transmission of microbial pathogens in water supplies. Good sanitation practices and adequate methods to dispose of human and animal excrement are thus first necessities. These establish the first barrier between contaminants such as fecal pathogens and the drinking water source. If surface waters are used as a source, appropriate filtration establishes a second barrier and should be considered wherever feasible. The resulting reduced turbidity generally enhances the effectiveness of all disinfection methods. However, if only one barrier is obtainable, WHO recommends that it should certainly be disinfection.

In the real world, the urban or rural poor in the developing countries have little or no control over their source water, nor over provision of adequate sanitation. In fact, it is common knowledge, and well documented by WHO, UNICEF and World Bank's Water and Sanitation monitoring reports that defecation in the open fields is routine practice for a large fraction of the poor in the developing countries. Ironically, so long as the large majority of the community practices open defecation, switching even to pit toilets confers little advantage to the small minority that might begin using them. The change is beneficial to all only when a large fraction of the community participates in the improved practice.

Recontamination of disinfected (or safe) water owing to poor storage practices or owing to dipping unwashed hands in the stored water can be a problem for a number of disinfection methods that leave no post-treatment residual protection (e.g. boiling, reverse osmosis, UV treatment, ozonation). However, small mouthed water storage vessels can effectively eliminate this route of recontamination. Mintz et al[9] report successful introduction into the Bolivian market of a 20-liter narrow-mouthed carrying and storage water vessel with a spigot for withdrawing water. The same vessel can also be used as the container in which a solution of chlorine bleach is added to disinfect the water in a few hours. The vessel sells for about $6, and the cost of the chlorine solution at the local dealer is about $0.40 per month for a typical local family (Robert Quick, personal communication, 2007).

Storage of surface waters in protected reservoirs or in impoundment lakes leads to considerable improvements in microbiological quality of water through predation, settling of bacteria attached to particulates, and the effect of solar UV in the near-surface layer of waters. Protection of the water source is also critical, which is often a challenge in areas where water is used for multiple purposes including watering cattle and washing, or where open human defecation is practiced in the vicinity.

For the urban and rural poor, such impoundment of their source water in protected reservoirs is beyond reach. For the urban poor, living in dense slums with hardly space for foot traffic between the crowded dwellings, space for a protected impoundment reservoir is unthinkable. Surprisingly, the same holds for the most of the rural poor, since they are usually without much (or any) land tenure, and have few opportunities to create or set aside protected reservoirs

Ground Water

A major alternative to impounding and/or filtering surface waters is to tap groundwater. Groundwater is naturally filtered through several meters of soil and rock, and is commonly free of protozoan cysts and larger parasites. It is also commonly free of significant suspended particles (i.e. turbidity), making subsequent disinfection treatment (e.g. chlorine, UV), if desired, more effective. Accessing such ground water requires drilling deep (e.g. 80 m) bore wells that tap deep aquifers containing old ground water that has little organic carbon and usually little biological contamination.

During the 1980s, UNICEF sponsored development of a rugged, inexpensive, and low-maintenance handpump, now known as UNICEF India Mark II. During the late 1990s the cost of a 5-inch-diameter well drilled by a private contractor to 60 meters in hard rock, cased to 10 meters, and fitted with an India Mark II handpump was US$1300. Remarkable advances are being achieved in India with installations of 100,000 deep borewell handpumps annually, with each handpump serving about 200 persons.[10] This publicly funded, impressive advance in drinking water access and quality has caused a massive shift in the last few decades for Indian rural populations from former overwhelming reliance on polluted surface sources to ground water. Average per capita investments required to deliver various kinds of improved or treated water in different parts of the world are reported in the WHO/UNICEF joint assessment of global water supply and sanitation.[11]

The 1990s exposed challenges related to ground water quality which could require introduction of new low-cost treatment options for ground water aquifers contaminated with high levels of arsenic or fluoride, high levels of total dissolved solids (TDS), or that have turned brackish owing to salt-water intrusion. An aggressive effort, installing shallow tube wells (instead of deep ones) in the Indian state of West Bengal and in Bangladesh has led to the previously mentioned calamity of mass arsenic poisoning.[8] UNICEF[12] estimates that 66 million Indians are exposed to toxic levels of fluoride in groundwater. In urban and peri-urban areas, sources of fecal pollution, such as pit toilets and leaks in city sewers, have largely contaminated local groundwater with pathogenic bacteria and viruses. The resulting ground water, while still free of turbidity, protozoan spores, and larger parasites, is contaminated with fecal bacteria and viruses, requiring disinfection.

Filtration

Rapid and slow filters effectively reduce turbidity of source water. High turbidity interferes with the effectiveness of disinfection by chlorine, ozonation, and UV. In addition to substantially removing turbidity, slow sand filters also permit large reductions in bacterial and viral contamination and remove larger biological contaminants (such as cryptosporidium, giardia, amoebae, parasite eggs, etc). Any given design of filter has inherent limits on the level of turbidity and total suspended solids (TSS, in units of mg/l) it can treat. If incoming water exceeds these design parameters, the filter may clog up rapidly and may produce filtrate (i.e. outlet water) with turbidity and TSS exceeding the intended design values. Water with high values for turbidity and TSS requires pretreatment before filtration. Such pretreatment can be obtained with flocculation followed by settling or filtration; however, this requires a much higher level of supervisory skills and supply of appropriate chemicals. With such chemical treatment, electrical charges on suspended fine solids in the water are neutralized and the coagulated particulates removed through settling or filtration.

Chlorination

Chlorine in various forms is the most common disinfectant used worldwide. The chlorine dose is measured in units of concentration times contact time. A chlorine dose of 2 mg/l and 30-minute contact time (in a chlorination holding tank) provides 99.9 percent disinfection of Giardia at $20\pm C$, turbidity of 1 NTU, and pH of 7.[13] The required chlorine dose for disinfection increases sharply with increasing turbidity, increasing pH, decreasing water temperature, and increasing concentrations of ammonia, hydrogen sulfide, Fe, and Mn. The chlorine doses needed over the full range of these water properties would differ by a factor larger than 10. With poor quality water (e.g. high turbidity, high pH), disinfection with chlorine may become impractical because the chlorine dose required may be so high that contact times of tens of hours are needed, or chlorine concentrations may exceed objectionable taste thresholds.

Conventional automated chlorine-dosing plants can apply the right amount of chlorine; however, they require highly trained operators, engineers, and repair and maintenance infrastructure available in and appropriate for only large urban populations. In many smaller communities in developing countries, various solid or liquid chemical forms of chlorine (e.g. bleaching powder [calcium hypochlorite, $Ca(OCl)_2$], or sodium hypochlorite (NaOCl), have been tried since they are safer to transport and handle than chlorine gas, and can be prepared locally. However, they have met with mixed success, primarily because the need for a trained operator drives up the unit costs of chlorine treatment rapidly as the scale of the treatment plant becomes smaller.

The major advantage of chlorine is its ability to leave a residual disinfection concentration in the water supply. Residual free chlorine is the available chlorine remaining in the water after a specified contact period, which can further disinfect small amounts of newly introduced biological contamination. The residual chlorine also suppresses re-growth of nuisance bacteria. A large infusion of pathogens and organic matter (as often happens in depressurized urban water pipes in metropolitan areas of developing countries), however, routinely overwhelms the protection provided by residual chlorine.

The primary disadvantage of chlorine is the necessity to maintain an appropriate supply chain of source chemicals to the water treatment location. Both liquid and powder bleach degrade over time with half lives of the order of weeks to months (depending on storage conditions). Cholera outbreaks have been reported in India when impassable roads blocked the chlorine supply chain during heavy monsoons. Another disadvantage is the need for a skilled and trained operator and a repair and maintenance infrastructure. For large systems (serving cities of 100,000 or more), chlorine disinfection costs are low, approaching about $0.02 per m^3 of water. As mentioned above, with small-scale systems, the costs rapidly increase, as does the impracticality of having skilled technical operators.

Boiling

Most of the world's 2 billion poor who lack access to safe drinking water, also don't have access (physical or economically affordable), to protected impoundment of source waters, slow sand filters, chlorine, and the technical know-how to use these methods. Owing to lack of access to modern scientific knowledge, they also commonly have other local explanations regarding the causes of diarrheal and other waterborne diseases. Boiling disinfects water very effectively and adequately, however, it is not generally appreciated that boiling a person's daily supply of safe water (say, 10 liters) requires about three times the fuel needed to cook a person's daily meals. Most of the world's poor rely on biomass for fuel (wood, twigs, agricultural waste, dry biomass of various kinds including cattle dung). This fuel is either collected by foraging or purchased in local markets. A three-fold increase in the fuel consumption for boiling water on a routine basis by those who don't have access to safe water will be either economically unaffordable (for purchased fuels), or environmentally difficult to support.

Furthermore, WHO estimates that there are 1.6 million excess deaths annually from adverse health effects of cooking smoke inhalation by the cooks and the young children that are kept in the proximity of the mothers as they cook.[14] A three fold increase in biomass burning to boil water will likely increase the smoke inhalation by a similar amount, and lead to a large increase in mortality.

UV Disinfection

Ultraviolet light in the wavelength range 240 to 280 nm has been known, for almost a century, to be germicidal. The germicidal effect occurs because the UV light causes severe damage to the DNA of the microorganisms. The UV dose is measured in microwatt seconds of UV energy (at or close to 260 nm) per sq. cm of water surface. DNA has an absorption peak at about 260 nm, well away from the absorption peaks of proteins and cell membrane; thus UV is able to penetrate the cell and disable the DNA quite effectively. Exposure to UV-C creates covalently bonded photo-adducts between adjacent pairs of Thiamine in the DNA's base sequence, disabling the DNA. Ordinary clear tap water has an extinction coefficient for UV-C of about 0.01/cm, allowing good penetration.

Appropriate filtration or sedimentation before UV treatment can remove larger pathogens (e.g., eggs of helminthes) and also reduce turbidity improving UV transmittance and reducing the shielding of microbial pathogens by particulate matter. A UV-C energy dose of 40 mJ/sq.cm. of water surface is adequate to meet disinfection criteria of WHO, which require a >5-log kill of *E. coli* in the water being treated.

Compared to boiling over a biomass cookstove of 12 percent efficiency, UV disinfection (delivered with a low-pressure mercury plasma discharge) can require 20,000 times less primary energy. In contrast to many of the chemical disinfectants, UV disinfection imparts no taste or odor to the water, and presents no risks from overdosing or formation of carcinogenic disinfection by-products. The very high sensitivity of DNA to UV light allows very short treatment time for the water. In contrast to chlorine (which requires contact times of 30–60 minutes), UV disinfects water in a few seconds. Because it does not have diseconomies of scaling down, the cost of disinfection per cubic meter remains about $0.05 even for small systems.

Since UV does not impart residual disinfection to the water, it is appropriate only under circumstances where the disinfected water will be protected from recontamination and end-users are educated about proper hygiene and water handling. Several thousand municipal water treatment systems using only UV-treatment for disinfection are in operation across Western Europe, delivering water through high quality and high integrity municipal pipe networks.

Enzyme mechanisms exist within several bacterial species that attempt to repair the damaged DNA (although in a slow and error-prone manner). A larger dose of UV energy can so severely damage the pathogens' DNA that they die before their repair mechanisms can restore the DNA. However, even at these energy doses, several non-pathogenic bacteria and algae can survive the treatment. While harmless, they can grow in the treated water (at rates controlled by dissolved nutrients, temperature, available light) and make the water eventually aesthetically unacceptable for drinking. Therefore, UV treatment by itself, at the minimum energy dose required by current standards, is not considered suitable for disinfection of drinking water intended for long-term storage.

Most UV systems use a linear UV lamp, enclosed within a cylindrical coaxial UV-transparent sleeve, submerged in water in a co-axial cylindrical UV-exposure chamber. Water flows axially in the annual space on the outside the quartz sleeve and receives the UV dose. Chemical fouling and biological film (particularly when the lamp is off and the water stagnant during hours of disuse) builds up on the sleeve surface over time. This fouling seriously impairs UV transmittance of the sleeve and necessitates periodic cleaning with acids and mechanical scrubbers. Depending on raw water quality, the cleaning may be needed every few weeks. This makes maintenance complex and expensive and puts it beyond the means of most rural communities. Another limitation of most UV systems is that their design requires the use of a pressurized raw water source (e.g. from a municipal tap). They are not useful for communities collecting water from handpumps and/or surface water sources (e.g. wells, rivers, or lakes).

THE ORIGIN OF UV WATERWORKS

In the summer of 1993, prompted by the outbreak of a mutant strain of cholera ("Bengal Cholera") against which there was no vaccine, we initiated a design effort for a low-cost, robust, and low maintenance device for drinking water disinfection. Using funds from the author's Pew Environmental Fellowship, and $10K each in seed funding from USAID and LBNL's Center for Building Science, we reviewed existing technologies, initiated a design effort, and held a workshop in Washington D.C. in November 1993 to address the issue.

We found that one could disinfect water with a UV dose of 40 mJ/sq.cm. at an attractively low cost of 0.5 cents per metric tonne of water. With a handsome 300% overdose UV energy to guard against malfunctions in the field (e.g., sags in supply voltage, unexpectedly lower UV transmittance of water, incomplete removal of turbidity, reduced UV output from lamps that are not replaced at end of their rated life, reflector degradation), the cost of disinfection is still only 1.5 cents per tonne. Since a drinking water needs for one person (at 10 L per day) is only 3.6 tonnes per year, it was amazing to compare this number to the annual death toll from diarrheal waterborne disease (estimated then at 5 million, mostly children below age 5).

Our goal was to disinfect communities' drinking water collected by hand from surface sources or with handpumps. The water entering the device might have a pressure of only a few cm of water column. Thus, the design was aimed to work with very low pressure drop (about 10 cm of water column), and also without any integrated filter. If filtering is necessary, we reasoned that it should be done outside the device, using a slow sand filter, or an in-line filter cartridge if one had a pressurized line. We circumvented the sleeve fouling problem with a design having a bare UV lamp supported below a reflector, above a free surface of water flowing in a shallow trough. Between the UV lamp and the water there are no solid surfaces on which the algae or chemical deposits can occur. We arranged the hydrodynamics of the flow through the trough to have as narrow a distribution of residence time for water parcels as feasible. We set the maintenance interval (simple cleaning of the *bottom* of the shallow trough) at six months. Our initial design was wholly of welded stainless steel sheet, consumed 40 watts, disinfected 30 liters per minute (lpm), and would cost about $900 at mature production volumes.

After a second workshop in Bhubaneshwar, India, in May 1994, we field-tested this design at several sites in India, including at Bhupalpur, Uttar Pradesh, in 1994-95. The Indian communities informed us that the flow capacity of the device was far higher than necessary – they had no water sources that came anywhere close to 30 lpm – and the device was too bulky and costly. In response, we developed a compact device sacrificing efficiency for size and weight. The flow rate was dropped to 15 lpm, the device is much more compact and has a substantially lower manufacturing cost at mature production volumes. The cost of disinfection climbed to 5 cents per tonne of water – still affordable to those making less than a dollar a day. The unit is designed to treat water with a UV extinction coefficient equal to the average effluent from a U.S. municipal sewage treatment plants. (Water with higher extinction coefficients will absorb more UV energy and thus have reduced disinfection performance unless the flow rate is correspondingly reduced). This design effort, completed in December 1995, exhausted all of the meager project funding. In June of 1996, LBNL concluded a license agreement with WaterHealth International for this technology, selecting it from among more than a dozen applicants for the license from around the world. Since that time, the design has seen several improvements from WaterHealth International's own redesign efforts for reducing production costs, improving product reliability, ease of manufacture and field maintenance. However the basic design remains unchanged from the schematic shown in Figure 1.

185

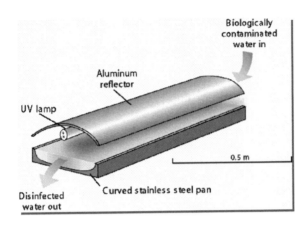

FIGURE 1. Schematic of UV Waterworks interior.

Following the device's receipt of two awards in 1996 (Popular Science's Best of What's New, and Discover Award for the Best Invention of the year in Environment category), the Department of Energy (DOE) made funds available for the first field testing of the 15 lpm device in South Africa. The field test was incorporated into the then ongoing U.S.-South Africa Binational Commission, Sustainable Energy Committee. The Natural Resources Defense Council (NRDC) partnered with LBNL in the field testing, as did the South African Center for Essential Community Services.[15] The first of these field tests was conducted outside Durban at the "Lily of the Valley," a privately-run hospice for HIV-positive orphan-infants, abandoned by HIV-positive mothers who could not care for them. The hospice had no protected water source, and the ground water on which it relied had about 4000 colony forming units (CFUs) of coliform bacteria, including 100 CFUs of thermotolerant coliforms (also commonly called fecal coliforms), per deciliter. Water leaving the unit showed no detectable coliforms (except for brief intervals – see below), during weekly testing throughout the duration of the trial. At the end of the year-long trial, the unit was donated to the hospice, which continued to use it until an year later when the local water utility found resources to extend a pressurized water line to the facility.

The one lesson learned from the South Africa field trials was the need to have a fail-safe unit, so that no water could emerge from the outlet of UV Waterworks if the electricity had been switched off, or if the UV-bulb had failed. The design under testing relied on a safe-viewing port, which would show the glow of the UV-lamp. However, the South Africa experience showed that occasionally the electricity would get turned off, and the unit continue to be used owing to poor communication between the technical staff and the hospice nurses. In all subsequent units, a solenoid valve controls the flow of raw water into UV Waterworks. In case of any reduction in the lamp output, or drop in UV-transmittance of water, the solenoid valve shuts off water supply to the device. The electronics and the solenoid valve are now mounted on top of the unit, as shown in Figure 2, and raise the total power consumption to about 50W.

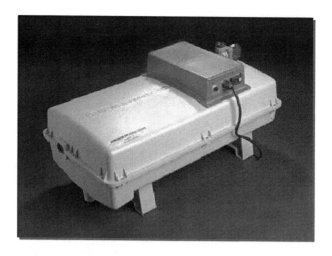

FIGURE 2. UV Waterworks showing the electronics piggyback box, and the solenoid valve (behind the piggyback box) on the top right of the main enclosure. The safe-viewing port is visible on the left face of the main enclosure. Water discharges from the port on the bottom left surface.

ENERGY AND CARBON IMPLICATIONS OF UV WATERWORKS

The UV-C lamps produces light with an electrical discharge through a low-pressure mercury plasma, and is electrically identical to a common fluorescent lamp. In terms of optics, the only difference is that the glass envelope of the fluorescent lamp, opaque to UV, is replaced with quartz, and the phosphor lining on the inside of the envelope is not needed, since we need the UV-C photons to come out. About 40% of the electrical energy in the low pressure mercury discharge gets converted into light (the rest comes out as heat), and of this light, about 95% is delivered at 253.7 nm, close to the 260nm absorption peak of the DNA. Since only a very small energy dose is needed to disinfect water adequately (recall: only 40mJ/sq.cm. at the water surface), and we pay for grid electricity at about 10 cents per kWh, the cost of disinfection is very small. It remains very small even at the current price of about 50 cents per kWh for photovoltaic electricity (after including all storage and power conditioning costs).

Energy Implications

The realistic alternative for most poor community residents in the developing countries would be to boil their drinking water on a biomass cookstove. Typical biomass cookstoves have an overall energy efficiency of about 12% (it is only about 6% for a three stone fire), measured as the chemical energy in the fuel that is transferred as heat into the pot. Since a hard boil for one minute is adequate to disinfect water at sea level, we can compare the energy in the fuel (biomass) needed to disinfect a liter of water by boiling, to the energy in the coal that is used to generate the electricity that would be used to disinfect the same amount of water in the UV Waterworks.

Starting from an initial temperature of 20 °C, to boil 1 kg water for 1 minute using a biomass stove of 12% efficiency takes about 3.3 MJ of biomass energy. Using UV Waterworks powered with electricity from a 33% efficient combined-cycle coal-fired power plant and transmission losses of 12% requires 690 J of thermal energy. Compared this way, UV Waterworks uses about 5000 times less primary energy (i.e., thermal energy in the coal at the powerplant) than the primary energy input to disinfect the same amount of water by boiling over a biomass cookstove.

Of course, in reality, most of the poor in the developing countries do not boil their drinking water –from either lack of relevant knowledge, or inability to afford the fuel to daily boil their drinking water. As real incomes slowly rise, and education and knowledge spreads, one can reasonably expect that members of these communities will seek to boil their drinking water unless cheaper means of obtaining safe water are available. Thus, apart from its immediate public-health benefit, UV Waterworks could quell what is currently a large suppressed demand for energy for boiling drinking water. This view is consistent with the position expressed by the Gold Standard (www.cdmgoldstandard.org), a Switzerland-based non-profit foundation supported by major international environmental non-profits that support a certification program for "clean" carbon savings under the Clean Development Mechanism (CDM) defined under the Kyoto protocols.

Carbon Implications

Estimates of avoided CO_2 emissions from the use of UV Waterworks require an estimate of the baseline. Assuming that the baseline represents emissions from disinfection by boiling water over a biomass cookstove. However, the CO_2 emissions are different depending on whether the boiling is fueled with renewable biomass ("RB") or non-renewable biomass ("NRB"). RB refers to biomass obtained through sustainable harvesting. Under steady-state conditions, the carbon in the biomass is emitted as CO_2 during combustion, and is recaptured by the growing biomass source through photosynthesis. However, combustion in a biomass cookstove also leads to non-CO_2 greenhouse gases (GHG) emissions such as methane, NOx, CO, etc. Contribution of these non-CO_2 GHG emissions from combustion of biomass can be quite significant. We estimate that with a RB fired cookstove, the emissions are 282 grams of CO_2-equivalent per kg of water, and with an NRB fired cookstove the emissions are 766 g of CO_2-equivalent per kg of water.[16] GHG emissions from using UV Waterworks are 100 milligrams of CO_2-equivalent per kg of water treated, so can be ignored in the calculations below.

Operated 20 hours a day, one UV Waterworks treats annually 7300 tonnes of water. Each tonne of treated water, assuming that it replaced boiling on an RB fired cookstove, avoids of 282 kg of CO_2-equivalent GHG emissions. Thus, each UV Waterworks working at 20 hours a day potentially avoids 2000 tonnes of CO_2-equivalent GHG emissions.

CURRENT STATUS

The notion that provision of charitable or government aid and expertise from the developed world by itself can solve the problems of the global poor is being seriously questioned, as fifty years of top-down development have failed to close the gap between the rich and the extremely poor.[17] Closing these gaps is likely to require not only philanthropic and governmental intervention, but also development of markets that can reach underserved segments with affordable goods and services.[18] WaterHealth International (WHI, www.waterhealth.com) is a for-profit corporation founded in 1995 to develop and market affordable water purification and disinfection systems, with a strategic goal of addressing the underserved needs of a large segment of the world's population without access to clean drinking water. WHI obtained an exclusive worldwide license for the UV Waterworks technology from the University of California Lawrence Berkeley National Laboratory.[19]

WHI soon realized that what communities want is not an individual component or a particular technology, but simple, convenient, and affordable access to safe drinking water. WHI now offers a turnkey system that includes a small building that houses the UV Waterworks system packaged with pumps, filters, controls, and water storage and distribution mechanisms, and includes service delivery and quality maintenance programs, as well as investments in education on health and hygiene in communities served by WHI systems. In India, Ghana and the Philippines, WHI's rural community water systems—"WaterHealth Centers"— make safe water available to the rural poor, charging them user fees for filtering and disinfecting the local water. These systems are designed to be financially viable and socially sustainable by working closely with local village-level governance structures and non-governmental agencies on the one hand, and financial institutions that extend credit for the system purchase on the other hand. In effect each WaterHealth Center is a village-owned micro-utility that is set up with financing and comes with a service and maintenance contract.

FIGURES 3a, 3b: A WaterHealth Center in Afuaman, Ghana (2008).

WHI is an example of a private-public partnership that uses a private sector approach to address a critical need with public and private components. In India, WHI partners with several local non-governmental organizations (NGOs) that provide outreach and support to the villages in which WHI works. WaterHealth's installed capacity to provide daily safe water had grown from several thousand in 2005 to over one million people in the Spring of 2008. At the 2007-2008 costs relevant to India (labor rates, civil works costs, costs for tanks, pumps, customs duty etc.), WHI was able to offer triple-filtered, carbon-filtered and UV-treated water to the local rural communities for a price of 0.2 US cent per liter of water!

CONCLUSIONS

While the global water market continues to surge (estimated at US$ 400 billion in 2007), it leaves behind about 2 billion people who lack access to safe drinking water. Innovation in technology, and in public-private partnership that turns the technology into an affordable offering of safe drinking water to the rural poor has been demonstrated on a small but still impressive scale in India. WaterHealth's installed capacity for delivering safe water is growing – sustainably – as measured by financial, social and environmental metrics. Much more (and more rapid) progress needs to take place, but the key point is that this is no longer charity!

ACKNOWLEDGEMENTS

This work was performed at Lawrence Berkeley National Laboratory, University of California, under contract DE-AC02-05CH11231. The author is grateful to several colleagues at LBNL who graciously and generously provided encouragement, ideas, loaned laboratory space and equipment. Funding support from USAID, LBNL's Technology Transfer office, DOE's Policy Office, Joyce Mertz Gilmore Foundation, Rockefeller Foundation, support in the form of the Pew Fellowship in Conservation and the Environment to the author, International Network of Resource Information Centers, and in-kind donations of UV lamps from GE and Philips are all gratefully acknowledged. Useful review comments on the manuscript from Philip Price and staff of WaterHealth are gratefully acknowledged.

REFERENCES

1. World Health Organization, *Measuring Child Mortality* (2007). Retrieved February 6, 2008, from http://www.who.int/child_adolescent_health/data/child/en/index.html
2. United Nations Development Program (UNDP), *Human Development Report 2006*. Retrieved March 10, 2008, from http://hdr.undp.org/en/reports/global/hdr2006/.
3. W.M. Reilly, "U.N. Calls for More Clean Water," United Press International (10 November 1996).
4. World Health Organization, *Guidelines for Drinking Water Quality*, Vols. 1–3. Geneva, Switzerland, WHO (1996–2004).
5. J. De Zuane, *Handbook of Drinking Water Quality* (2nd ed.), New York, Van Nostrand (1997).
6. World Health Organization, *Water, sanitation, and health: Facts and figures* (2004). Retrieved 6 February 2008, from http://www.who.int/water_sanitation_health/publications/facts2004/en/index.html
7. J. Briscoe, "Poverty and water supply: how to move forward," *Finance Dev.* 29(4), 16–19 (1992).
8. West Bengal and Bangladesh Arsenic Crisis Information Centre (1998). Retrieved February 4, 2008, from http://bicn.com/acic/

9. E.D. Mintz, F.M. Reiff and R.V. Tauxe, "Safe water treatment and storage in the home: a practical new strategy to prevent waterborne diseases," *JAMA* 273(12), 948–53 (1995).

10. R. Talbott, "Rural water supply and sanitation program in India—goals, roles and innovation," *Proc. 23rd WEDC Conf*erence (1–5, September 1997), Durban, South Africa. Loughborough Univ., UK, Water, Eng. Dev. Cent. (WEDC).

11. WHO/UNICEF Global Water Supply and Sanitation Assessment 2000 Report, Retrieved May 2008, from http://www.who.int/water_sanitation_health/monitoring/globalassess/en/.

12. UNICEF, *Water. Environment and Sanitation - Universal sanitation gains momentum* (2006). Retrieved February 6, 2008. from http://www.unicef.org/india/wes_1359.htm

13. Environmental Protection Agency, *Guidance Manual for Compliance with the Filtration and Disinfection Requirements for Public Water Systems Using Surface Water Sources.* Washington, DC: Environmental Protection Agency Office (1991).

14. World Health Organization, *Fuel for Life: household energy and health (2006).* Retrieved May 20, 2008, from http://www.who.int/indoorair/publications/fuelforlife.pdf

15. A. Drescher, A. Gadgil, D. Greene, N. Kibata and P. Miller, "UV Disinfection of drinking water: tests in South Africa," *Providing Safe Drinking Water in Small Systems: Technology., Operations, and Economics,* Proceedings of symposium under same title, May 10–13, Washington, DC (1998). Sponsored by NSF-Int. PAHO & WHO. Washington, D.C. Lewis Publishers.

16. A.J. Gadgil, D.M. Greene and A.H. Rosenfeld, "Energy-efficient Drinking Water Disinfection for Greenhouse Gas Mitigation" *Proceedings* 1998 ACEEE Summer Study "Energy Efficiency in a Competitive Environment" Asilomar, Pacific Grove CA (23-28 August 1998).

17. K. Danaher (ed.), *Fifty Years is Enough: The Case Against the World Bank and the International Monetary Fund.* United States: Global Exchange (1994).

18. J. Novogratz, "Meeting Urgent Needs with Patient Capital," *MIT Innovations* (12 June 2007). Retrieved February 7, 2008, from
http://www.mitpressjournals.org/doi/abs/10.1162/itgg.2007.2.1-2.19.

19. C.G. Reuther, "Brighter light better water." *Environmental Health Perspectives* 104(10), 1046–48 (1996).

Saving Energy and Improving Air Quality in Urban Heat Islands

Hashem Akbari

Heat Island Group
Environmental Energy Technologies Division
Lawrence Berkeley National Laboratory
Berkeley, CA 94720

Abstract. Temperatures in urban areas have increased because solar energy is more strongly absorbed by additional roofs and pavements. Downtown Los Angles is now 2.5 Kelvin warmer than in the 1930s, which requires 1–1.5 GWe more electricity to cool buildings on summer days, costing an extra $100 million/year. Cool roof and pavement materials with increased reflectivity of 0.25 can lower surface temperatures by 10 K. If Los Angles urban temperatures could be reduced by 3 K, ozone concentrations could be reduced considerably.

INTRODUCTION

World energy use is the main contributor to atmospheric CO_2. In 2002, about 7.0 giga metric tons of carbon (GtC) were emitted internationally by combustion of gas, liquid, and solid fuels. Increasing use of fossil fuel and deforestation together have raised atmospheric CO_2 concentration some 25% over the last 150 years. According to global climate models and measurements, these changes in the composition of the atmosphere have already begun raising the Earth's average temperature. If current energy trends continue, these changes could drastically alter the Earth's temperature, with unknown but potentially catastrophic physical and political consequences.

Of all electricity generated in the United States, about one-sixth is used to air-condition buildings. The air-conditioning use is about 400 tera-watt-hours (TWh), equivalent to about 80 million metric tons of carbon (MtC) emissions, and translating to about $40 billion (B) per year. Of this $40 B/year, about half is used in cities that have pronounced "heat islands." The contribution of the urban heat island to the air-conditioning demand has increased over the last 40 years and it is currently at about 10%. Metropolitan areas in the United States (e.g., Los Angeles, Phoenix, Houston, Atlanta, and New York City) have typically pronounced heat islands that warrant special attention by anyone concerned with broad-scale energy efficiency.

The ambient air is primarily heated through three processes: direct absorption of solar radiation, convection of heat from hot surfaces, and man-made heat (exhaust from cars, buildings, etc.). Air is fairly transparent to light; the direct absorption of solar radiation in atmospheric air only raises the air temperature by a small amount. On a sunny day, typically

CP1044, *Physics of Sustainable Energy, Using Energy Efficiently and Producing It Renewably*
edited by D. Hafemeister, B. Levi, M. Levine, and P. Schwartz
© 2008 American Institute of Physics 978-0-7354-0572-1/08/$23.00

about 90% of solar radiation reaches the Earth's surface and then is either absorbed or reflected. The absorbed radiation on the surface increases the surface temperature. And in turn the hot surfaces heat the air. This convective heating is responsible for the majority of the diurnal temperature range. The contribution of man-made heat (e.g., air conditioning, cars) is very small, compared to the heating of air by hot surfaces, except for the downtown high-rise areas.

Modern urban areas have darker surfaces or lower "effective" albedo and relatively less vegetation than their more natural surroundings, which affects urban climate, energy use, and thermal environmental conditions. Dark roofs, for example, heat up more than their more reflective counterparts and thus raise the summertime cooling demands of buildings. Collectively, on a neighborhood scale, dark surfaces and reduced vegetation warm the air over urban areas, contributing to urban heat islands. On a clear summer afternoon, the air temperature in a typical city can be as much as 2.5 Kelvin (K) higher than surrounding rural areas. In hot cities in the U.S., peak urban electric demand rises by 2-4% for each 1 K rise in daily maximum temperature above ambient air temperatures of 15-20°C.

Temperatures in cities are generally increasing. An analysis of summertime monthly maximum and minimum temperatures between 1877-1997 in downtown Los Angeles clearly indicated that maximum temperatures are now about 2.5 K higher than in 1930 (Fig. 1 and 2).[i] Minimum temperatures are about 4 K higher than in 1880. A California study analyzing the average urban-rural temperature differences for 31 urban and 31 rural stations from 1965-1989 showed that urban temperatures have increased by about 1 K (Fig. 3).[ii] This trend in increasing temperatures in urban areas is typical of most U.S. metropolitan areas and observed in many other cities across the world.[ii] Summertime urban heat islands can exacerbate demand for cooling energy. Note that this is above and beyond what is believed to be the global warming trend. Since most people live in cities, they would experience the effects of both global warming *and* urban heat islands.

Increasing urban ambient temperatures results in increased system-wide electricity use. In the Los Angeles Basin, the heat-island-induced increase in power consumption of 1-1.5 GW can cost rate-payers $100 million per year (see Fig. 4). In the United States, additional air-conditioning use from increased urban air temperature comprises 5-10% of urban peak electric demand at a direct cost of several billion dollars per year. Since cooling-demand on hot summer days is the cause of peak demand for electricity, the electric utilities have installed additional capacity to compensate for the heat-island effects.

Besides increasing system-wide cooling loads, summer heat islands increase smog production. Smog production is a highly temperature-sensitive process. In the Los Angeles Basin, at daily maximum temperatures below 22°C, maximum ozone concentration is typically below the California standard [90 parts per billion (ppb)]; at above 32°C, practically all days are (see Fig. 5).

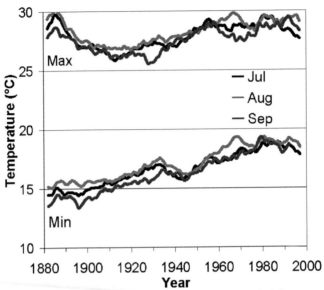

FIGURE 1. Ten-year running average summertime monthly maximum and minimum temperatures in Los Angeles, California (1877-2004). The ten-year running average is calculated as the average temperature of the previous 4 years, the current year, and the next 5 years. Note that the maximum temperatures have increased about 2.5 K since 1920. During the same period, the minimum temperature is also increased by about 3 K. (Source: Akbari[iii])

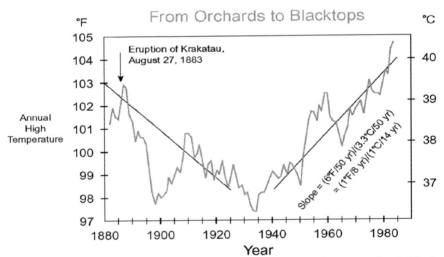

FIGURE 2. Ten-Year running–average maximum annual temperatures in Los Angeles, California (1877–1997). The ten-year running average is calculated as the average temperature of the previous 4 years, the current year, and the next 5 years.

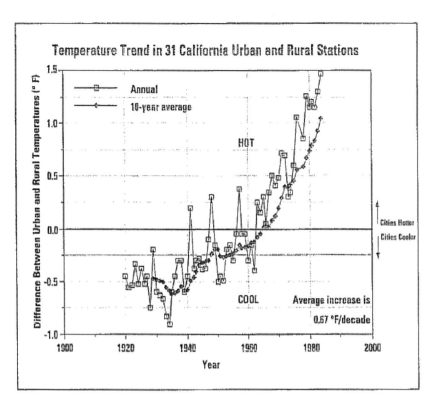

FIGURE 3. Warming trend in California Urban areas. Since 1940, the temperature difference between urban and rural meteorological stations has shown an increase of about 0.67 F per decade. Note that during 1920-1960, cities were actually cooler than suburban areas, probably because of relatively more vegetation in urban areas. (Source: Akbari et al.[ii]).

Summer heat islands increase citizens discomfort and heat wave related mortalities. According to the U.S. Centers for Disease Control and Prevention, over the past 20 years, more Americans were killed by heat than by hurricanes, lightning, tornadoes, floods, and earthquakes combined. Within a five-day period, the 1995 Chicago heat wave killed between 525 and 726 people, depending on the method used for determining which deaths were attributable to the high temperatures. In the heat wave of 1980, some 1,250 Americans died. A heat wave in summer of 2003 in India killed at least 1,200 people. Most tragic is the death of between 10,000 to 15,000 people who died in France's scorching heat wave in August 2003. Many of the victims were elderly people living in poorly designed houses or apartments that were not air-conditioned.

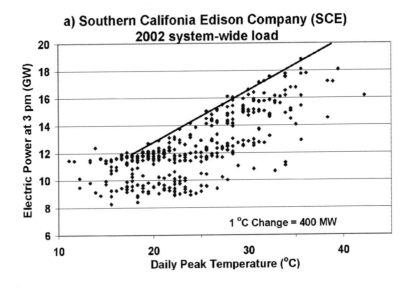

a) Southern Califonia Edison Company (SCE) 2002 system-wide load

1 °C Change = 400 MW

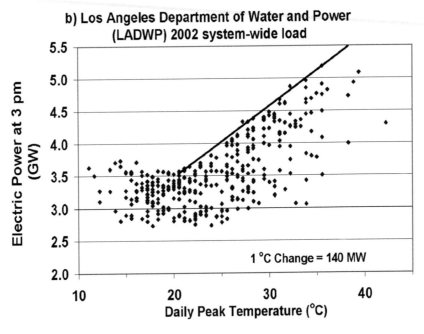

b) Los Angeles Department of Water and Power (LADWP) 2002 system-wide load

1 °C Change = 140 MW

FIGURE 4. Daily peak utility electric power demand vs. daily peak air temperature. The increased summertime temperatures cause increased cooling requirements. In Los Angeles Basin (primarily served by Southern California Edison and Los Angeles Department of Water and Power), we estimate that about 1-1.5 GW of power are used to compensate the heat island effect. This increased power adds about $100,000 per hour ($100 million a year) during summer days to the utility customers' electricity bills.

**Ozone concentration measured at
Los Angeles, W Flint Street, 2002**

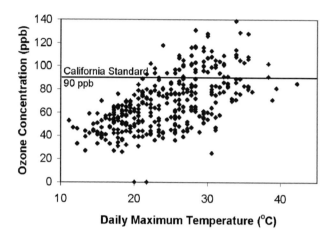

**Ozone concentration measured at
Los Angeles, North Main Street, 2002**

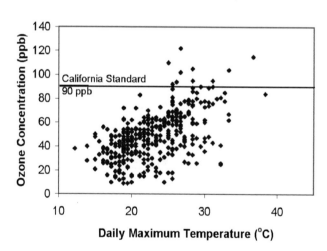

FIGURE 5. Daily maximum ozone concentration vs. daily maximum temperature in two locations at Los Angeles. The impact of the heat island is also seen in smog. The formation of smog is highly sensitive to temperatures; the higher the temperature, the higher the formation and, hence, the concentration of smog. In Los Angeles at temperatures below 22°C, the concentration of smog (measured as ozone) is below the California standard. At temperatures of about 32°C practically all days are smoggy. Cooling the city by about 3°C would have a dramatic impact on smog concentration.

It is important to note that heat island is a direct result of urbanization that creates an urban fabric consisting mostly of roofs, paved surfaces (roads, driveways, parking lots), and less vegetation (trees, lawns, bushes, shrubs). Understanding and quantifying the fabric of a city is an important first step in analyzing and designing implementation programs to mitigate urban heat islands. Of particular importance is the fraction of each surface type within an area. An accurate characterization of the urban surfaces will also allow a better estimate of the potential for increasing solar reflectance of urban surfaces (roofs, pavements) and increasing urban vegetation. This would in turn provide more accurate modeling of the impact of heat-island reduction measures on ambient cooling and urban ozone air quality.

In four studies, Akbari et al.[iv], Akbari and Rose[v,vi] and Rose et al.[vii] characterized the fabric of Sacramento CA, Salt Lake City UT, Chicago IL, and Houston TX, using high-resolution aerial digital orthophotos covering selected areas in each city. Four major land-use types were examined: commercial, industrial, transportation, and residential. These orthophotos were analyzed to estimate the fraction of each major land use type (defined as urban fabric) and to estimate the land-use land-cover (LULC) in each city (Figure 6). Although there were differences among the fabrics of these four metropolitan areas, some significant similarities were found.

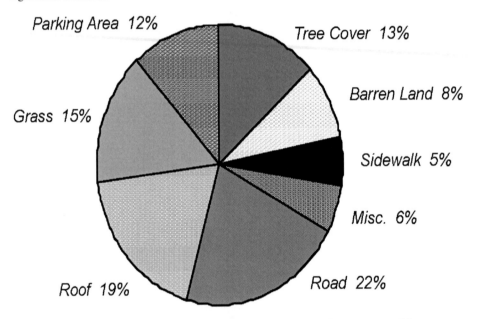

FIGURE 6. The Land Use/Land Cover (LULC) percentages for Sacramento, CA.

Of approximately 800 km² of urban area in Sacramento, about 49% was residential; in Salt Lake City about 59% of the 620 km²; in Chicago about 53% of 2,520 km², and in Houston about 56% of 3,430 km². The fraction of industrial, transportation, and mixed urban land-uses in these four cities varied only by a few percent. For the entire metropolitan area, the percentage of the total roof areas, as seen from above the canopy was about 19% in the Sacramento and Salt Lake City metropolitan areas, 25% in Chicago metropolitan area, and 21% in Greater Houston. The percentage of paved areas ranged from 29% to 39%, vegetated areas 29% to 41%, and other areas 10%-40%. Under the canopy, the roof area ranged from 20% to 25%, paved surfaces 29% to 45%, vegetated areas 20% to 37%, and other areas 9% to

15%. In residential areas, the percentage of the total roof areas, as seen from above the canopy, ranged from 19% to 26%, paved surfaces 25%-26%, vegetated areas 39%-49%, and others 4%-16%. Under the canopy, roof area ranged from 20% to 27%, paved surfaces 24% to 32%, vegetated areas 33% to 47%, and other areas 6% to 17%.

MITIGATION TECHNOLOGIES

Cool roofing material, urban shade trees, and cool pavements are three effective technologies to cool summer heat islands. We estimate that by full implementation of these mitigation measures the cooling demand in the U.S. can be *decreased by 20%*. This equals to about 40 TWh/year in savings, worth over $4B per year by 2015 in cooling-electricity savings alone. If smog reduction benefits are included, savings could total to over $10B/year. Achieving these potential savings is conditional on receiving the necessary federal, state, and local community support. Scattered programs for planting trees and increasing surface albedo already exist, but the initiation of an effective and comprehensive campaign would require an aggressive agenda.

Cool Roofs

When dark roofs are heated by the sun, they directly raise the demand for cooling for the buildings beneath those roofs. For highly absorptive (low-albedo) roofs, the surface/ambient air-temperatures difference may reach 50 K, while for less absorptive (high-albedo) surfaces (e.g. white-coated roofs), the difference can be only about 10 K.

Hot roofs also heat the outside ambient air, thus indirectly increasing cooling demand of neighboring buildings. We have simulated the effect of urban-wide application of reflective roofs on cooling-energy use and smog in many metropolitan areas.[viii,ix,x] We estimate roof albedos can realistically be raised by 0.30 on average, resulting in a 1-2.5 K cooling at 3pm (on a sunny August day). This temperature reduction reduces building cooling-energy use even further. Other benefits of light-colored roofs include a potential increase in the roofs useful life.

Direct Energy Benefits of Cool Roofs

Field studies in California and Florida have demonstrated direct cooling-energy savings in excess of 20% upon raising the solar reflectance of a roof to 0.6 from a prior value of 0.1–0.2. Energy savings are particularly pronounced in older houses that have little or no attic insulation, especially if the attic contains the air distribution ducts. Akbari *et al.* observed cooling-energy savings of 46% and peak power savings of 20% achieved by increasing the roof reflectance of two identical portable classrooms in Sacramento, California.[xi] Konopacki *et al* documented measured energy savings of 12–18% in two commercial buildings in California.[xii] In a large retail store in Austin, Texas, Konopacki and Akbari documented measured energy savings of 12%.[xiii] Akbari documented energy savings of 31–39 Wh/m²/day in two small commercial buildings with very high internal loads, by coating roofs with a white elastomer with a reflectivity of 0.70.[xiv] Parker *et al.* measured an average of 19% energy savings in eleven Florida residences by applying reflective coatings on their roofs.[xv] Parker *et al.* also monitored seven retail stores in a strip mall in Florida before and after applying a high-albedo coating to the roof and measured a 25% drop in seasonal cooling energy use.[xv,xvi] Hildebrandt *et al.* observed daily energy savings of 17%, 26%, and 39% in an office, a museum and a hospice, respectively, retrofitted with high-albedo roofs in Sacramento.[xvii]

Akridge reported energy savings of 28% for a school building in Georgia after an unpainted galvanized roof was coated with white acrylic.[xviii] Boutwell and Salinas showed that an office building in southern Mississippi saved 22% after the application of a high-reflectance coating.[xix] Simpson and McPherson (1997) measured energy savings in the range of 5–28% in several quarter-scale models in Tucson AZ.[xx]

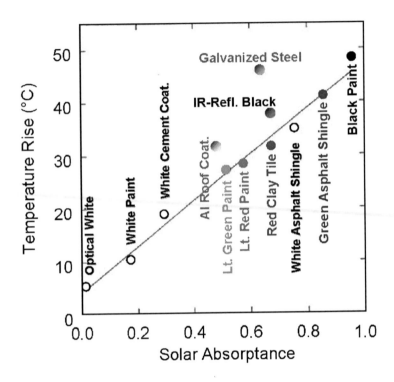

FIGURE 7. Temperature rise (surface temperature minus air temperature) of various roofing materials measured at peak solar conditions. All samples were insulated on the back and the measurements were made at low wind speed.

Computer simulations of cooling energy savings from increased roof albedo in residential and commercial buildings have also been documented by many studies, including Konopacki and Akbari, Akbari et al., Parker et al., and Gartland et al.[xxi,xxii,xv,xxiii] Konopacki et al. estimated the direct energy savings potential from high-albedo roofs in eleven U.S. metropolitan areas (see Figure **8**).[xxiv] The results showed that four major building types account for over 90% of the annual electricity and monetary savings in the U.S.: pre-1980 residences (55%), post-1980 residences (15%), and office buildings and retail stores together (25%). Furthermore, these four building types account for 93% of the total air-conditioned roof area. Regional savings were found to be a function of three factors: energy savings in the air-conditioned residential and commercial building stock; the percentage of buildings that were air-conditioned; and the aggregate regional roof area. Metropolitan-wide annual savings from the application of cool roofs on residential and commercial buildings were as low as $3M in the heating-dominated climate of Philadelphia and as much as $37M for Phoenix and $35M in Los Angeles.

The results for the 11 Metropolitan Statistical Areas (MSAs) were extrapolated to estimate the savings in the entire United States. At 8¢/kWh, the value of U.S. potential nationwide net commercial and residential energy savings (cooling savings minus heating penalties) exceeds $750 million per year.[xxv] The study estimates that, nation-wide, light-colored roofing could produce savings of about 10 TWh/yr (about 3.0% of the national cooling-electricity use in residential and commercial buildings), an increase in natural gas (heating) use by 26 GBtu/yr (1.6%), and a decrease in peak electrical demand of 7 GW (2.5%) (equivalent to 14 power plants each with a capacity of 0.5 GW).

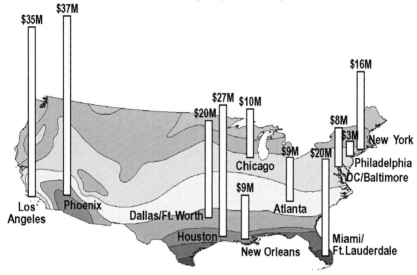

FIGURE 8. Estimated energy-saving potentials of light-colored roofs in 11 U.S. metropolitan areas. About 10 residential and commercial building prototypes in each area are simulated. Both savings in cooling and penalties in heating are considered. The estimated saving potentials is about $175M (1997 energy prices) per year for the 11 cities. Extrapolated national energy savings is about $0.75B per year. (Source: Konopacki et al. [xxiv])

Indirect Energy and Smog Benefits of Cool Roofs

Indirect effects require that a large fraction of the urban area be modified to produce a change in the local climate. To date, results have been attained only by computer simulations. Using the Los Angeles Basin as a case study, Taha examined the impacts of using cool surfaces (cool roofs and pavements) on urban air temperature and thus on cooling-energy use and smog.[xxvi,xxvii] In these simulations, Taha estimates that about 50% of the urbanized area in the L.A. Basin is covered by roofs and roads, the albedos of which can realistically be raised by 0.30 when they undergo normal repairs. This results in a 2 K cooling at 3 p.m. during an August episode. This summertime temperature reduction has a significant effect on further reducing building cooling-energy use. The annual savings in Los Angeles are estimated at $21M.[xxviii]

Taha also simulated the impact of urban-wide cooling in Los Angeles on smog—predicting a reduction of 10–20% in population-weighted smog (ozone).[xxvii] In L.A., where smog is especially serious, the potential savings were valued at $104M/year. Table 1 also shows the present value (PV) of all future savings associated with installation of cool roofs.

Table 1. Energy savings, ozone reduction, and avoided peak power resulting from use of Cool Roofs in the Los Angeles Basin (Source: Rosenfeld et al. [xxviii]).

Benefits	Direct	Indirect	Smog	Total
1 Cost savings from cool roofs (M$/yr)	46	21	104	171
2 Δ Peak power (GW)	0.4	0.2		0.6
3 Present value per 100 m^2 of roof area ($)	153	25	125	303

Cool-Colored Roofing Materials

Suitable cool *white* materials are available for most products, with the notable exception of asphalt shingles; cooler *colored* (nonwhite) materials are needed for all types of roofing, especially in the residential market. Coatings colored with conventional pigments tend to absorb the invisible "near-infrared" (NIR) radiation that bears more than half of the power in sunlight (see Error! Reference source not found.). Replacing conventional pigments with "cool" pigments that absorb less NIR radiation can yield colored coatings that look the same to the eye but have higher solar reflectance. These cool coatings lower roof surface temperature, reducing the need for cooling energy in conditioned buildings and making unconditioned buildings more comfortable.

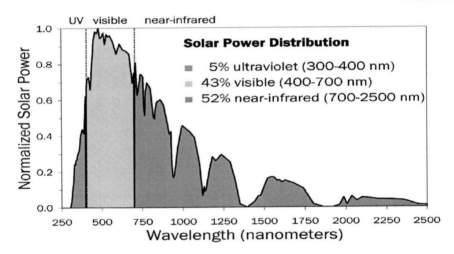

FIGURE 9. Peak-normalized solar spectral power; over half of all solar power arrives as invisible, "near-infrared" radiation.

According to *Western Roofing Insulation and Siding* magazine, the total value of the 2002 projected residential roofing market in 14 western U.S. states (AK, AZ, CA, CO, HI, ID, MT, NV, NM, OR, TX, UT, WA, and WY) was about $3.6 billion (B).[xxix] We estimate that 40% ($1.4B) of that amount was spent in California. The lion's share of residential roofing expenditure was for fiberglass shingle, which accounted for $1.7B, or 47% of sales. Concrete and clay roof tiles made up $0.95B (27%), while wood, metal, and slate roofing collectively represented another $0.55B (15%). The value of all other roofing projects was about $0.41B (11%). We estimate that the roofing market area distribution was 54–58% fiberglass shingle,

8-10% concrete tile, 8–10% clay tile, 7% metal, 3% wood shake, and 3% slate.

Suitable cool *white* materials are available for most roofing products, with the notable exception (prior to March 2005) of asphalt shingles. Cool nonwhite materials are needed for all types of roofing. Industry researchers have developed complex inorganic color pigments that are dark in color but highly reflective in the near infrared (NIR) portion of the solar spectrum. The high near-infrared reflectance of coatings formulated with these and other "cool" pigments—e.g., chromium oxide green, cobalt blue, phthalocyanine blue, Hansa yellow—can be exploited to manufacture roofing materials that reflect more sunlight than conventionally pigmented roofing products.

Cool-colored roofing materials are expected to penetrate the roofing market within the next few years. Preliminary analysis suggests that they may cost up to $1/m^2$ more than conventionally colored roofing materials. However, this would raise the total cost of a new roof (material plus labor) by only 2-5%.

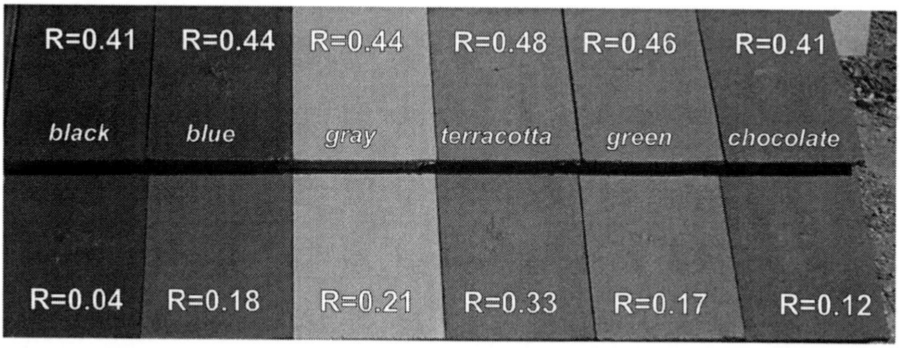

FIGURE 10. Palette of color-matched cool (top row) and conventional (bottom row) roof-tile coatings developed by industrial partner American Rooftile Coatings. Shown on each coated tile is its solar reflectance, R.

Cool Pavements

Urban pavements are made predominantly of asphalt concrete. The advantages of this smooth and all-weather surface type for vehicles are obvious, but some associated problems are perhaps not so well appreciated. Sunlight on dark asphalt surfaces produce increased heating. An air-temperature increase, in turn, increases cooling-energy use in buildings, and can accelerate smog formation. The albedo of fresh asphalt concrete pavement is about 0.05: the relatively small amount of black asphalt coats the lighter-colored aggregate. As an asphalt concrete pavement is worn down and the aggregate is revealed, albedo increases to about 0.10 to 0.15 (the value of ordinary aggregates). If a reflective aggregate is used, the long-term albedo can be higher.

Furthermore, the temperature of a pavement affects its structural performance; cooler pavements last longer in hot climates. Reflectivity of pavements can improve visibility at night and can reduce electric street-lighting demand. Street lighting is more effective if pavements are more reflective, increasing safety as a result. Despite concerns that, in time, dirt will darken light-colored pavements, experience with cement concrete roads suggests that the light color of the pavement can actually persist after long usage.

The challenge is to develop cool pavements that are economical and practical. Measurements show that an increased albedo lowers pavement temperature. The data of Fig.

11 clearly indicate that significant modification of the pavement temperature can be achieved: a 10 K decrease in temperature for a 0.25 increase in albedo.

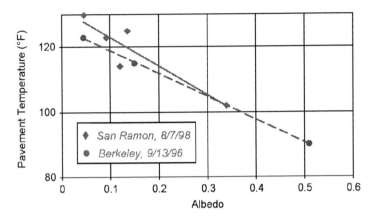

FIGURE 11. The dependence of pavement surface temperature on albedo. Data in Berkeley, California were taken at about 3 pm, on new, old, and light-color coated asphalt pavements. The data from San Ramon, California were taken at about 3 pm on 4 asphalt and 1 cement concrete (albedo = 0.35).

Energy and Smog Benefits of Cool Pavements

Cool pavements affect energy use and air quality through lowered ambient temperatures. Lower temperature has two important effects: 1) reduced demand for electricity for air conditioning and 2) decreased production of smog (ozone). Rosenfeld *et al.* (1998) estimated the cost savings of reduced demand for electricity and of the externalities of lower ozone concentrations in the Los Angeles Basin.

Simulations for Los Angeles (L.A.) Basin indicate that a reasonable change in the albedo of the city could cause a noticeable decrease in temperature. Taha predicted a 1.5K decrease in temperature of the downtown area.[xxviii] The lower temperatures in the city are calculated based on the assumption that all roads and roofs are improved. From the meteorological simulations of three days in each season, the temperature changes for every day in a typical year were estimated for Burbank, typical of the hottest 1/3 of L.A. basin. The energy consumptions of typical buildings were then simulated for the original weather and also for the modified weather. The differences are the annual energy changes due to the decrease in ambient temperature. The result is a city-wide annual saving of about $71M, due to combined albedo and vegetation changes. The kWh savings attributable to the pavement are $15M/yr, or $0.012/m²-yr. Analysis of the hourly demand indicates that cooler pavements could save an estimated 100 MW of peak power in L.A.

The simulations of the effects of higher albedo on smog formation indicate that an albedo change of 0.3 throughout the developed 25% of the city would yield a 12% decrease in the population-weighted ozone exceedance of the California air-quality standard.[xxvi] The estimated annual cost to the residents of L.A. because of air quality related medical costs and lost work time is about $10 B.[xxx] [xxx] The greater part of pollution is particulates, but the ozone contribution averages about $3 B/yr. Assuming a proportional relationship of the cost with the amount of smog exceedance, the cooler-surfaced city would save 12% of $3 B/yr, or $360M/yr. As above, we attribute about 21% of the saving to pavements. Rosenfeld *et al.*

value the benefits from smog improvement by altering the albedo of all 1250 km^2 of pavements by 0.25 saves about \$76M/year (about \$0.06/m^2 per year).[xxviii]

Shade Trees

The beneficial effects of trees are both direct in shading of buildings and indirect in cooling the ambient air (urban forest). Trees can intercept sunlight before it warms buildings and cool the air by evapotranspiration. In winter, trees can shield buildings from cold winds. Urban shade trees offer significant benefits by reducing building air-conditioning, and lowering air temperature, thus improving urban air quality (reducing smog). Savings associated with these benefits vary by climate and region and, over a tree's life, can reach up to \$200/tree. The cost of planting and maintaining trees can vary from \$10-500/tree. Tree-planting programs can be low-cost, offering savings to tree-planting communities. The choice of tree species is also important. Low-emitting drought-resistance trees are typically recommended.

The issue of direct and indirect effects also enters into our discussion of atmospheric pollutants. Planting trees has the direct effect of reducing atmospheric CO_2 because each individual tree directly sequesters carbon from the atmosphere through photosynthesis. However, planting trees in cities also has an indirect effect on CO_2. By reducing the demand for cooling energy, urban trees indirectly reduce emission of CO_2 from power plants. Akbari *et al* showed that the amount of CO_2 avoided via the indirect effect is considerably greater than the amount sequestered directly.[xxxi] Similarly, trees directly trap ozone precursors (by dry-deposition, a process in which ozone is directly absorbed by tree leaves), and indirectly reduces the emission of these precursors from power plants—by reducing combustion of fossil fuels and hence reducing NOx emissions from power plants.

SUMMARY AND CONCLUSIONS

Most urban areas are warmer than their surrounding rural areas. The temperature difference between urban and rural areas is commonly referred to as urban heat islands. With the rapid expansion of cities in the last five decades, heat islands are growing and are affecting the world's ever-increasing urban population. Increasing urban ambient temperatures raise building cooling energy use, worsen the urban air quality, and reduce citizens' comfort. Cool surfaces (cool roofs and cool pavements) and urban trees can have a substantial effect on urban air temperature and hence can reduce cooling-energy use and smog. In the United States, it is estimated that about 20% of the national cooling demand can be avoided through a large-scale implementation of heat-island mitigation measures. This amounts to 40 TWh/year savings, worth over \$4B per year by 2015 in cooling-electricity savings alone. Once the benefits of smog reduction are accounted for, the total savings could add up to over \$10B per year.

Achieving these potential savings is conditional on receiving the necessary governmental and local community support. Scattered programs for planting trees and increasing surface albedo already exist, but to start an effective and comprehensive campaign would require a more aggressive agenda. Much of the fundamental work to promote heat-island mitigation measures are already in place. The American Society for Testing of Materials (ASTM) has developed standards for measurement of solar reflectance of roofing and pavement materials. The Cool Roof Rating Council (CRRC) has been organized to measure, rate, and label the solar reflectance and thermal emittance of roofing materials. Many industrial leaders have introduced cool roofing materials on the market. In contrast, the development of cost-effective solutions for cool pavement has been very slow. The cool roofs criteria and standards are incorporated into the Building Energy Performance Standards of ASHRAE (American Society of Heating Refrigeration, and Airconditioning Engineers), California Title 24 building code, and the California South Coast's Air Quality Management Plans. Many field projects have demonstrated the energy benefits of cool roofs and shade trees. The South Coast Air Quality Management District and the United States Environmental Protection Agency (EPA) now recognize that air temperature is as much a cause of smog as NO_X or volatile organic compounds. In 1992, the EPA published a milestone guideline for tree planting and light-colored surfacing. Many countries have joined efforts in developing heat-island-reduction programs to improve urban air quality. The efforts in Japan are of quite notable interest.

Trees can potentially reduce energy consumption in a city and improve air quality and comfort. These potential savings are clearly a function of climate: in hot climates, deciduous trees shading a building can save cooling-energy use, in cold climates, evergreen trees shielding the building from the cold winter wind can save heating-energy use. Trees also improve urban air quality by lowering the ambient temperature and hence reducing the formation of urban smog, and by dry deposition to absorb directly gaseous pollutants and PM10 from the air. Low-emitting trees should be considered in designing a tree-planting program, so that volatile organic compounds emitting trees would not undermine our efforts. Finally, a major cost of a tree-planting program is that associated with planting and maintaining by tree professionals. The cost of water consumption of trees in most climates is small compared to planting and maintenance costs. It is quite possible to design a low-cost tree-planting program that utilizes and employs the full voluntary participation of the population.

Pavements cover a surprisingly large fraction of a city's surface and typically are among the darkest and hottest surfaces. There are well-accepted methods of creating lighter-colored pavements, such as chip-seals using whiter aggregate. The difficulty in implementing cooler pavements is in taking a long-term and city-wide view of the situation. Most often, the decision about pavements is made on the basis of initial cost, without regard for the shortened lifetime of hot pavements or the heat-island effects. When these are taken into account, the life-time costs of cooler pavements may be lower for many kinds of roads.

REFERENCES

[i] H. Akbari, M. Pomerantz, and H. Taha, "Cool surfaces and shade trees to reduce energy use and improve air quality in urban areas," Solar Energy, 70(3):295-310, 2001.

[ii] H. Akbari, S. Davis, S. Dorsano, J. Huang, and S. Winnett (editors), *Cooling Our Communities: A Guidebook on Tree Planting and Light-Colored Surfacing*, U.S. Environmental Protection Agency, Office of Policy Analysis, Climate Change Division, 1992.

[iii] H. Akbari,, "Opportunities for saving energy and improving air quality in urban heat islands," Chapter 2 in *Advances in Passive Cooling*, M. Santamouris (Ed.) Earthscan Publications Ltd. London, U.K. 2007.

[iv] H. Akbari, L. S. Rose, and H. Taha, "Characterizing the Fabric of the Urban Environment: A Case Study of Sacramento, California," LBNL-44688, Lawrence Berkeley National Laboratory, Berkeley, California (December 1999).

[v] H. Akbari and L. S. Rose, "Characterizing the Fabric of the Urban Environment: A Case Study of Salt Lake City, Utah," LBNL-47851, Lawrence Berkeley National Laboratory, Berkeley, California (February 2001).

[vi] H. Akbari and L. S. Rose, "Characterizing the Fabric of the Urban Environment: A Case Study of Chicago, Illinois," LBNL-49275, Lawrence Berkeley National Laboratory, Berkeley, California (October 2001).

[vii] L. S. Rose, H. Akbari, and H. Taha, "Characterizing the Fabric of the Urban Environment: A Case Study of Greater Houston, Texas," LBNL-51448, Lawrence Berkeley National Laboratory, Berkeley, California (January 2003).

[viii] H. Taha, S.C. Chang, and H. Akbari, "Sensitivity of the Houston-Galveston meteorology and ozone air quality to local perturbations in surface albedo, vegetation fraction, and soil moisture: Initial modeling results." Report prepared for the Global Environment and Technology Foundation, Center for Energy and Climate Solutions, March. Report No. LBNL-47663. Lawrence Berkeley National Laboratory Berkeley, CA, 2001.

[ix] H. Taha, S.C. Chang, and H. Akbari, " Meteorological and air-quality impacts of heat island mitigation measures in three U.S. cities," Report No. LBL-44222, Lawrence Berkeley National Laboratory Berkeley, CA, 2000.

[x] H. Taha, S. Douglas, J. Haney, A. Winer, M. Benjamin, D. Hall, J. Hall, X. Liu, and B. Fishman, "Modeling the ozone air quality impacts of increased albedo and urban forest in the South Coast Air Basin," Lawrence Berkeley Laboratory Report LBL-37316, Berkeley, CA, 1995.

[xi] H. Akbari, S. Bretz, D. Kurn, and J. Hanford, "Peak Power and Cooling Energy Savings of High-Albedo Roofs," *Energy and Buildings* 25:117-126, 1997.

[xii] S. Konopacki, H. Akbari, L. Gartland, and L. Rainer, "Demonstration of Energy Savings of Cool Roofs," Lawrence Berkeley National Laboratory Report LBNL-40673. Berkeley, CA, 1998.

[xiii] S. Konopacki, and H. Akbari, "Measured Energy Savings and Demand Reduction from a Reflective Roof Membrane on a Large Retail Store in Austin," Report number LBNL-47149. Berkeley, CA: Lawrence Berkeley National Laboratory, 2001.

[xiv] H. Akbari, "Measured energy savings from the application of reflective roofs in 2 small non-residential buildings," *Energy*, 28:953-967, 2003.

[xv] D.S. Parker, J.R. Sherwin, and J.K. Sonne, "Measured Performance of Reflective Roofing Systems in a Florida Commercial Buildings," ASHRAE Transactions, 104(1), American Society of Heating, Refrigeration, and Air Conditioning Engineers, Atlanta, Georgia, (January 1998).

[xvi] D. Parker, J. Sonne, and J. Sherwin, "Demonstration of Cooling Savings of Light Colored Roof Surfacing in Florida Commercial Buildings: Retail Strip Mall," Florida Solar Energy Center Report FSEC-CR-964-97. Cocoa, FL, 1997.

[xvii] E. Hildebrandt, W. Bos and R. Moore, "Assessing the Impacts of White Roofs on Building Energy Loads," *ASHRAE Technical Data Bulletin* 14(2), 1998.

[xviii] J. Akridge, "High-Albedo Roof Coatings - Impact on Energy Consumption," *ASHRAE Technical Data Bulletin* 14(2), 1998.

[xix] C. Boutwell and Y. Salinas, "Building for the Future—Phase I: An Energy Saving Materials Research Project," Oxford: Mississippi Power Co., Rohm and Haas Co. and the University of Mississippi, 1986.

[xx] J.R. Simpson and E.G. McPherson, "The Effect of Roof Albedo Modification on Cooling Loads of Scale Residences in Tucson, Arizona," *Energy and Buildings*, 25:127-137, 1997.

[xxi] S. Konopacki and H. Akbari, "Simulated Impact of Roof Surface Solar Absorptance, Attic, and Duct Insulation on Cooling and Heating Energy Use in Single-Family New Residential Buildings,"

Lawrence Berkeley National Laboratory Report LBNL-41834. Berkeley, CA, 1998.

[xxii] H. Akbari, S. Konopacki, C. Eley, B. Wilcox, M. Van Geem and D. Parker, "Calculations for Reflective Roofs in Support of Standard 90.1," *ASHRAE Transactions* **104**(1):984-995, 1998.

[xxiii] L. Gartland, S. Konopacki, and H. Akbari, "Modeling the Effects of Reflective Roofing," *Proceedings of the ACEEE 1996 Summer Study on Energy Efficiency in Buildings* **4**:117-124. Pacific Grove, CA, 1996.

[xxiv] S. Konopacki, H. Akbari, S. Gabersek, M. Pomerantz, and L. Gartland, "Cooling Energy Saving Potentials of Light-Colored Roofs for Residential and Commercial Buildings in 11 U.S. Metropolitan Areas," Lawrence Berkeley National Laboratory Report LBNL-39433, Berkeley, CA, 1997.

[xxv] H. Akbari, S. Konopacki, and M. Pomerantz, "Cooling energy savings potential of reflective roofs for residential and commercial buildings in the United States," *Energy*, **24**, 391-407, 1999.

[xxvi] H. Taha, "Modeling the impacts of large-scale albedo changes on ozone air quality in the South Coast Air Basin," Atmospheric Environment, 31(11):1667-1676, 1997.

[xxvii] H. Taha, "Modeling the Impacts of Increased Urban Vegetation on the Ozone Air Quality in the South Coast Air Basin," Atmospheric Environment, 30(20):3423-3430, 1996.

[xxviii] A.H. Rosenfeld, J.J. Romm, H. Akbari, and M. Pomerantz, "Cool Communities: Strategies for Heat Islands Mitigation and Smog Reduction," *Energy and Buildings*, **28**, pp. 51-62, 1998.

[xxix] Western Roofing, Online at http://WesternRoofing.net , 2002.

[xxx] J.V. Hall, A.M. Winer, M.T. Kleinman, F.M. Lurmann, V. Brajer and S.D. Colome, "Valuing the Health Benefits of Clean Air," *Science*, **255**: 812-817, 1992.

[xxxi] H. Akbari, A. Rosenfeld and H. Taha, "Summer Heat Islands, Urban Trees, and White Surfaces," *ASHRAE Transactions* 96(1) (1990), American Society for Heating, Refrigeration, and Air Conditioning Engineers, Atlanta, Georgia.

Standby Energy Use in California Homes

Alan Meier

Environmental Energy Technologies Division
Lawrence Berkeley National Laboratory
Berkeley, CA 94720

Abstract. Many electrical devices in homes continue to draw power while switched off or not actively performing their primary function. These devices include familiar appliances, such as televisions, microwave ovens, computers, set-top boxes, mobile phone chargers, and video and audio components but also less obvious devices like dishwashers, tankless water heaters and smoke detectors. The energy use of these devices while in their low-power modes is now about 980 kWh/year (or 112 watts) per home in California, corresponding to about 13% of the state's total residential electricity use in 2006. If treated as a separate end use, low-power mode energy use is the fourth largest residential end use. About half of the electricity in the electronics end use is consumed in the low-power modes.

THE INCREASING IMPORTANCE OF STANDBY ENERGY USE

In the mid 1990s, researchers noticed the rising amount of electricity consumed by appliances that were either switched off or not performing their primary function (Meier, Rainer et al. 1992; Sandberg 1993). This electricity consumption was often needed to power internal clocks, infrared receivers (for remote controls), digital displays, or other electronic components. In other cases, the standby power use simply energized power supplies and circuits that did nothing. "Standby power" is the best-known term used to describe this phenomenon but many other terms are used including, phantom loads, leaking electricity, and off-mode power consumption. Any product with a remote control, external power supply, continuous display (including a sole LED), or a soft-touch keypad will have standby energy use.

Researchers estimated that standby energy use was responsible for as much as 11% of residential electricity use in Australia, 10% in Japan, 5% in the United States, and over 5% in some European countries (International Energy Agency 2001). In the late 1990s, Energy Star developed specifications to reduce standby energy use in televisions and VCRs. At about the same time European manufacturers voluntarily agreed to cut standby in those products, and Japan established an aggressive program to reduce standby in all products. Considerable progress has been made in reducing the standby energy use in some products, especially in televisions, computers, and external power supplies (Roth and McKenney 2007). Power-saving innovations

CP1044, *Physics of Sustainable Energy, Using Energy Efficiently and Producing It Renewably*
edited by D. Hafemeister, B. Levi, M. Levine, and P. Schwartz
© 2008 American Institute of Physics 978-0-7354-0572-1/08/$23.00

include the switch-mode power supply, lower-power displays, and improved techniques for power management. However, the number of products with standby power use began to grow rapidly; manufacturers wanted every product to have memory, keep track of the time, display information, or be able to act upon receiving a signal from a handheld remote control, a computer, or (in the case of a television set-top box) from a service provider. The savings achieved in some product types were offset by the increased number of new products with continuous power consumption; as a result, the total electricity consumed by products while not being used appears to be climbing.

As electrical products became more sophisticated, and performed more services while "off", the problem of standby energy use expanded to cover a range of low-power modes that were nevertheless clearly different in function from "active". There now exists a wide range of services performed while the products are not actively switched *on* ranging from *nothing* (beyond contributing waste heat to the space) to ensuring uninterrupted communications links. The common feature of all these modes, however, is that they are not in an *active* mode, that is, performing a primary function (the television showing an image, the PC computing, the coffee maker brewing coffee, or the subwoofer pumping out music). These modes have many names—sleep, hibernate, ready, etc.—and considerable confusion exists about terminology. The International Electrotechnical Commission (IEC) initially defined "standby power" as the lowest power mode of a product while connected to the mains (IEC 2005). However, this definition is undergoing revision and the likely outcome will be that "standby" will be used in a generic sense to describe all low-power modes. In this paper, we have adopted the expansive definition.[1]

MEASURING AND ESTIMATING STANDBY ENERGY USE

It is useful to understand the methods for measuring and estimating standby energy use because these procedures illustrate some of the technical aspects related to reducing standby. A test procedure to measure standby power for a particular mode was developed by the IEC (IEC 2005). Many products operate in one mode for essentially all the time and, if periods of activity are brief, the annual energy consumption is roughly equal to the product of power measurement and the time interval. If energy consumed during the active mode is significant, then this consumption must be subtracted. More complicated products operate in several low-power modes so the measurement procedure may need to be repeated for each stable mode. Standby *energy* use is more difficult to obtain because the time the product resides in each mode must be measured. In most cases, the energy use of a product is estimated. The procedure begins by identifying each mode and the functionality associated with it. For each standby mode, the following information must be measured or estimated:

---

[1] To avoid confusion, several researchers have adopted the term "lopomo" to describe all of the non-active modes. (Lopomo is a contraction of "Low-power mode".)

- The average power use (in watts, W); and
- The fraction of time the product resides in that mode.

The standby energy use of a single product is the sum of the product's energy use in each low-power mode. This can be written as:

$$\text{Annual standby energy use (in kWh/year)} = 8.76 \cdot \sum_{\text{all low power modes, i}} P_i \cdot U_i \qquad (1)$$

where P_i is the Power (in watts) and U_i is the Usage (in terms of a fraction of time), both for mode i. The term 8.76 converts the watts into kWh/year. If the *active* mode is included, then the formula predicts a product's total energy use.

The *national* standby energy consumption of the product is calculated by multiplying the single unit's energy consumption by the number of homes in the country and the fraction of homes with that product (the "saturation"):

$$\text{Standby energy use in the United States} = N \cdot S \cdot 8.76 \cdot \sum_{\text{all low power modes, i}} P_i \cdot U_i \qquad (2)$$

where N is the number of homes in in the United States and S is the saturation of that product type. The saturation typically ranges from zero to two though it can exceed two for products like televisions, smoke detectors, and ceiling fans. The saturation must also take into account the fact that some designs within a product type may not have any low-power modes while others do. For example, some washing machines rely on electromechanical controls (and have no standby energy) while most modern machines rely on electronic controls (which do have standby energy use).

In practice, estimating standby energy use is extraordinarily complex owing to the large number of products, modes, and operating hours. The data must be stitched together from a variety of sources because no single group can collect all the necessary data on modes, power, and operating practices. An estimate of standby energy use in California homes is described below based on a 2008 study (Meier, Nordman et al. 2008).

STANDBY ENERGY USE IN CALIFORNIA HOMES

To accurately estimate standby energy use, data on over 170 different types of electricity-consuming products were collected. Some products were familiar, such as VCRs, microwave ovens, and computers, but others were not previously carefully investigated, such as tankless water heaters, controls for heating systems, toaster ovens, modems, and compact audio systems. Each product had one or more low-power modes and some had more than four standby modes.

The identification of modes and their power use mostly relied on measurements of products in 75 homes, plus intensive measurements in eight homes (Nordman 2004). Even though about 2000 products were measured, other data sources were still required to fill in the gaps, some from other countries, notably Australia (Energy

Efficient Strategies 2006). Usage patterns were based on a telephone survey of 306 homes, detailed analyses of long-term measurements, and data and technical literature about specific products. Data on saturations of products came from a telephone survey, utility surveys, and estimates by trade associations. TABLE 1 shows the key sources for data on power, usage, and saturation.

TABLE 1. Major sources of data to estimate California standby energy use.

Power	Usage	Saturation
Spot measurements in 75 homes	Telephone survey of 306 homes	Telephone survey of 306 homes
Earlier measurements in eight houses	Time series measurements in 50 homes	In-home survey of 75 homes
LBNL measurements of builder-installed miscellaneous energy	Surveys (e.g., Nielsen)	Appliance saturation survey organized by utilities
Other in-house measurements	Magazine and journal articles (typically dealing with one product)	California Lighting and Appliance Saturation Survey (CLASS)
Measurements from other countries (principally Australia, Denmark, UK, Germany, New Zealand)	Australian case studies	Trade journals
Product technical specifications	Other technical reports	Trade associations
Magazine and Journal articles		U.S. DOE Residential Energy Consumption Survey (RECS)

It is generally recognized that most consumer electronics have standby energy use; however standby is appearing in unexpected places, too. Most gas-fired tankless water heaters, for example, draw a few watts continuously. Building codes now require mains-powered smoke detectors instead of battery-powered models; each of these units draws about 0.5 W and a large home can easily have eight smoke detectors.

Based on the data collected and calculations described earlier, California's standby energy use is about is about 982 kWh/year or equivalent to a constant draw of 112 W. This corresponds to roughly 13% of 2006 residential electricity use. The key results are summarized in TABLE 2. Some differences exist between homes in California and the rest of the nation but these results are roughly applicable to the United States. A first-order estimate for standby energy use in the average US home is one thousand kilowatt-hours per year. However, the US percentage is somewhat lower, roughly 9% vs. 13%, because the average national average residential electricity consumption—

the denominator—is higher (about 11,000 kWh/year vs. 7350 kWh/year) (Energy Information Administration 2005).

TABLE 2. Estimated California standby energy and power use in 2006.

	Per Home		California	
	Power	**Energy/year**	**Power**	**Energy/year**
Standby energy	112 W	982 kWh	1.29 GW	11,300 GWh
Lowest mode	54 W	470 kWh	0.617 GW	5,410 GWh
Total residential electricity use	840 W	7,350 kWh	9.6 GW	84,500 GWh
Number of products that constantly draw power	44		506 million	

About half of the standby energy occurs while the products are in their lowest power mode. The lowest power mode has been the target of numerous voluntary programs and regulations (IEC 2005; California Energy Commission 2007; Energy Star 2008). Several studies have shown that manufacturers have made considerable progress in reducing power consumption in this mode; however, these results indicate that energy consumption in the lowest power mode still represents a large fraction of total standby energy use. Consumers cannot lower the energy consumption of these products without physically disconnecting them from the outlet or by inserting a switchable power strip. The remaining energy use occurs at higher modes and functionality where, in principle, users could reduce energy use by enabling power management or other features.

The average California home—and probably the average American home—contains about 44 products that draw power all the time. This estimate includes the visible products like televisions, VCRs, and computers but also often overlooked devices such as doorbell transformers, ground fault circuit interrupt (GFCI) outlets, and smoke detectors. The saturations of these built-in devices are certainly expanding as a result of code requirements for new homes and through remodeling of existing homes.

The ten product types responsible for the largest standby energy use are listed in TABLE 3. Together, these product types account for roughly 40% of total standby energy use. Two forms of set-top boxes dominate the list. Other product types within the video category, televisions and DVRs, also rank among the highest users of standby energy.

TABLE 3. Products with highest standby energy use in California.

Product Type	Average Standby Electricity Use Per Home		Fraction of Total Standby Energy Use
	(kWh/year)	(W)	
Set-top box, satellite	58	7	6%
Set-top box, digital cable	50	6	5%
Television, CRT	47	5	5%
Video, DVR	38	4	4%
Audio minisystem	38	4	4%
Computer, desktop	37	4	4%
Receiver (audio)	33	4	3%
Phone, cordless	33	4	3%
Air conditioning, central	27	3	3%
Oven, microwave	26	3	3%
Total of top 10	**387**	**44**	**40%**

This ranking should be interpreted with caution because some product types might have ranked higher if similar product types had been combined. New products, such as digital photo frames and digital television adaptors—both so new that no homes had them in 2006—can appear in a large number of homes very quickly. New, energy-intensive features, such as hard disks in set-top boxes (DVRs), have gained market share with unexpected speed. Other products, such as VCRs, are disappearing (though usually not as rapidly as competing devices are appearing.) As a result the list is constantly changing.

Many of these products fall into a growing category of devices whose primary function is related to information processing, storage, and transmission. This is evidence of the energy implications of the "information age." One feature of this category is the large fraction of energy consumed while not performing their primary function. About half of the electricity consumed by products in this category occurs while in their standby modes.

If treated as a separate end use, standby energy use is the fourth largest residential end use in California. Standby energy use is likely to grow as consumers purchase more products with low-power modes and as a consequence of firm trends and building codes. Most new consumer electrical products have a standby power mode. An increasing fraction will link to networks, which may further increase power consumption while not active. Building codes requiring greater use of ground fault interrupt circuits (GFCI) and mains-wired smoke and carbon-monoxide detectors will also contribute to larger standby energy use. The consumption per device is small, but the number of devices involved is enormous. Standby energy use in California's homes will soon exceed 132 W (up from 112 W) as these trends continue.

Prospects for Reducing Standby Energy Use

From a physical perspective, a large fraction of standby energy use is unnecessary and could be avoided. Two design strategies could achieve reductions. First, more efficient components must be substituted. The power supply, that is, the component responsible for transforming mains power (115 or 220 V AC) to low voltage direct current, has received the greatest attention (and made the greatest progress). In the last decade, the traditional "linear" power supply has been replaced by "switching" power supplies with significantly higher conversion efficiencies and lower no-load losses.

Second, the product needs to automatically shift into the lowest possible functional mode. To accomplish this goal, power management, that is procedures to ensure that only components whose functionality is actually required are energized, must be implemented. Considerable energy savings potential exists in this aspect. The chips in some devices (such as set-top boxes) often have power management capabilities but which are not fully enabled. In other cases, products are prevented from "powering-down" by requirements imposed by a network. Communications protocols designed without consideration of energy use force devices to maintain higher levels of functionality than are actually necessary. In some cases, implementation of network communication protocols may increase standby energy use because they will allow computers, printers, set-top boxes, powered speakers and other products to shift from an active mode to standby. The net result is desirable, however, because the product's total energy use falls. The technologies already applied to mobile telephones and other portable devices demonstrate the impact of careful application of efficient components and power management. They often accomplish the same functionality but with power consumption levels one hundredth of fixed products.

The switchable "power strip", that is, an extension cord with multiple outlets and an independent switch, offers further energy-saving opportunities. Initially the switch on the power strip simply completely cut off power to the product (or products) without physically unplugging it. However, new designs include features that permit much more sophisticated control. A master-slave arrangement allows the power strip to sense when the user shuts down one product (such as a computer) and automatically disconnects closely related products (display, printer, speakers, etc.) Other designs respond to remote controls, motion sensors, and other external information. The power strip works best for clusters of products used in a coordinated fashion, such as computers, audio, or video equipment. The combined standby power use of these clusters can easily exceed 10 W, so the savings can be substantial. The power strip provides a crude form of power management in the absence of effective power management and communications protocols in the individual products; in the future many of these products will accomplish the same task without additional equipment.

The principal obstacle to greatly reducing standby energy use is economic; manufacturers are unwilling to invest in the improvements that would allow these savings. Some of these costs must also be borne by the industry as a whole in the form of development of common operating standards and communications protocols. The energy performance of mobile products, such as mobile telephone and ipods, demonstrate the potential savings when these investments are made.

CONCLUSIONS

Standby energy use is now a common feature of electrical appliances. Nearly all new products draw power continuously. This power consumption typically maintains a low level of functionality, such as powering an LED or an infrared sensor but, in many cases, serves no function. A first-order estimate for standby energy use in the average US home is one thousand kilowatt-hours per year, corresponding to roughly 9% of residential electricity use. The average California home—and probably the US home—contains at least forty products constantly drawing power. About half of the energy use arises from products related to information processing, storage, and transmission.

A large fraction of standby energy use is unnecessary but reductions are hindered by manufacturers reluctance to increase production costs. At the same time, industry-wide solutions are necessary to allow networked products to reliably power-down when their services are not actually required.

REFERENCES

California Energy Commission, *2007 Appliance Efficiency Regulations*. Sacramento, CA, State of California (2007).

Energy Efficient Strategies, *2005 Intrusive Residential Standby Survey Report*, Warragul (2006). Prepared for the Australian/NZ Ministerial Council on Energy.

Energy Information Administration, "Residential End-Use Consumption of Electricity 2001," (24 May 2005). Retrieved July 17, 2008, from http://www.eia.doe.gov.

Energy Star, "Energy Star Home Electronics," (2008). Retrieved July 16, 2008, from www.energystar.gov.

International Electrotechnical Commission, *International Standard 62301: Household electrical appliances - Measurement of standby power*, IEC, Geneva (2005).

International Energy Agency, *Things That Go Blip in the Night: Standby Power and How to Limit It*, IEA, Paris, France (2001).

A. Meier, B. Nordman, et al, *Low-Power Mode Energy Consumption in California Homes*, California Energy Commission, Sacramento, CA (2008).

A. Meier, L. Rainer, et al, "The Miscellaneous Electrical Energy Use in Homes," *Energy - The International Journal* 17(5), 509-518 (1992).

B. Nordman, *Developing and Testing Low Power Mode Measurement Methods*, California Energy Commission, Sacramento, CA (2004).

K. Roth and K. McKenney, *Energy Consumption by Consumer Electronics in U.S. Residences*, TIAX, Cambridge, MA (2007).

E. Sandberg, *Electronic Home Equipment — Leaking Electricity*, The Energy Efficiency Challenge for Europe, Rungstedgard, Denmark, European Council for an Energy Efficient Economy (1993).

Infrared Technology Trends and Implications to Home and Building Energy Use Efficiency

FLIR Systems
70 Castilian Drive
Santa Barbara, California 93117

Abstract. It has long been realized that infrared technology would have applicability in improving the energy efficiency of homes and buildings. Walls that are missing or are poorly insulated can be quickly evaluated by looking at the thermal images of these surfaces. Similarly, air infiltration leaks under doors and around windows leave a telltale thermal signature easily seen in the infrared. The ability to view, evaluate and quickly respond to these images has immediate benefits in addressing and correcting situations where these types of losses are occurring. The principle issue that has been limiting the use of infrared technology in these applications has been the lack of availability and accessibility of infrared technology at a cost point suited to this market. The emergence of low cost microbolometer based infrared cameras, not needing sensor cooling, will greatly increase the accessibility and use of infrared technology for *House Doctor* inspections. The technology cost for this use is projected to be less than $1 per inspection.

INFRARED TECHNOLOGY TRENDS

Modern infrared systems have been driven by a number of factors over their development. These include the physics of thermal radiation, the physics of infrared detectors, political, military, and more recently, the influences of a large commercial market. Systems once reserved for high dollar military and government customers are increasingly becoming common place and are finding homes in a wide variety of commercial and civilian markets. As this trend continues, infrared systems will become common place for use in the field of energy engineering and ensuring efficient energy use.

A Bit of Infrared History

In the year 1800, Sir William Herschel, Royal Astronomer to King George III of England wrote "There are rays coming from the sun... invested with a high power of heating bodies, but with none of illuminating objects... The maximum of the heating

CP1044, *Physics of Sustainable Energy, Using Energy Efficiently and Producing It Renewably*
edited by D. Hafemeister, B. Levi, M. Levine, and P. Schwartz
© 2008 American Institute of Physics 978-0-7354-0572-1/08/$23.00

power is vested among the invisible rays"[1]. He had discovered infrared radiation. However, it wasn't until 1900 that Max Planck[2], a theoretical physicist at the University of Berlin, wrote the equations now known as Planck's law. Planck's law describes the emission of photons as a function of wavelength for an object at a given temperature. Higher temperature objects emit their peak radiation at shorter wavelengths while cooler objects have peak emission at longer wavelengths. Planck's law expressed in radiant photon emittance is shown in Figure 1 below. It can be seen that the peak emission in the infrared for a 300 Kelvin terrestrial environment (room temperature is about 295 Kelvin) is at approximately 10um and that the emitted infrared radiation falls off steeply at shorter wavelengths. Most infrared systems have been developed for 300 Kelvin terrestrial imaging. The radiant photon emittance is

$$Q = \frac{2\pi c}{\lambda^4} \cdot \frac{1}{e^{\frac{hc}{\lambda kT}} - 1}$$

where λ, k , T and c are the wavelength, Boltzmann's constant, temperature, and the speed of light, respectively.

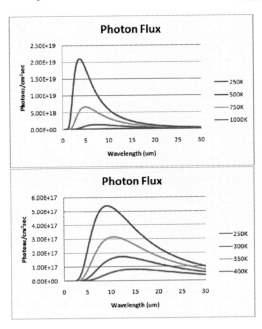

FIGURE 1. Radiant photon emittance as a function of object temperature and wavelength (note that the upper and lower graphs differ only in the temperature ranges plotted)

A second influence for the infrared technology development is that the atmosphere is mostly opaque to infrared radiation and is only transparent at certain wavelengths. This can be seen below in Figure 2, which shows typical transmission for infrared radiation over a 5km path at sea level. Here a number of transmission bands can be seen, several short wavelength bands at 1 to 2.5um (SWIR), a mid wavelength band at 3um to 5um (MWIR), and a long wavelength band at 8um to 12um (LWIR). For infrared radiation to be a useful tool there needs to be adequate radiation. For the 300 Kelvin source, this indicates wavelengths longer than 3um. Secondly, there needs to be adequate transmission through the atmosphere to support the imaging of these wavelengths. Two of the bands, the LWIR band, and the MWIR band (to a lesser degree), fall conveniently close to the peak region of emission for a 300 Kelvin object. Since these bands are well situated to transmit radiated infrared light from 300 Kelvin objects, they have largely been the focus for infrared detector developments.

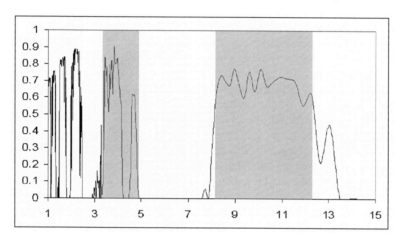

FIGURE 2. Transmission vs. wavelength (um) for atmospheric transmission of a 5km path at Sea Level (US standard Model)[3]. The 3 to 5 micron region is the midrange IR band (MWIR), and the 8 to 12 micron region is the long-wavelength IR band (LWIR).

There are many types of detectors that have been developed to detect signals in these bands and they conveniently fall into to two principle detector categories; 1) Detectors that detect infrared photons through quantum means, and 2) Detectors that detect by absorbing incident infrared radiation, changing temperature in response, and by exhibiting a change in a measurable property in response to their change in temperature.

The first category of detector generally operates on the principle of a photon exciting an electron from the valance band to the conduction band in a semiconductor material. These are typically photoconductive or photovoltaic types of detectors and are often implemented in narrow bandgap binary or tertiary material systems. For example photovoltaic silicon detectors have a bandgap of 1.1eV and can detect radiation to about 1um. A suitable infrared detector such as photovoltaic HgCdTe capable of the detection of 10um photons may have a bandgap of about 0.1eV. FLIR typically uses photovoltaic InSb with a band gap of 0.25eV in the MWIR band and microbolometers, as described in more detail below, in the LWIR band. The narrow bandgap is necessary to allow the longer wavelength infrared photon to exhibit adequate energy to free a valance band electron to move to the conduction band. The longer the wavelength of the infrared photon, the lower the energy bandgap needed. The band gap energy is $E = hc/\lambda$, where h is Planck's constant. One of the problems associated with this type of detection system is that thermal excitation of carriers can be adequate to overcome the narrow energy band gaps. This can present the problem of excess conduction band current (dark current) that can easily swamp out any signals detected through IR photon detection. As such, it is necessary to cool these detectors, often to cryogenic temperatures, to allow the photon generated current to dominate the thermal current and to achieve optimal detector performance. The requirement for cryogenic cooling of the infrared detector leads to the need for a vacuum, a dewar, a cryogenic cooler, and adds substantial cost and complexity to the system.

Much of the early development of infrared technology was driven by military system needs and the desire to achieve a very high level of detection performance. Since quantum detectors offered a higher degree of detection performance, much of the early funding and development was directed in this area. This has largely driven infrared technology down the path of quantum detectors requiring a vacuum dewar, cooler, with the associated additional cost and complexity. This has played a large role in slowing the adoption of infrared technology into many commercial applications.

The second principle type of detector that has been developed operates by the detector absorbing incident infrared radiation, changing temperature in response, and providing a means to measure this change, typically by a change in its resistance. These are referred to as bolometers. Unlike the quantum detectors, the bolometers may operate at room temperature. In the late 1980's employees[4] at the Honeywell Corporation invented a way to take the concept of the traditional bolometer and miniaturize it for high volume low cost manufacturing. This variation of the bolometer is referred to as a microbolometer or Honeywell microbolometer.

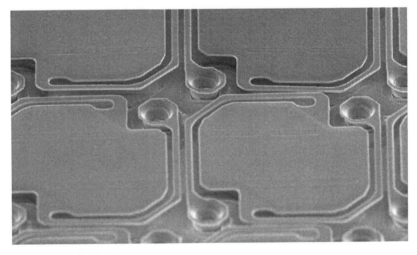

FIGURE 3. Scanning Electron Microscope image of several 38um x 38um microbolometer array elements.

220

Figure 3 above, shows a small portion of a microbolometer array. Each microbolometer, 38um x 38um pixels are shown, has two micro fine legs that suspend the main body of the microbolometer over an advanced integrated circuit. While the integrated circuit may utilize the same type of semiconductor manufacturing technology used in commercial micro circuits, the microbolometer is created through manufacturing techniques know as Micro-Electro-Mechanical Systems (MEMS). The suspended structure contains multiple layers and forms an optical cavity for maximizing photon absorption at a desired wavelength. The microbolometer heat capacity and thermal isolation are engineered and selected to provide a thermal time constant consistent with thermal imaging applications. While not providing the sensitivity of the quantum detectors, microbolometer arrays offer substantial advantages in systems. There is no need for a cryogenic cooler, and the size, power, and cost can be significantly improved, greatly increasing their utility for *House Doctor* inspections.

FIGURE 4. Packaged microbolometer array ready for infrared camera integration (3cm x 3cm).

Shown above in Figure 4, is a vacuum packaged microbolometer array. Total package size is about 3cm x 3cm and is less than 1cm thick. The FLIR units of this type have a sensitivity of about 0.05 Kelvin for room temperature objects. Electrical interface leads, vacuum housing, and optical interface window are visible. This component represents a substantial advancement in infrared technology for commercial applications. A lens element, video processing electronics, battery, and housing are added to complete the infrared camera.

Transition from Military to Commercial Markets

Probably the easiest way to visualize the changes that are occurring in the infrared industry is to look at what has already taken place in the Global Positioning System Technology (GPS) market. In the late 1970s and early 1980s, GPS technology was primarily funded, developed, and used by the military. Pull from commercial markets and a desire from the government to see reduced system cost led to a "dual use" phase for the technology starting in the late 1980s. Today, while there are still strong military needs, the commercial market has dominated the development and sales for GPS technology.

Figure 5. Evolution of Infrared System Technology,
House Doctor FLIR B-Series Camera is shown right.

The same factors are at work in the infrared markets. Infrared technology has made three distinct transitions. The first, as in the case for GPS technology, was supported by the military. These are systems largely designed, developed, and deployed to meet specific military needs. Accompanying this are extensive development phases, high development costs, high production costs, and limited production volumes to dilute these expenses. While these systems have been expensive, the value of these to the military is undeniable and the costs can be justified. Again, as in the case for GPS technology, commercial market pull accompanied by the desired for lower cost systems, lead to a dual-use phase in infrared technology beginning in the 1990s. Here, systems were designed and conceived for possible use in military and commercial markets, thus increasing production volumes, further offsetting development costs, and reducing system price. The final phase, which has just taken hold in the infrared market, is the commercial phase. The FLIR B-Series camera, shown above in Figure 5, was conceived, developed, marketed and is being produced based on commercial market needs.

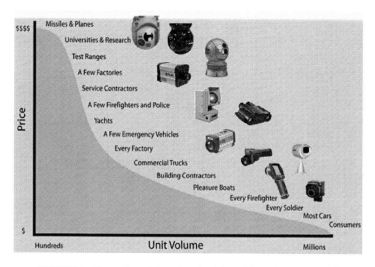

FIGURE 6. Illustration of the price elasticity of the infrared market.[5]

As the manufacturing volumes increase, costs come down and markets not previously accessible to the technology become available. This has a snow ball effect as the new markets further increase manufacturing volumes further reducing cost. Figure 6 above illustrates the elasticity of the infrared markets. For the upper left part of the chart, military and research applications purchase low volume high price systems for high added value applications. For the consumer applications, few will spend 10s or 100s of thousands of dollars for an infrared camera to troubleshoot thermal issues in their home. Few professional *House Doctors* would spend these amounts; however, when the cost of a useful infrared camera reaches 10s or 100s of dollars the market will be substantial.

FIGURE 7. Average selling price for *House Doctor* suited infrared cameras over the past ten years.

223

Figure 7 shows the average selling price for a series of infrared cameras that would be well-suited for the field of energy engineering and home and building diagnostics. These systems have seen more than a tenfold reduction in price while becoming more portable and easier to use. Common features today include;

- Stores 1,000 Radiometric JPEG Images
- Razor-Sharp, Thermal Image Quality
- Long, 7 Hour Battery Life
- Ultra-Portable, Weighs Just 1.2 lbs
- Built-in Laser LocatIR™, Removeable SD Memory Card
- Includes New, Powerful QuickReport Software
- Easy-to-Operate, Large 3.5 Full Color LCD
- Dew Point, Insulation and Color Alarm Features

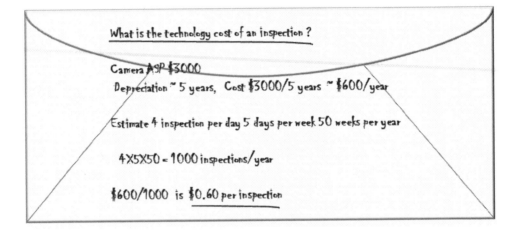

What is the technology cost of an inspection ?

Camera ASP $3000
Depreciation ~ 5 years, Cost $3000/5 years ~ $600/year

Estimate 4 inspection per day 5 days per week 50 weeks per year

4X5X50 = 1000 inspections/year

$600/1000 is $0.60 per inspection

Utility of Infrared Technology

Objects have two basic properties when it comes to emitting infrared radiation. The first, Planck's Law, was discussed earlier and is illustrated in Figure 1 above. The warmer an object is, the higher the emitted radiation and the shorter the wavelength of peak emission. For example, an object at 300 Kelvin will have peak emission at approximately 10um and at 750 Kelvin will have peak emission at about 5um. This is the case for objects that are perfect emitters. A perfect emitter is something that can be thought of as a perfectly black object, in that it absorbs and emits perfectly. This is an area in physics where the law of conservation of energy says that a good emitter must be a good absorber and a poor absorber (a reflector) must be a poor emitter. This effect is referred to as the emissivity of a material. A high emissivity material is a good emitter and good absorber and a low emissivity material is the converse, a poor emitter and poor absorber.

224

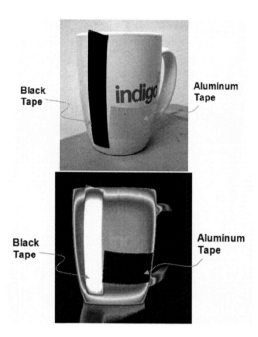

FIGURE 8. Visible and infrared image of a warm coffee cup showing the effects of emissivity[6]

Figure 8 illustrates the effects of temperature and emissivity in thermal images. On the top and bottom are images of a warm coffee cup. The upper image is a conventional visible image acquired with a digital camera, and on the bottom is an infrared image. The infrared image is acquired by integrating the radiated photons with a sensor as described above and then processing the signals from the sensor to a video signal. As such, all photons within the infrared wavelength band contribute to the integrated signal and any information associated with the wavelength of these photons is lost in the process. In this way, infrared "color" information is lost and the resulting image can be thought of in the same way as a black and white photograph in the visible spectrum. It is common practice in the infrared camera video processing electronics, to add back color to increase the dynamic range of the image to a cameras color display. This process is akin to taking a black and white image and arbitrarily mapping specific shades of grey to specific colors. In the infrared image shown in Figure 8 white is mapped to white, black is mapped to black and the grey scales in between are mapped through a rainbow of colors. While the inverse would be just as valid, the image above is shown as a "white hot" image.

Again, looking at the thermal image of the coffee cup, it can be noticed that the lip of the cup and the handle are at "cooler" temperatures. These regions of the image are representative of Planck's Law at work. Notice that the cup has a strip of black tape and aluminum tape on the surface. Because the tape is thin and attached to the surface of the cup, very little actual temperature difference is present in the surface of the cup and the surface of the tape. For the infrared image however, striking differences in emitted radiation from the cup in these regions is observed. The color black generally represents the absence of reflected light as such is a good absorber and consequently a good emitter. For the region of the cup where the black tape is present a higher signal is present. In contrast, the aluminum tape, somewhat reflective, is a poor absorber and a poor emitter rendering a lower emitted signal. The black tape has a higher emissivity than the cup, which in turn, has a higher emissivity than the aluminum tape.

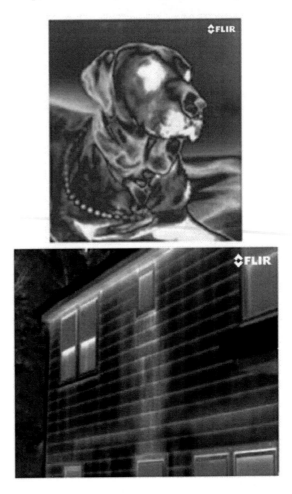

FIGURE 9. Infrared images of the IR dog Jake and Allen's house.

When we look at an infrared image, the different shades of grey or colors may therefore be due to signal mapping, temperature variations, emissivity variations, or as is the case most often, a combination of all three of these elements. It is probably worth noting here that there can be much more to the issue of signal mapping than can be addressed in this paper. This is an area that can be very important for those wanting to make measurements that have known and calibrated results (see the subject of radiometric images for more information[7]).

Infrared thermal imaging is an excellent tool for detecting moisture, liquids, and other fluids that effect either temperature or infrared radiation transmission. While these features are not as much of a factor for building heat loss analysis, they can be useful in identifying additional issues. These effects can further complicate interpreting an infrared image. For example, looking at the infrared home image of Figure 9, the location of the wall studs can be seen from the outside of the house through the siding and a difference in the quality of insulation can be observed between the studs. Heat loss in the windows is also apparent, however, many other features are present in the image that are somewhat complex and are representative of combinations of the elements discussed previously. These are all areas where skill and experience become a factor in understanding what is being observed.

Infrared Technology for Building Engineering and Diagnostics

There are a number of areas that infrared technology becomes useful in building energy efficiency:

1) Thermal Loss
2) Air Infiltration
3) Heat flow and Stack effect
4) Material inconsistency and or damage

Thermal Loss. Figure 10 shows the heat loss in a ceiling section due to a piece of missing insulation. This type of infrared image is probably the most commonly used in building diagnostics as the images are often easily understood and quickly show where to fix problems. In most buildings, it is desirable to control the interior temperature within a certain range to provide comfort for the occupants. Exterior temperatures may be below or above interior temperature. In situations where a temperature difference exists between the interior and exterior, the interior surface temperatures will change temperature based on the heat capacity of the materials and the thermal resistance between the material and the outside temperature. Ideal thermal imaging situations are present when there is an interior to exterior temperature difference and the conditions are in a steady state. Under these conditions performance aspects of the building, such as wall R-Value can be extracted from the infrared images. It should be appreciated that situations can exist where it is cooler in the morning and warmer in the afternoon on the exterior of a building. These situations produce thermal crossover where the direction of heat flow from the interior to the exterior can change direction during the day. Imaging through these periods of change can produce difficult to interpret results. In addition interior to exterior isothermal situations do not produce thermal images containing useful information on heat leaks or loss. As such, attention to time of day and weather conditions is required to obtain the best images for direct heat loss measurements.

FIGURE 10. Thermal imaging showing thermal difference due to missing insulation with a temperature difference of about 5 K.

Air Infiltration. Air can leak into a building at times when there is a temperature difference between interior and exterior and constitute an energy loss mechanism. The traditional method of measuring the area of infiltration leakage for a building has been to employ a "Blower Door" to pressurize the building and to fit the expression:

$$\mathcal{V} = c\Delta P^m$$

The coefficient c is called the flow coefficient and it corresponds to the nominal air flow value extrapolated for a building pressure of 1 Pa. Using a blower door to provide air flow into a building and measuring the pressurization of the building at several known flow rates, the area of leakage for the building can be estimated. Figure 11 below shows a modern blower door for small building for home infiltration measurements. The door expands to seal an interior exterior surface of the building and a large calibrated fan or blower is used to provide known flow into the building. The operator is seen making pressurization measurements.

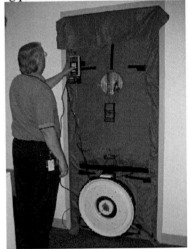

FIGURE 11. Dr. Robert Madding, Director of the FLIR Infrared Training Center (ITC™), operates a "Minneapolis Blower Door"[8]

The pressurization of the building results in interior air flowing out openings or defects in the construction to the lower pressure environment. As discussed, the blower door offers the ability to calculate the leakage area, however, finding the leaks is a bit of a detective effort. The use of a smoke stick or smoke pen as shown in Figure 12, enables one to see the path of air flow and to find areas where the interior air is escaping from the building.

FIGURE 12. Investigative engineer uses smoke pen to visualize air infiltration[9]

Infrared imaging has become an important tool in the area of air infiltration and can produce thermal images illustrative of the leaks. Additionally, for situations where a temperature difference exist between interior temperature and exterior air temperature, a blower door can be used to depressurize a building for the purpose of drawing air in through the leakage area. This process provides increased temperature contrast for the areas surrounding the leakage point. Figure 13 shows a window and a door and the effects of depressurization and the resulting thermal contrast generated in the leakage areas. Air leaks around the door and window are very visible and while detectable with a smoke stick, the infrared imagery provides a fast and accurate view of them.

Not all infiltration leaks are as easily diagnosed with a smoke stick. Figure 14 shows a log cabin that was analyzed for air leaks by Professional Investigative Engineers Amhaus & Fronapfel. A significant leakage area was discovered in high, difficult to reach areas using the combination of infrared thermal imaging and blower door depressurization.

FIGURE 13. Thermal image of Air infiltration around window and exterior door.[10]

FIGURE 14. Blower door depressurization of home is used to expose infiltration issues in construction.[11]

Heat Flow and Stack Effect – The high sensitivity of thermal imaging can be applied to many other aspects of building engineering and diagnostics. Temperature differences of < 50mK can be measured on floor, wall, and ceiling surfaces providing information for computer modeling and thus supporting a wide variety of diagnostic applications. Thermal effects due to air flow, interior thermal gradients, or updraft effects can be measured and used to diagnose these problems.

Material Inconsistency and or Damage – An additional and advantageous element of the thermal imaging is the ability to distinguish material differences in heat capacity, thermal conductivity, and emissivity. For example, walls built from different building materials are often easily distinguished in the infrared while they may look identical in the visible spectrum. Often these differences can help to piece together the story of how something was build, repaired, and where different materials may have been used.

SUMMARY

Infrared thermal imaging technology, once the purview of high end military products, is now being used for beneficial uses in the field of building design, engineering and energy use diagnostics. Progress with microbolometer technology has greatly advanced IR technologies for energy conservation. No longer must thermal cooling be considered and cameras with greatly reduced weight and cost are available. Thermal images are providing pictures of heat loss through walls, insulation, and air infiltration. The combination of using a blower door for depressurization and an infrared imager to capture the enhanced infiltration is providing images allowing fast accurate evaluation, diagnostics and correction for a buildings leakage area. Although progress is being made, the emergence of very low cost, handheld infrared imagers will soon allow a range of home diagnostics uses not previously practical. Microbolometers at today's prices of $3000 can reduce the equipment cost for a *House Doctor* inspection to less than $1 per inspection. As this technology reaches the price point where contractors, consumers, and home owners will purchase, easy to see and understand problems may be more rapidly identified, diagnosed, and resolved.

ACKNOWLEDGMENTS

We would like to acknowledge the FLIR Infrared Training Center (ITC) for providing technical information and infrared images for this field of use. Additionally, we would like to thank and acknowledge the information and images supplied by ITC InfraMation participants Edward L. Fronapfel P.E., Eric G. Amhaus Professional Investigative Engineers Inc., Jack M. Kleinfeld, P.E., Kleinfeld Technical Services Inc., Larry Steinbronn, CIE, CBST, Advanced Restoration Services Inc, Raphael Danjoux, ITC, FLIR Systems and Karl Grimnes, National Institute of Technology, Oslo, Norway.

REFERENCES

[1] R. Hudson Jr., "Infrared Systems Engineering", *John Wiley & Sons*, pg 3 (1969).

[2] H. Kragh, "Max Planck: the reluctant revolutionary," *Physics World* (2000).

[3] M. Nussmeier, simulations, *FLIR Systems*, Santa Barbara, CA.

[4] Cole, *"Microstructure Design for a High IR Sensitivity"*, US Patent 5,286,976 (7 Nov. 1988).

[5] J. Frank, VP Product Strategy, FLIR Systems, Santa Barbara California.

[6] Photographs by Tanya Stahlbusch, *FLIR Systems*, Santa Barbara, CA

[7] FLIR Infrared Training Center, FLIR Systems, Inc. has extensive information on the subject.

[8] The Energy Conservatory, Minneapolis, MN, http://www.energyconservatory.com.

[9] E.L. Fronapfel and J.M. Kleinfeld, *"Analysis of HVAC System and Building Performance Utilizing IR, Physical Measurements and CFD Modeling,"* Proc. InfraMation (2005).

[10] D. Raphael, ITC, FLIR Systems and K. Grimnes, National Institute of Technology, Oslo, Norway.

[11] E.G. Amhaus and E.L. Fronapfel, *"Infrared Imaging and Log Construction Thermal Performance,"* Proc. InfraMation (2005).

SESSION C

ENERGY USE BY AUTOMOBILES

The Race for 21st Century Auto Fuels

Alex Farrell, Adam Brandt and Sam Arons

Energy and Resources Group
University of California at Berkeley
Berkeley, CA 94720

Abstract. Will automobiles be predominately propelled with petroleum or other fuels during the 21st century? This paper carries out a comparison of five automotive technologies (efficiency, fossil, biofuel, electricity and hydrogen) under four basic criteria (infrastructure, vehicles, resources, and environment).

Economic growth brings more and faster travel. As nations become wealthier, their citizens travel further. An increase in per capita income by a factor of 10 from $3000 to $30,000 increases auto travel from 2000 miles/year to about 15,000 miles/year (Schafer 2005). Thus, a factor of 10 increase in income increases travel by a factor of seven. As China, India and other nations become wealthier, it will be difficult to constrain their desire for travel.

It is widely expected that petroleum production will not be able to fulfill future demand from automobiles. Conventional oil production will peak in the near future (some argue it has already peaked). The 1956 predictions by M. King Hubbert that US oil production in the lower-48 states would peak between 1966 and 1971 at 9 million barrels per day (Mbbl/d) were essentially correct (Deffeyes 2001). His method does not apply to all regions and reservoirs in detail, and making detailed predictions using his method is problematic (Brandt 2007). For example, his method did not consider the effects of higher oil prices and enhanced-recovery technology, and it also did not consider other resources such as tar sands and synthetic fuels. These concerns aside, his general conclusions are irrefutible: at some point petroleum production will drop and this will clash with increased demand for oil by the developing world.

Some 20 years after peak, we might expect oil production to fall short of projected demand by nearly 50% (Hirsch 2005). The inelastic demand for fuels will result in significantly escalated energy costs at this point (some even argue for prices in the hundreds of dollars per barrel (Leeb and Leeb 2004)). At this point, the *oil transition* will be well underway (Farrell and Brandt 2006). We define the oil transition as the transition to substitutes for conventional petroleum, be they fossil-based, biologically-based, or based on other energy carriers such as electricity. This transition will bring intertwined strategic, economic, and environmental concerns.

CP1044, *Physics of Sustainable Energy, Using Energy Efficiently and Producing It Renewably*
edited by D. Hafemeister, B. Levi, M. Levine, and P. Schwartz

On the environmental side, we are confronted with the issue of climate change. Greenhouse gas emissions must be greatly reduced if atmospheric carbon dioxide concentration is to be constrained to 450 ppm. It is generally assumed that this means greater reductions in the developed world in the beginning, but expanded to the developing nations before very long. And, of course, there are other environmental issues beyond that of climate change, such as land use, water use, and air pollution.

From the economic and strategic perspectives, one key feature is the rate of transition from oil to oil substitutes. At what rate will innovation and investments in new technologies be made to replace conventional petroleum? This rate will affect the likelihood of price spikes and vulnerability to supply interruptions. And while security risks from petroleum dependence may be growing, they have not yet motivated large–scale substitution with other fuels.

METHOD TO EXAMINE THE RACE FOR 21st CENTURY FUELS

Understanding the possible fuels of the future is key to developing effective policies to address carbon emissions from transport (Farrell, Sperling et al., 2007). We examine four different fuels (fossils, biofuels, electricity and hydrogen). In addition, as a "fifth fuel," we include increased end-use efficiency. The matrix of FIGURE 1 displays the five fuels as columns. The rows contain the four areas of concern (infrastructure, vehicles, resources, and environment). The plus signs (advantages), minus signs (disadvantages) and question marks (yet to be determined) indicate the tentative standings of the contenders. We support the arguments for these judgments below.

Increased Auto Energy Efficiency

On a cost basis, increased efficiency is the clear winner. However, efficiency improvements can only provide so much reduction in fuel use, and alternative energy sources must still be developed. This is especially true in the current climate, where rapid global economic growth ensures that savings from efficiency are quickly offset by increased demand in developing countries. The oil embargo of 1973-74 strengthened the political will of the Congress, resulting in the Corporate Average Fuel Economy (CAFE) standards during the fall of 1975. Before the oil embargo, automobile fuel economy was generally between 10 and 15 miles per gallon (mpg). CAFE effectively doubled fuel economy by increasing the standard to 27.5 mpg. Because it averages the fuel economy of the entire fleet of vehicles produced by a manufacturer, CAFE allows car companies to produce inefficient cars as long as this production is balanced with more efficient automobiles.[1]

---

[1] An excellent overview of CAFE is given by N.R. Council, *Effectiveness and Impact of Corporate Average Fuel Economy (CAFE) Standards*. Washington, D.C., National Academies Press (2002).

	Efficiency	Fossil	Biofuels	Electricity	Hydrogen
Infrastructure	+	+	+	+	−
Vehicles	+	+	+	−	−
Resources	−	+	−	+	+
Environment	+	−	?	?	?

Note: These are rough, subjective judgments that depend on how various fuels are produced, and they will change with innovation. 36

FIGURE 1. The score card comparing the entrants in the race for automobile fuels of the 21st century.

Figure 2 shows EPA data on fuel economy and vehicle performance (EPA 2007). Note that it took about a decade for light duty vehicles to obtain the 27.7 mpg fleet average, due to the pace of increases mandated by CAFE. After this target was reached and further increases in CAFE were halted, efficiency began to be "traded away" for increased mass, power, and acceleration. Part of this is because light duty vehicles have lost ground while sport utility vehicles (SUV) and light truck sales increased.

Figure 2 illustrates that engineering has allowed the return of vehicle weight to pre-CAFE levels and a 60% increase in horsepower, while slipping some 10% in fuel economy. US (and California) vehicle efficiency lags behind world standards. Japan and the European Union have set standards at about 50 mpg for the 2010–12 time window. China has a standard of about 35 mpg for the year 2008.

Higher vehicle efficiency should be our first priority, if for no other reason than the monetary savings that come with efficiency (see Figure 3). The issue of safety must be explored. (See chapter on *Safe Automobiles* by Tom Wenzel and Marc Ross in this book.) But, efficiency is not the entire answer for three fundamental reasons:

•Efficiency is not a source of energy, fuels will still be needed.

•Inherent tradeoffs in vehicle performance and cost will eventually emerge.

•U.S. automakers and unions are poorly positioned to compete in the efficient auto sector.

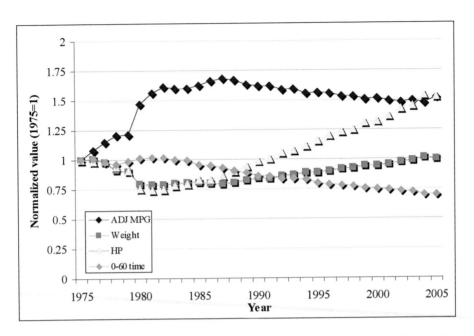

FIGURE 2. US auto and truck fuel economy, horsepower, weight and 0 to 60 mph acceleration times (1975–2005). Values normalized in 1975. Values in 1975 were: MPG = 13.1 mi/gal; Weight = 4060 lbs; HP = 137 HP; 0-60 time = 14.1 sec. Source: (EPA 2007)

	Average fuel economy improvement	Net savings (3 years, no discount)
Subcompact car	12%	$200
Midsize car	20%	$350
Large car	27%	$1,500
Small SUV	25%	$1,500
Large SUV	42%	$1,300
Large Pickup	38%	$1,100

FIGURE 3. National Academy of Sciences projected savings from fuel economy improvements (Council 2002). Notes: Gasoline price of $1.5 per gallon was used, diesel and hybrid drivetrains ignored. Since the bench mark cost used was $1.50/gallon, the savings will be increased by at least a factor of two.

238

THE REIGNING CHAMPION: FOSSIL FUELS

Petroleum-based automobiles utilize a large existing infrastructure. This infrastructure needs expansion (additional refineries and exploration) if it is to meet future demand, but its familiarity make this less problematic than for our other fuels. Storage facilities are in place, both private and governmental. These could be expanded to protect further against oil supply shocks or politically-based disruption of supply. The fossil-fuel resource base is very large, but it will be under severe pressure from increased demand.

Substitutes for Conventional Petroleum (SCP)

Because a significant fraction of conventional oil reserves are nationalized, the large international oil companies lack access to new oil fields. This is driving oil production towards more abundant, low-quality resources. The transition to these fossil-based substitutes for conventional petroleum (SCPs) has begun (Brandt and Farrell 2007).

For the purposes of this paper, SCPs include oil produced using enhanced oil recovery (EOR, used in depleted or low-quality heavy conventional oil deposits), deep-water extraction, synthetic crude oils produced from tar sands (~200 Gbbl barrels of reserves, +1 Tbbl resource in place), coal-based synthetic fuels (often called CTL synfuels), and oil shale. Figure 4 shows that these technology/resource combinations represent massive amounts of oil-equivalent, potentially dwarfing what has already been extracted (Brandt and Farrell 2007). Two things are of key importance with regard to this figure: first, these substitutes for oil emit more carbon dioxide than by conventional oil production and consumption; and second, while they are significantly more expensive than conventional oil extraction, the cost is far less than the present market value of $120 / bbl (May, 2008).

For each resource type in Figure 4, the quantity of liquid hydrocarbon fuels that could be produced is plotted on the horizontal axis, accounting for losses in conversion. The dark portion of each resource is a conservative estimate of resource availability (typically, reserves), while the lighter portion represents less certain resources. The monetary and GHG "costs" are plotted for each resource on the vertical axis, given in dollars per barrel (year 2000 $/bbl, top) and carbon emissions in grams of carbon equivalent emitted per mega-joule of refined product (gCeq./MJ, bottom). The vertical dimension for each segment of the curve represents the range of variability or the uncertainty associated with the implications of utilizing each resource.

EOR typically requires injection of materials into the reservoir, often carbon dioxide or thermal energy (e.g. steam) (Lake 1989). About half of US EOR is steam-based, while half is CO_2 based (Moritis 2006). Low-quality heavy oils must often be chemically upgraded and often cleaned of impurities.

Synthetic crude oil (SCO) is produced from the Alberta tar sands using a number of techniques. Production techniques include mining-based or in-situ production of bitumen in the tar sand deposit. Mining-based approaches dig up and slurry the bitumen/sand mixture, and remove the bitumen from the sand with heat and detergents. In situ processes use steam to extract bitumen with the sand left in place. Upgrading the bitumen to produce refinery-ready synthetic crude oil occurs using fluid or delayed coking processes, followed by hydrotreating. This produces refinery-ready synthetic crude of about 32 °API, 0.1 wt% sulfur and 500 ppm nitrogen (Gray 1994).[2]

Synthetic liquid fuels (e.g. synthetic diesel) can be produced in a two step process. First, a syngas comprised mainly of CO and H_2 is created through catalysis (in the case of gas-to-liquids, or GTL) or gasification and reformation (in the case of coal-to-liquids, or CTL). The syngas is then converted into liquid fuel using the Fischer-Tropsch (FT) catalytic process.[3] CTL synfuels have higher GHG emissions than GTL synfuels because of the higher proportion of carbon per unit hydrogen in coal. Note that in FIGURE 4, the CTL and GTL quantities are theoretical maxima because they assume all gas and coal are used as feedstock for SCPs and none for other purposes. We ignore methane hydrates due to a lack of peer-reviewed data, but these resources are potentially very large.

Lastly, synthetic crude oil can be extracted from oil shale, a sedimentary rock that contains a solid hydrocarbon-like substance. The oil shale resource endowment is large (Dyni 2006), but processing is costly – both financially and environmentally. The standard approach to oil shale production is to mine and crush the rock, and then heat it in a retort, releasing synthetic oil and gas (Bartis, LaTourrette et al. 2005). This process requires more capital, energy, and water than conventional oil production and has higher GHG emissions. A new in situ process developed by Shell Oil may reduce these challenges, but it is still in the development stages (Brandt 2007).

One concern with SCPs is the amount of energy returned per unit of energy invested in extraction and refining of fuels. Although these SCPs requires more energy investment than historic values for conventional oil, the energy return on investment (EROI) from SCP is still generally favorable (Cleveland 2005). US coal production has an EROI of over 80, so conversion to liquid fuels should not doom this resource to a negative energy return. The prospects for oil shale seem somewhat more dubious because it has a lower energy density than coal. A Shell executive recently claimed that their process had an EROI of 3.5 based on direct energy inputs (Shell 2006).

SCP technologies may lead to significant environmental damage. Using GHG emissions as a proxy, the potential environmental effects from production of SCPs could be quite large, possibly twice those of conventional oil production per unit of fuel delivered. A partial solution is carbon capture and storage (CCS), which could sequester additional upstream GHG emissions (see chapter on CCS by Larry Myer) (Intergovernmental Panel on Climate Change 2005). However, carbon capture and sequestration would only go so far: emissions due to fuel combustion would remain. For some SCPs, the difference would be small (e.g. 10%-20% reductions for tar sands, EOR and GTLs) but for others, CCS could significantly lower total emissions (reductions up to 50% for CTLs and oil shale). The large endowment of coal resources suggests that emissions would still be large even if CCS were employed.

---

[2] These values are the average of Syncrude and Husky Oil SCO product.
[3] Note also that the FT process can be used to produce synthetic hydrocarbons from biological feedstocks.

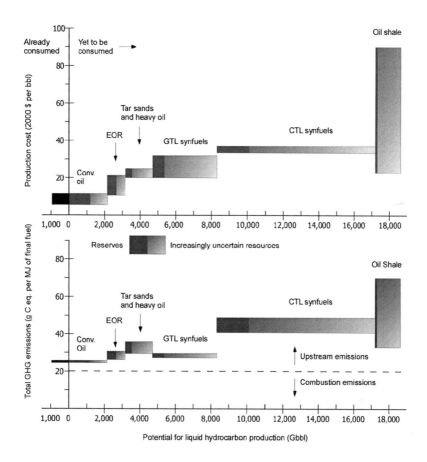

FIGURE 4. Global supply of liquid hydrocarbons from all fossil resources and associated costs in dollars (top) and GHG emissions (bottom). EOR is enhanced oil recovery, GTL and CTL are gas- and coal-derived synthetic fuels. The lightly shaded portions of the graph represent less certain resources. GHG emissions in the lower figure are separated into fuel combustion (downstream) and production and processing (upstream) emissions by a dashed line. Results are based on costs and conversion efficiencies of current technologies available in the open literature Gas hydrates are ignored due to a lack of reliable data. The GTL cost estimates assume a range of $0.5 to $2 per MBTU. See (Brandt and Farrell 2007) for details.

GTL and CTL production processes have similarities to hydrogen production (Yamashita and Barreto 2005), for which the cost of adding CCS has been estimated as an additional 5%-30% of production costs (Intergovernmental Panel on Climate Change 2005). Thus, total costs for most of the available resources would likely remain below $50 per barrel even with CCS. Note, however, that the prospects of large scale carbon storage are not assured (Wilson, Johnson et al. 2003), and other environmental issues would remain to be addressed even if CCS were used.

GHG emissions have no market value today, so SCPs are currently being produced without CCS. This phenomenon is not captured in current forecasts of GHG emissions, so actual emissions may be worse than "business as usual" scenarios (Intergovernmental Panel on Climate Change 2005). Given the expense involved and the realities of the market, government policies to internalize the cost of GHG emissions will be needed to induce CCS (the exception being CCS used for enhanced oil recovery, which will likely be developed as part of EOR through CO_2 injection).

Overall, FIGURE 4 shows that the oil transition is not a shift from abundance to scarcity: fossil fuel resources abound. Rather, the oil transition is shift from high quality resources to lower-quality resources that have increased risks of environmental damage, as well other risks.

BIOFUELS

The chapter on *The Science of Photons to Fuels* by Steve Chu in this book describes the basic science and approaches to obtaining considerably more fuel than corn feedstock presently yield. Some new distribution infrastructure will be needed for biofuels, especially for ethanol, which must be splash-blended with gasoline at the terminal and cannot be carried in existing pipelines. Biofuels generally offer good energy storage, although biodiesel (fatty acid methyl ester) can be prone to spoilage, especially when produced from certain feedstocks such as algae (Chisti 2007). It is possible to create bio-based hydrocarbons (e.g., Fischer-Tropsch diesel) that completely eliminate infrastructure and storage issues, however these are not commercially viable at this time (Hamelinck, Faaij et al. 2004). Current vehicles need little to no change to use existing biofuels (Lave, Griffin et al. 2001).

There is a limited resource base for biofuels from corn feedstock (Jones, O'Hare et al. 2007). Any use of large amounts of land will have uncertain environmental effects (Farrell, Plevin et al. 2006). The rapid increase in ethanol production for automobiles is shown in FIGURE 5 (Kammen, Farrell et al. 2008). Note that the price of petroleum has risen markedly, at the same time that the ethanol fuel production has also risen. The success of biofuels has relied on subsidies and mandates. It is a small industry that, until recently, was quite profitable and growing rapidly. Record high corn prices driven largely by ethanol demand, as well as infrastructure limitations, have reduced profits and slowed the pace of growth (Kotrba and McElroy 2008; Taylor and Murphy 2008). Ten billion gallons per year of ethanol corresponds to about 0.2 billion barrels/year of petroleum on an energy basis. This corresponds to about 0.5 Mbbl/day, or about 2% of US petroleum usage, a significant portion of which is an additive to gasoline to prevent preignition, a service originally provided by lead additives.

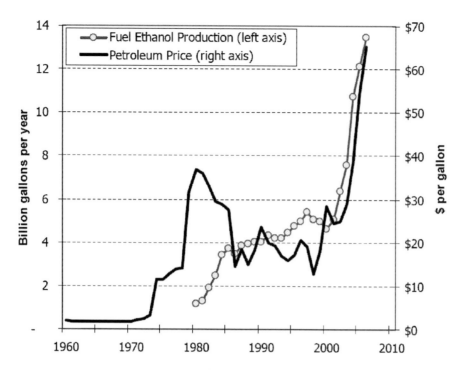

FIGURE 5. Petroleum price ($/barrel) and fuel ethanol production (billion gallons/year). Sources: EIA, BP, RFA.

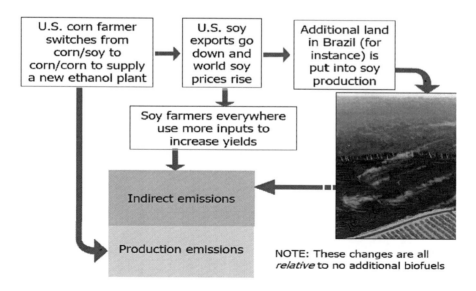

FIGURE 6. Indirect increase of green house gases from increased corn production for ethanol.

Land use is central to most environmental and social effects of biofuels. There are direct effects from soil erosion, fertilizer runoff, and biodiversity loss. There are also indirect environmental effects caused by displacing land use through the global market competition between energy and food production. For example, the displacement of soybean acreage in the US to increase corn production has reduced the global supply of soybeans and raised their price. This price increase can result in the clearing of new land somewhere for increased soy production, see FIGURE 6, and thus can result in increased GHG emissions (Searchinger, Heimlich et al. 2008).

In the long-term, biofuels may be produced from feedstocks that do not compete with food production. The technologies that could greatly strengthen the production of biofuels include ligno-cellulosic fermentation, gasification and synthesis, fast pyrolysis, and algae (again, see chapter by Steve Chu). FIGURE 7 displays some of the pathways for biofuel production (Farrell and Gopal 2008).

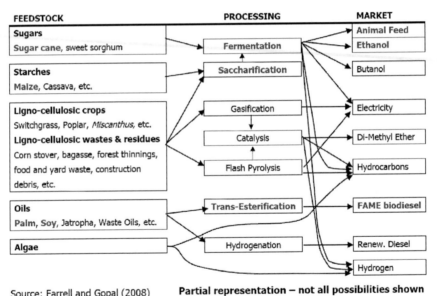

FIGURE 7. Many possible biofuel production pathways exist.

ELECTRICITY

The electric car is evolving. If lithium batteries can be made viable and inexpensive, the electric automobile that uses off-peak power will blossom. See chapter in this book on batteries by Venkat Srinivassen. Other issues are as follows:

- Little new distribution infrastructure is needed in the US for vehicles powered by electricity at least at first, but the grid is aging, which can cause problems. Spare generating capacity currently on the grid would allow millions of electric and/or plug-in hybrid vehicles to charge in California without the need for new power plants (Lemoine et al., 2008), especially if they charge at night. This result should apply to other state or regional grids as well.
- Electric motors use electrical energy with very high efficiency, much more efficiently than combustion engines use petroleum fuels (on order 90% as compared to 20% for IC engines). If electricity is generated from fossil fuels, we must also consider the efficiency of producing the electricity from the power plants in the first place, and transmitting it over the electricity grid to the vehicles. Because electricity plants are larger and designed to handle higher temperatures than the combustion engines in vehicles, they can achieve higher thermodynamic efficiency (close to 50% for co-generation plants). Because power plants can achieve such high efficiencies, even when transmission and distribution losses (about 8-9%; Wang, 2001) are taken into account, electric vehicles' overall efficiency is still greater than those achievable by conventional IC engines.
- Energy storage batteries are presently expensive and too short-lived, so electric vehicles are expensive. Improved technologies (such as nanophospate batteries) are being vigorously developed, but uncertainty remains.
- The environmental effects from electricity production depend strongly on which fuel/technology combination is used (see Table 2).

The hybrid (HEV) has a 15% smaller IC engine than a traditional car. It shuts down when in the idle mode and it uses regenerative breaking when slowing down. The battery is kept within a narrow range around 50% charge in order to prolong battery life (see work by Andy Frank at UC Davis).

The Plug-in Hybrid Electric Vehicle (PHEV) may enter the competitive marketplace in less than a decade. The traditional hybrid electric vehicle (HEV) essentially is a PHEV with a very small range since its battery is very small. The Plug-in Hybrid Electric Vehicle (PHEV) engine downsized by 33% compared to the conventional car. A larger lithium-ion battery is used to give it a pure electric range for, perhaps, 20-40 miles, without petroleum use. It is charged from the grid, preferably at night. The batteries are discharged more deeply than for the HEV, raising the requirements for battery quality. A battery pack might cost $10,000 initially and perhaps $5,000 with volume production. Manufactures of the advanced Li ion batteries claim battery life of 10 years, the life of the car. However, this remains to be proven over the next 10 years. Many manufacturers are developing, or have announced plans to develop, PHEVs.

Currently, fuel savings do not offset PHEVs' additional capital cost at today's battery prices (see Table 1; Lemoine et al., 2008), but PHEVs can reduce GHG emissions substantially, depending on the carbon intensity of the electricity source (see Table 2). However, because of today's high battery prices, the GHG emissions reductions can be very expensive on a cost/tonne-CO_2eq basis ($1,000-$2,000/tonne in some cases). Because of these high costs, the best thing we can do to achieve a large amount of emissions in the most cost-effective manner is to replace/convert conventional SUVs with/to hybrid, electric, or plug-in versions as soon as possible. This is because conventional SUVs are so inefficient in comparison to smaller compact cars that any gain in efficiency is disproportionately greater for SUVs (Arons et al., 2008).

Tables 1 examines the break-even points for PHEV, in comparison to both a conventional compact vehicle (CV) and a hybrid vehicle (HEV), under varying electricity and gasoline prices (note that insurance and O&M are ignored). The table shows that at today's prices (around $3/gal and $0.10/kWh), driving a PHEV allows you to save about $2,000 in fuel costs over the vehicle lifetime in comparison to a CV, and about $1,000 in comparison to an HEV. However, battery prices are in the $10,000 range (or $5,000 at scale), so PHEVs' fuel savings do not compensate for the additional capital cost of purchasing the vehicle in the first place. Put differently, the battery costs would have to drop to around $300-$500/kWh from their current level of over $1,000/kWh in order for PHEVs to become cost-effective (Lemoine et al., 2008).

TABLE 1. The discount rate is r, 16% n = 12-year vehicle lifetime. Use r to get NPV of savings. Divide by battery size (about 5 kWh) to get $/kWh breakeven battery costs.

Gasoline price	$2/gal		$3/gal		$4/gal	
Annual PHEV fuel savings						
Elec. Price ($/kWh)	CV	HEV	CV	HEV	CV	HEV
$0.05	$294	$155	$471	$264	$649	$373
$0.10	$231	$93	$409	$202	$587	$311
NPV of PHEV fuel savings (n=12, r=16%)						
Elec. Price ($/kWh)	CV	HEV	CV	HEV	CV	HEV
$0.05	$1,525	$807	$2,450	$1,372	$3,375	$1,938
$0.10	$1,201	$483	$2,216	$1,048	$3,051	$1,614
Breakeven battery costs ($/kWh, n=12, r=16%)						
Elec. Price ($/kWh)	CV	HEV	CV	HEV	CV	HEV
$0.05	$298	$277	$479	$472	$600	$666
$0.10	$235	$166	$416	$360	$597	$555

Source: Lemoine et al (2008)

The values in Table 2 are calculated using the GREET model (Wang, 2001), updated with lifecycle GHG emission data for power plants (Pacca and Horvath, 2002). When driving in gasoline mode, hybrids (HEVs) and PHEVs reduce GHGs/mile in comparison to conventional vehicles (CVs) because of their higher efficiency (in part due to regenerative braking), and when driving in electric mode they can reduce GHGs substantially, especially if charging from a clean energy grid (such as the average California grid) or a clean energy source (such as wind power). Note that for compact cars, a PHEV20 (a plug-in hybrid with a battery big enough to drive 20 miles in electric mode) reduces about 30% of GHGs/mile (300 gCO_2e/mi to 200), whereas for an SUV, the percentage reduction is about 45% (600 gCO_2e/mi to 350). This result underscores the fact that, taken SUVs' existence as a given, the best way to quickly and cost-effectively reduce automotive GHG emissions with electricity is to convert or replace IC SUVs with electric/hybrid/PHEV versions (Arons et al., 2008).

TABLE 2. Greenhouse gas emissions for conventional, HEV and PHEV cars.

		Energy sources		
	Gasoline	US avg.	CA avg.	Wind
Compact Car				
CV	294			
HEV	225			
PHEV20	211	199	116	1
PHEV60	203	198	115	1
Sport Utility Vehicle				
CV	605			
HEV	401			
PHEV20	375	346	202	2
PHEV60	367	329	192	2

Source: Arons et al (2008)   Units: gCO2e/mi.

HYDROGEN

The hydrogen car, and particularly hydrogen storage, is discussed in detail in the chapter by Jan Herbst. The hydrogen car is an electric car where electricity is generated from hydrogen by an onboard fuel cell. The source of the hydrogen for the car can be either from electricity or natural gas reformation. There are reasons to be skeptical about the near-term prospects for hydrogen cars (Keith and Farrell 2003). The hydrogen car has the following issues that must be resolved before it can be commercialized (Farrell, Keith et al. 2003):

• Compared to an electric car, the additional conversion processes can introduce losses in efficiency.
• Major new distribution infrastructure is needed for hydrogen.
• Energy storage is bulky and expensive
• Fuel cell vehicles are currently very expensive

- The resource base is very large if hydrogen is obtained from natural gas, but there will be energy losses if electrolysis is used to obtain hydrogen.
- Environmental impacts will depend on which processes are used.

CONCLUSION

The matrix of five technologies (columns) and four evaluation criteria (rows) has been filled out in Figure 1. In brief, we summarize the results as follows:

- *Efficiency* is most cost effective and least damaging, but external energy is still needed to run automobiles.

- *Fossil* driven autos have infrastructure, existing vehicles and resources (if you allow for synfuel production). Fossil fuels damage the earth and create climate change. The production of fossil synfuels will be more damaging than the present conventional oil production.

- *Biofuels* are commercially competitive in Brazil but require significant subsidies to be competitive in the US. The existing infrastructure and vehicles can accommodate biofuels. However, the use of biofuels is limited by the amount of cropland available, and the small relative gain in energy. Too much use of land for ethanol would damage crop land over the long haul. Cellulosic feedstocks such as switchgrass could someday be useful.

- *Electricity* infrastructure is mature and capable, but the grid is aging and needs improvement. If the fraction of electric cars is limited, it would be possible to use the off-peak power of base-loaded plants, but at some point dedicated new power plants would have to be constructed. A key challenge is the lifetime, size, and expense of the lithium ion battery. There are sufficient resources to make electricity for cars, so the challenge is matching the size (and expense) of the battery with the necessary range and refueling speed society demands.

- *Hydrogen* pipelines and gas stations do not exist and would have to be created. The hydrogen car would need greatly improved storage capacity and cheaper fuel cells. Hydrogen can be produced from electrical power, but with an efficiency loss. The issue of the environment is uncertain since it depends on which method is used to obtain the hydrogen.

In our view the oil transition (or the race for 21st century fuels) brings long-term environmental concerns that loom larger than economic or security threats. This is because tradeoffs between these three types of impacts have a strong potential to be resolved by accepting increased environmental damage in order to avoid economic or security risks. The global petroleum industry has begun to recognize this interaction, but strategies to deal with them have not yet emerged.

248

ACKNOWLEDGMENT

Rich Plevin provided assistance with the biofuels portion of this paper.

REFERENCES

J.T. Bartis, T. LaTourrette, et al, *Oil Shale Development in the United States: Prospects and Policy Issues. RAND: Infrastructure, Safety and Environment*, Santa Monica, CA, RAND, 68 (2005).

A.R. Brandt, *Converting Green River oil shale to liquid fuels with the Shell in situ conversion process: energy inputs and greenhouse gas emissions*, Berkeley, CA, Energy/Resources Group (2007).

A.R. Brandt, "Testing Hubbert," *Energy Policy* 35(May), 3074-3088 (2007).

A.R. Brandt and A. E. Farrell, "Scraping the bottom of the barrel: CO_2 emission consequences of a transition to low-quality and synthetic petroleum resources," *Climatic Change* 84(3-4), 241-263 (2007).

Y. Chisti, "Biodiesel from microalgae," *Biotechnology Advances* 25(3), 294-306 (2007).

C.J. Cleveland, "Net energy from the extraction of oil and gas in the United States," *Energy* 30(5), 769-782 (2005).

N.R. Council, *Effectiveness and Impact of Corporate Average Fuel Economy (CAFE) Standards*, Washington, D.C., National Academies Press (2002).

K.S. Deffeyes, *Hubbert's Peak: The Impending World Oil Shortage*, Princeton, Oxford, Princeton University Press (2001).

J.R. Dyni, *Geology and resources of some world oil-shale deposits*, Reston, Virginia, US Geological Survey, US Department of the Interior, 42 (2006).

EPA, *Light-duty automotive technology and fuel economy trends: 1995-2007*, Washington, D.C., Environmental Protection Agency (2007).

A. Farrell and A. Gopal, "Bioenergy research needs for heat, electricity, and liquid fuels," *MRS Bulletin* 33, 373-380 (April 2008).

A. Farrell, R.J. Plevin, et al, "Ethanol can contribute to energy and environmenal goals," *Science* 311, 506-508 (27 January 2006).

A. Farrell, D. Sperling, et al, *A Low-Carbon Fuel Standard for California: Part 1 - Technical Analysis*, Berkeley, CA, Institute for Transportation Studies (2007).

A.E. Farrell and A.R. Brandt, "Risks of the oil transition," *Envir. Research Letters* 1(1) (2006).

A.E. Farrell, D.W. Keith, et al, "A strategy for introducing hydrogen into transportation," *Energy Policy* 31(13), 1357-1367 (2003).

M.R. Gray, *Upgrading petroleum residues and heavy oils*, New York, Marcel Dekker (1994).

C.N. Hamelinck, A.P.C. Faaij, et al, "Production of FT transportation fuels from biomass; technical options, process analysis and optimisation, and development potential," *Energy* 29(11), 1743-1771 (2004).

R.L. Hirsch, *The Shape of World Oil Peaking*, *Learning From Experience* 13 (2005).

Intergovernmental Panel on Climate Change, *IPCC Special Report on Carbon dioxide Capture and Storage*, Cambridge, University of Cambridge Press (2005).

A.D. Jones, M. O'Hare, et al, *Biofuel Boundaries: Estimating the Medium-term Supply Potential of Domestic Biofuels*, Berkeley, CA, Institute for Transportation Studies, U. California Berkeley, 36 (2007).

D.M. Kammen, A. Farrell, et al, *Energy and Greenhouse Impacts of Biofuels: A Framework For Analysis*, Berkeley, CA, Institute for Transportation Studies, University of California, Berkeley, 31 (2008).

D.W. Keith and A. E. Farrell, "Rethinking Hydrogen Cars," *Science* 301, 315-316 (18 July 2003).

R. Kotrba and A.K. McElroy, "Back to Reality," *Ethanol Producer Magazine* (2008).

L.W. Lake, *Enhanced oil recovery*, Upper Saddle River, NJ, Prentice Hall (1989).

L.B. Lave, W.M. Griffin, et al, "The Ethanol Answer to Carbon Emissions," *Issues In Science and Technology* 18(2), 73-78 (2001).

S. Leeb and D. Leeb, *The oil factor*, New York, Warner Business Books (2004).

D. Lemoine, D.M. Kammen, et al, "An innovation and policy agenda for commercially competitive plug-in hybrid electric vehicles," *Environmental Research Letters* 3 (014003): 10 (2008).

G. Moritis, "CO2 injection gains momentum," *Oil & Gas Journal* 104(15), 37-+ (2006).

A. Schafer, "Transportation, energy, and technology in the 21st Century," *GCEP Advanced Technology Workshop*, Stanford, CA, Global Climate and Energy Project, Stanford University (2005).

T. Searchinger, R. Heimlich, et al, "Supporting Online Materials for: Use of U.S. Croplands for Biofuels Increases Greenhouse Gases Through Emissions from Land Use Change," *Science* 319, 1238-1240 (29 February 2008).

Shell, *Oil Shale Test Project, oil shale research and development project (Plan of operation, submitted to Bureau of Land Management)*, Shell Frontier Oil and Gas Inc. (2006).

T. Taylor and R. Murphy, "Managing Through Tough Times in Ethanol Production," *Ethanol Producer Magazine* (2008).

E.J. Wilson, T.L. Johnson, et al, "Regulating the ultimate sink: Managing the risks of geologic CO_2 storage," *Environmental Science & Technology* 37(16), 3476-3483 (2003).

K. Yamashita and L. Barreto, "Energyplexes for the 21st century: Coal gasification for co-producing hydrogen, electricity and liquid fuels," *Energy* 30(13), 2453-2473 (2005).

The Relationship between Vehicle Weight/Size and Safety

Tom Wenzel[a] and Marc Ross[b]

[a]Lawrence Berkeley National Laboratory
1 Cyclotron Road, 90R4000
Berkeley, CA 94720
[b]University of Michigan
Ann Arbor, MI 48109

Abstract. Light-duty vehicles account for about 20% of US CO_2 emissions. However, new vehicle fuel economy standards have not been significantly tightened since they were first enacted three decades ago. A historical impediment to imposing tougher fuel economy standards has been the long-standing perception that reducing the mass of a car or truck would make it more dangerous to its occupants in a crash. One often hears that this perception is dictated by "simple physics:" that, all else being equal, you are at greater risk in a lighter vehicle than in a heavier one. Our research on driver fatality risk has found that, when it comes to vehicle safety, all else is never equal. Vehicle mass is not the most important variable in determining occupant safety, not even in frontal crashes between two vehicles. You are at no greater risk driving an average car than you are driving a much heavier (and less fuel efficient) truck-based SUV. And larger and heavier truck-based SUVs and pickups impose enormous risks on car occupants. We summarize the most recent research on the interplay between vehicle weight, size and safety, and what the implications are for new state and federal standards to reduce vehicle CO_2 emissions.

INTRODUCTION

Motor vehicles on roads and highways account for roughly 20% of total CO_2 emissions in the US; however, standards to reduce motor vehicle CO_2 emissions, and improve fuel economy, have not been substantially changed since they were first enacted in 1978. As a result, over time vehicle manufacturers have used new technologies that could have reduced fuel consumption to increase vehicle speed, acceleration, and weight. In addition, the popularity of "light" trucks (pickups, sport utility vehicles, and minivans) which get worse fuel economy and are subject to less stringent standards than cars, has resulted in the average on-road vehicle consuming the same amount of fuel per mile driven as twenty years ago. Because the number of vehicles registered and miles traveled continue to increase, large reductions in fuel consumption per mile driven would be needed if the transportation sector is going to contribute to the effort to reduce greenhouse gas emissions and slow global warming.

One of the historical impediments to increasing fuel economy standards has been the long-standing argument that reducing vehicle mass to improve fuel economy will inherently make vehicles less safe. This flies in the face of several recent studies that indicate that there are many fuel economy technologies that, if implemented widely,

CP1044, *Physics of Sustainable Energy, Using Energy Efficiently and Producing It Renewably*
edited by D. Hafemeister, B. Levi, M. Levine, and P. Schwartz
© 2008 American Institute of Physics 978-0-7354-0572-1/08/$23.00

would result in substantial improvements in fuel economy. One of the most promising routes to increasing fuel economy is mass reduction. However, the argument that "simple physics" dictates that lighter vehicles are less safe than heavier vehicles continues to sway many regulators, policy makers and many in the general public. We have spent the last several years examining the research underlying this argument, recent research that challenges it, and our own analyses. We conclude that the research claiming that lighter vehicles are inherently less safe than heavier vehicles is flawed: that other aspects of vehicle design are more important to the on-road safety record of vehicles. Moreover, it doesn't matter what analyses of historical data on vehicle mass and safety tell us; the use of new designs and light-weight, high-strength materials in motor vehicles has the potential to create new vehicles that are simultaneously lighter, with higher fuel economy, and safer. Widespread use of such materials may put to rest the tired argument that lighter vehicles are less safe than heavier ones.

TECHNOLOGIES TO IMPROVE VEHICLE FUEL ECONOMY AND SAFETY

A recent workshop of experts on vehicle safety and fuel economy, including representatives from auto manufacturers, academia, materials industry, and NGOs, found that there is little, if any, trade-off between improvements in fuel economy and in safety in light-duty motor vehicles, particularly if priority is given to both of these goals (ref. 1). A reduction in vehicle mass is an important technique for improving fuel economy, but certainly not the only one.

Figure 2 shows the most recent analysis of the cost of specific technologies to increase the fuel economy of light-duty vehicles. These technologies, such as torque converter lockup, aggressive shift logic, variable valve timing, variable valve lift, camless valve actuation, engine off at idle, cylinder deactivation, electric power steering and water pump, continuously variable transmission, turbocharging, as well as reductions in tire rolling resistance, air drag, engine friction, and mass, can increase fuel economy by over 50%, and are cost effective at current gas prices ($3 per gallon). Use of new power-trains (such as hybrid electric, plug-in hybrid electric, homogenous charge compression ignition, and fuel cells) or fuels (such as clean diesel or other low-carbon fuels) were not considered in the analysis; as gas prices climb even higher, more fuel economy technologies become cost effective.

Conversely, strategies can be employed to increase vehicle safety, with little effect on fuel economy, other than a small increase in vehicle mass. Electronic stability control, in which the vehicle computer automatically brakes individual wheels to inhibit a dangerous skid, better seat belts, and stronger roofs and pillars, and vehicle-to-vehicle communication, all can make vehicles more safe.

Although many technologies exist to simultaneously improve both vehicle fuel economy and safety, auto manufacturers may turn to perhaps the least costly and easiest method to meet higher fuel economy regulations: reducing vehicle mass.

Vehicle mass is a major factor in fuel economy. In response to federal fuel economy standards for new vehicles first enacted in 1975, the heaviest (over 4,000 lbs) cars were virtually eliminated from the new vehicle fleet, from 46% of sales in

1975 to 9% in 1980. These heavier vehicles remained a small part of the new vehicle fleet until the late 1980s, when the sales of heavier light trucks, many used as substitutes for cars, began to increase. By 2003 the fraction of light trucks over 4,000 lbs (32%) was approaching the level of heavy car sales in 1975 (40%) (Figure 2).

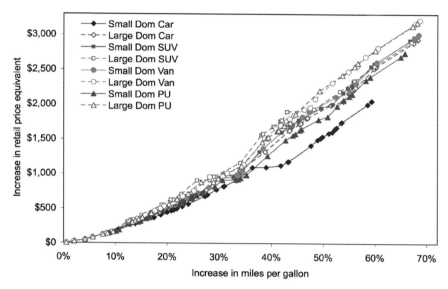

FIGURE 1. Fuel economy cost curves for domestic vehicles, medium term (2013-2018; reference 2).

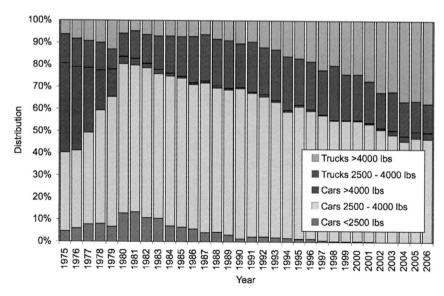

FIGURE 2. Distribution of new light-duty vehicle sales, by year, type, and EPA weight (curb weight plus 300 lbs; reference 3).

If the mass of a typical modern car were reduced 10%, the fuel economy would be increased 3% to 8%. (The larger increase applies if the engine size is reduced to keep acceleration performance the same.) The mass of vehicles, and especially the heavier pickups and SUVs, can be reduced substantially with little effect on vehicle size, by down-weighting components, replacing most of the body-on-frame designs of trucks with unibody designs used in cars, and increased use of light materials (such as high-strength steel, light metals such as aluminum and magnesium, and fiber-reinforced plastics).

However vehicle mass has just one *intrinsic* property critical to safety that is independent of design and materials: in a collision with another vehicle or a roadside object, the lighter of the pair is more strongly decelerated. Depending on the details of the crash, the stronger deceleration may create a greater risk of injury or death to an occupant of the lighter vehicle. But this additional risk is relatively small compared to what frequently happens in a serious crash: a) intrusion of the other vehicle or roadside object into the passenger compartment of the vehicle, b) rollover of the vehicle, or c) failure of the restraints to keep the occupants away from contact with hard interior surfaces.

The key issue to prevent intrusion is the strength of the passenger compartment and the height and stiffness of the collision partner. It is practical to use stronger materials and more compatible designs to reduce casualties in two-vehicle collisions. Mass is not *intrinsic* to any of this; for example, light honeycomb or fiber-reinforced materials can sever the *historical* connection between mass and strength.

The key issue to prevent rollover is to lower a vehicle's center of gravity. Although SUVs and pickups are more likely to roll over than passenger cars, the height of a vehicle's center of gravity, and not mass, determine the propensity of a vehicle to roll over. The propensity to rollover can be reduced by lowering the center of gravity and/or by increasing the trackwidth (the width of the vehicle between the tires). Electronic stability control (ESC) is a new technology that provides automatic braking separately at the four wheels to inhibit rollovers from occurring, and should substantially reduce rollovers. However if a rollover occurs, the crush resistance and performance of the roof will affect whether a belted occupant will be injured, and whether the occupants are belted will affect whether they are ejected (ejection is likely to result in serious to fatal injury).

Finally restraints (both safety belts and air bags) and interior padding provide important protection to occupants in all types of crashes. Side curtain air bags, which reduce head contacts with windows, are becoming more prevalent, and an increasing number of these systems can be triggered in a rollover. Advanced seat belts, with pretensioners and load limiters, are being incorporated in many models. Under research are four-point seatbelts, which would hold occupants in position in side-impact crashes better than today's three-point lap/shoulder belts. Improved restraints would also better control the deceleration in crashes with roadside objects, and thus further minimize the historical relationship between vehicle mass and safety.

Although vehicle mass is not intrinsic to improving occupant safety, currently safety technologies, such as ESC, curtain side airbags, and advanced seat belts, tend to be included in heavier, and more expensive, car models.

NOTHING SIMPLE ABOUT VEHICLE SAFETY

While lighter vehicles are at a disadvantage in crashes with heavier ones, all else being equal, a general reduction in vehicle weight across all vehicle types would not have a significant impact on safety. This is not only because the physics of the situation concerns only the relative masses of colliding vehicles, but also because a substantial majority of casualties in motor vehicle crashes are unrelated to the masses of the vehicles involved.

It is generally thought that driver factors are the most important causes in motor vehicle crashes, followed by environmental factors and vehicle factors (GAO). Driver behavior and environmental factors contribute greatly to whether a serious crash occurs; however, vehicle design dominates the type and extent of injuries once a crash has occurred. Whether a vehicle occupant is wearing a safety belt properly also is an important factor, as are the occupants age and physical condition.

Many consumers seek information on safety characteristics of different vehicles, and consult numerous published sources on safety equipment installed, and results of vehicle crash tests when deciding which vehicle model to purchase. How well a vehicle avoids potential crashes is based on how well it handles and brakes; how well a vehicle protects its occupants once a serious crash occurs may be indicated by laboratory crash tests using instrumented dummies. The National Highway Traffic Safety Administration regularly conducts crash tests and rates vehicle models under its New Car Assessment Program; the Institute for Highway Safety also conducts tests of different types of crashes, and publicizes their results. Consumer Reports publishes a Safety Assessment rating, which is a combination of NHTSA NCAP test results and results of handling and braking tests conducted by Consumers Union.

However, these tests provide only limited information, as no test procedure can replicate all of the conditions that occur on the road. Therefore researchers turn to real-world data on crashes and fatal crashes to shed light on how driver and environmental conditions affect whether crashes occur, and how well safety devices and vehicle design protect occupants once a serious crash has occurred. These analyses take the form of fatality or serious injury rates, per million registered vehicles or vehicle-miles, for specific types and makes/models of vehicles. Unfortunately, in these type of analyses it is difficult to determine the relative importance of driver behavior, environmental conditions, or vehicle factors in a fatality or injury rate for a specific class of vehicle. However, the most important driver and environmental factors can be analyzed to determine if they are biasing the overall result.

FATALITY RISKS

Rather than relying on crash tests in laboratory settings, we have used data on fatalities in the real world to assess whether heavier vehicles are indeed safer than lighter ones (ref. 4, ref. 5). We calculated the risk of driver fatality using the Fatality Analysis Reporting System, or FARS, the federal database of all fatalities occurring on public roadways. Fatality risk is simply the number of driver fatalities divided by the number of registered-vehicle-years, for each type of vehicle; registered-years are the number of vehicles registered as of January 2005, multiplied by the number of years each model year has been on the road (4.7 years for model year 2000 vehicles, 0.7 years for model year 2004 model year vehicles and so on). We did not develop this metric; it is regularly used by IIHS and other researchers. However, we were interested in exploring not only the risk to drivers of a certain type of vehicle, but also the risk that type of vehicle imposes on drivers of other vehicles. The sum of risk-to-driver and risk-to-driver-of-other-vehicles can be viewed as the total risk society bears by encouraging one type of vehicle over another. Usually analysts only present and discuss the risk to occupants of the vehicle in question, as if we as individuals or society at large don't care about the mayhem caused by aggressive vehicles and their drivers. In the following, we present both the risk-to-driver and risk-to-other-drivers.

Because these risks are calculated using data on actual crash fatalities, they account for both the risk of involvement in a serious crash, which is influenced by driver and environmental factors, as well as the risk of fatality once a serious crash has occurred, which is influenced by belt use, vehicle design, and driver frailty. As such, our use of the word "risk" can be taken to mean "risk as driven."

Our analysis of risk by vehicle type indicates that drivers are as safe in cars as in truck-based SUVs and pickups (Figure 3). Although the risk of fatality in a crash without a rollover is often lower in a SUV or pickup than in a car, their high center of gravity makes SUVs and pickups more susceptible to rollover crashes than a car. The recent trend of manufacturers producing crossover SUVs, which are lower and sometimes wider, and thus have a lower center of gravity, than conventional truck-based SUVs, has led to large reductions in the rollover fatality risk in SUVs. Among cars, large cars have only slightly lower risks than the midsize and the safer subcompact cars; this difference is within the statistical uncertainty of the risks, however. What really surprised us, however, is that there is a wide range in risk among different models of subcompact cars; in fact, the worst subcompact car model has a fatality risk three times that of the best subcompact cars. In Figure 3 we divide subcompact car models into a "low-risk" and a "high-risk" group.

On the other hand, in terms of risk to drivers of other vehicles, SUVs and pickups impose much higher risks than cars and minivans. This is not unexpected; most conventional SUVs were merely car-like cabins bolted onto the rigid steel frames of pickup trucks. These frames include two steel rods that can act like spears or fork tines when striking a car, overriding the bumper in a frontal crash, or punching through the door in a side impact (Figure 4). The combination of high and rigid steel structures make the designs of pickup trucks and truck-based SUVs incompatible with those of cars; researchers in the field refer to the "aggressivity" of such designs. Research on how to reduce light truck aggressivity towards cars includes whether truck fronts and bumpers should be redesigned, to make them lower and/or softer; whether car door sills should be raised to increase structural interaction with truck bumpers; or a combination of the two. Our analysis of car-based crossover SUVs, with unibody designs that don't have rigid frame rails, and have lower fronts, indicates that they are much less aggressive to other vehicles than the truck-based SUVs. As shown in Figure 3, the aggressivity of pickup trucks actually increases as the rated capacity of pickups increases (from compact pickups to one-ton pickups); the largest pickups are nearly six times more aggressive to other vehicles than the average car.

256

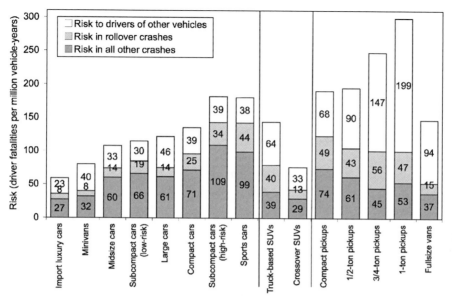

FIGURE 3. Risk-to-drivers, and risk-to-drivers-of-other-vehicles, by vehicle type; differences between vehicle types less than 10% are not statistically significant (reference 5).

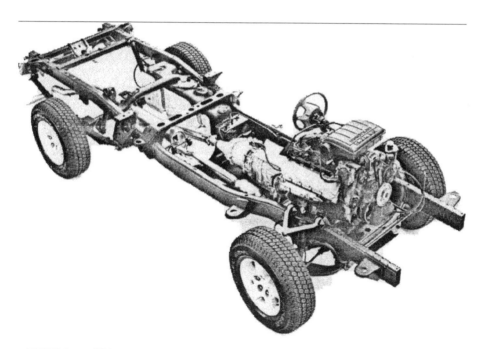

FIGURE 4. Stiff frame rails of pickups and truck-based SUVs (model year 2002 Dodge Ram 150 pickup truck).

VEHICLE DESIGN OR DRIVER BEHAVIOR?

Some of our results clearly have less to do with vehicle design, and more to do with who tends to drive a particular vehicle, and how. We used driver age and gender, as well as a measure of "bad" driving (based on alcohol or drug involvement, driving without a valid license or reckless driving in the current crash, as well as the driver's driving record in the last three years), to compare driver characteristics and behavior across vehicle types and models. The safest vehicles for their own drivers are minivans, probably because they are often used to transport children and their drivers tend to be more careful. On the other hand, the riskiest vehicles for their own drivers are sports cars, for obvious reasons. Minivans have the lowest fraction of young male drivers (4%) and "bad driver rating" (0.21), while sports cars have the highest (39%, 0.77). The cause of the extremely low risk of import luxury cars is not as obvious. Expensive luxury cars typically have the newest safety technologies, such as side curtain airbags and electronic stability controls; however, they also tend to have risky drivers (21% young males, 0.57 bad driver rating). The low risk of import luxury cars suggests that vehicle design and features can partially offset the risky nature of their drivers. On average, the high-risk subcompact cars have only slightly more risky drivers (23% young males, 0.57 bad driver rating) than the low-risk subcompact cars (21% and 0.49), which suggests that these low-price cars are intrinsically dangerous.

Drivers of truck-based SUVs and pickups are no different, at least in terms of age, gender, and driving history, from drivers of most cars; therefore, the differences we see in high rollover risk, and risk-to-others, in SUVs and pickups are probably not caused by their drivers. However, pickups tend to be driven on rural roads to a greater extent than cars. Rural roads are particularly dangerous, for a host of reasons: they tend to be designed for lower speeds, but speed limits are rarely enforced; oncoming traffic is not separated by a barrier, and shoulders are not protected by guardrails; they are not as well lit; they are further from trauma centers; etc. The more rural an area, the higher the fatality risk per vehicle (Figure 5). We calculated the population density of the county in which each crash occurred, and averaged over all vehicle types. As expected, pickup fatalities occur in more rural areas than fatalities in cars or SUVs (Figure 6). Analysis of California fatality risks by county indicates that, for all vehicle types, fatality risks both to drivers and to drivers of other vehicles increase as population density decreases (Figure 7). Therefore, some of the high risk to others that we calculate for pickups is the result of driving on dangerous rural roads, rather than the aggressivity of the design of pickups. SUVs tend to be driven on the same roads as cars.

To take the analysis one step further, we also calculated the two types of risk for individual makes and models, as shown in Figure 8. There are some interesting results for particular models. The lowest risk subcompact car model, the VW Jetta, has the most risky drivers (32% young males, 0.66 bad driver rating). The risks for models that are corporate twins (e.g. Ford Taurus and Mercury Sable, Chevrolet Cavalier and Pontiac Sunfire, etc.) differ somewhat, but are within the uncertainty of the estimates. In some cases there also are small differences in the drivers of each model, but these differences generally don't explain the differences in risk. The Hyundai Elantra was redesigned, with side curtain airbags added, in model year 2001; we see a 30%

reduction in risk to Elantra drivers before and after the redesign, which coincided with an improvement in its frontal crash test rating, from 3 to 5 stars. Similarly, the Ford Focus has a nearly 40% lower risk to its drivers than the model it replaced, the high-risk Escort. Because the average driver of these models likely did not change much in such a short time period, these reductions suggest that vehicle design has a large effect on safety.

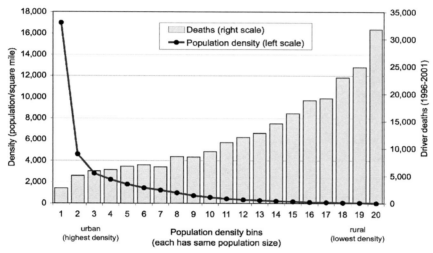

FIGURE 5. Driver deaths by population density (equal-population bins).

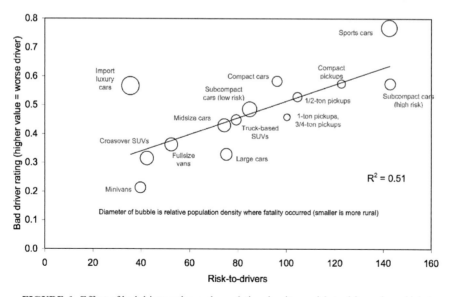

FIGURE 6. Effect of bad driver rating and population density on risk-to-drivers, by vehicle type (reference 5).

259

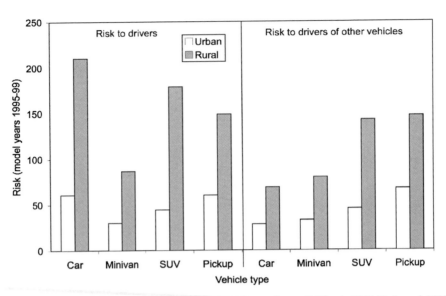

FIGURE 7. Risk to drivers and to others, by vehicle type and area, California 1995-99 through 2002 (reference 5).

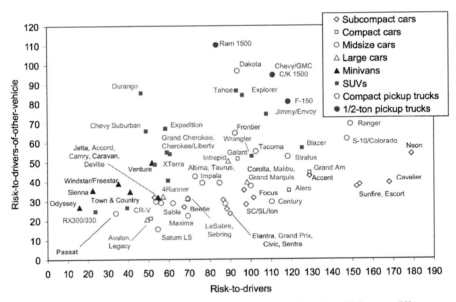

FIGURE 8. Risk-to-drivers in rollover crashes and all other crashes, by vehicle type; differences between models less than 20% are not statistically significant (reference 5).

So how does the fatality risk of car models correlate with their mass? Not as well as one might be led to believe (Figure 9). Other vehicle factors, such as vehicle nameplate (that is, Japanese/German vs. U.S. manufacturers), appear to have a stronger correlation with fatality risk than mass (Figure 10). And resale value after five years is much more strongly correlated with fatality risk than mass (Figure 11). This is more evidence that smart design can overcome any disadvantage a lower mass vehicle may impart.

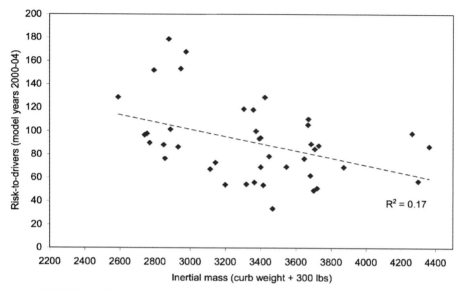

FIGURE 9. Risk-to-drivers and vehicle mass, for individual car models (reference 5).

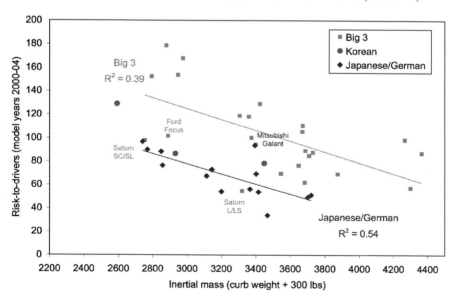

FIGURE 10. Risk-to-drivers and vehicle mass, for individual car models (reference 5)

261

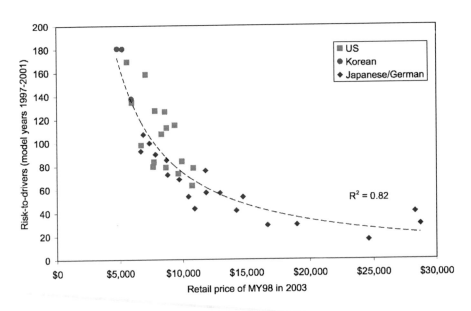

FIGURE 11. Risk-to-drivers and resale value, for individual car models (reference 5).

Even in crashes that are most affected by the disparity in vehicle mass, head on crashes between two cars, there is very little relationship between car mass and fatalities (Figure 12).

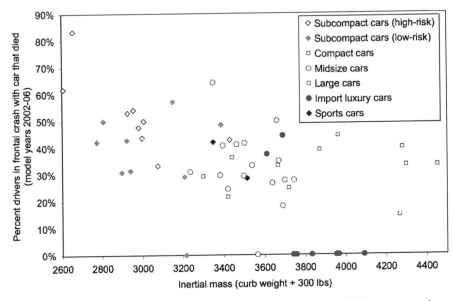

FIGURE 12. Percent of drivers in a fatal frontal crash with another car that died, by car type and mass.

Recall that in Figure 3 we split out conventional truck-based SUVs from the more recent car-based, or "crossover", SUVs. Crossover SUVs have lower frontal heights and masses than truck-based SUVs, resulting in not only lower risk in rollover crashes but also lower risk to drivers of other vehicles (Figure 13). In addition, crossover SUVs have 17% higher fuel economy for a given interior volume than truck-based SUVs (Figure 14).

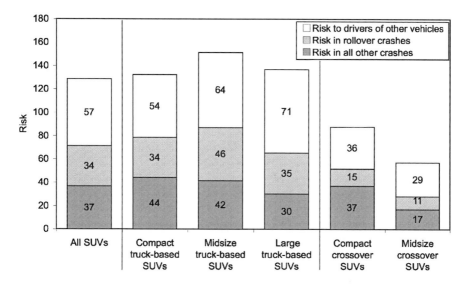

FIGURE 13. Risks in truck-based and crossover SUVs.

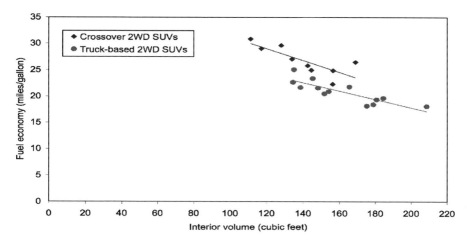

FIGURE 14. Fuel economy and interior volume of 2005 truck-based and crossover SUV models (2WD versions, reference 5).

263

CONCLUSION

For too long the conventional wisdom that lighter vehicles are less safe has helped stymie important efforts to increase new vehicle fuel economy and reduce greenhouse gas emissions. While vehicle fuel economy can be dramatically improved with no or little reduction in vehicle mass, with thoughtful design, lighter and smaller vehicles can be made as safe as larger, heavier ones.

NHTSA recently adopted revised fuel economy standards for light trucks, which allow larger trucks (defined by their exterior length times width "footprint") to have lower fuel economy than smaller trucks. This size-based standard is preferable to a weight-based standard some were proposing; however, NHTSA adopted the size-based standard because it is clinging to the myth that smaller trucks are less safe than larger, heavier ones. We and others have shown that this myth is unfounded; aside from specific designs, center of gravity (in rollovers) and frontal height and stiffness (in crashes with objects), and not footprint, are the important safety variables for occupant protection. And our analysis shows that larger (as measured by rated capacity) pickups impose greater risk on other vehicles than smaller pickups. NHTSA's size-based standard is not necessary to protect truck drivers, and will jeopardize car drivers; and it will not provide the dramatic reductions in CO_2 emissions needed to slow global warming. As the agency responsible for regulating both fuel economy and safety, NHTSA can do better on both fronts by adopting fuel-economy standards independently from safety standards. For safety standards, the nation clearly needs rules that take into account not just the dangers to a vehicle's occupants, but also the risk that a vehicle poses to others on the road.

The 2008 Energy Independence and Security Act does not explicitly call for a size-based standard, but continues to allow separate fuel economy levels for cars and trucks. By continuing to treat cars and light trucks differently, the proposed fuel economy regulation, as well as current safety standards, encourage consumers to purchase gas-guzzling and aggressive pickups and SUVs. Steps should be taken to apply the same standards equally to all light vehicles. Requiring that all light-duty vehicles meet the same stringent fuel-economy standards would increase the manufacturing costs of large pickup trucks and SUVs substantially. But some sticker shock here would be beneficial, because it would discourage consumers from purchasing a big vehicle unless they really needed one. Those people who truly require a large truck, say to pull a trailer or to haul cargo for a business, could be helped by offering them appropriate tax incentives (as are available now for vehicles weighing more than 6,000 pounds).

That change alone would improve highway safety for society as a whole. Other gains would come from appropriate regulations on vehicle compatibility and aggressivity, tougher roof-crush standards and improved seat-belt technology, which together would have a greater, direct effect on occupant safety than tinkering with fuel economy regulations.

ACKNOWLEDGMENTS

This work was supported in part by the Energy and the William and Flora Hewlett Foundations, through the U.S. Department of Energy under Contract No. DE-AC02-05CH11231.

REFERENCES

1. D. Gordon, D.L. Greene, M.H. Ross and T.P. Wenzel, *Sipping Fuel and Saving Lives: Increasing Fuel Economy without Sacrificing Safety: A report informed by an October 3, 2006 Experts Workshop on Simultaneously Improving Vehicle Safety and Fuel Economy through Improvements in Vehicle Design and Materials*, Washington DC: the International Council on Clean Transportation (2007).
2. Energy and Environmental Analysis, Inc., *Technologies to Reduce Greenhouse Gas Emissions from Light-duty Vehicles, Draft Final Report prepared for Transport Canada*, Arlington VA, Energy and Environmental Analysis, Inc. (2006).
3. R. Heavenrich, *Light-Duty Automotive Technology and Fuel Economy Trends: 1975 Through 2005*, EPA 420-S-05-0001, Washington DC, U.S. Environmental Protection Agency (2005).
4. T.P. Wenzel and M. Ross, "The effects of vehicle model and driver behavior on risk," *Accident Analysis and Prevention* 37, 479-494 (2005).
5. T. Wenzel and M. Ross, *Increasing the Fuel Economy and Safety of New Light-Duty Vehicles: White paper prepared for The William and Flora Hewlett Foundation Workshop on Simultaneously Improving Vehicle Safety and Fuel Economy Through Improvements in Vehicle Design and Materials*, LBNL-60449, Berkeley CA, Lawrence Berkeley National Laboratory (2006).

The Science of Photons to Fuel

Steven Chu

Lawrence Berkeley National Laboratory
Berkeley, CA 94720

Abstract. Transportation consumes 28% of US energy, and 60% of that is consumed by personal transportation. Because liquid fuels have high energy density, they will be the dominant fuel, until batteries have improved enough to support plug-in cars on an economic basis. Fifty million acres of energy crops plus agricultural wastes can produce roughly half of all of current US consumption of gasoline. Although ethanol from corn has received much attention as a possible substitute for gasoline, other biofuels feedstocks such as perennial grasses and agricultural wastes have greater potential for a much more environmentally friendly substitute for oil. The advantages of grasses over food crops such as corn include higher yield for given water and nutrient inputs; lower soil depletion and fertilizer run-off pollution. The major challenge in biofuels production from perennials is to improve the efficiency of conversion of the plant material to fuel. This paper describes some of the research that is being done to make biofuels from cellulose.

Helios Project and Energy Biosciences Institute

Biomass, mankind's long-standing energy source, has the potential of supplying significant amounts of renewable energy if efficiencies for the conversion to fuels can be substantially improved. Corn is currently the major source of bio-ethanol in the US today, but the current overall ethanol production that includes all energy inputs consumes nearly as much fossil energy than it replaces. Moreover, the net CO_2 emitted produce ethanol is only a slight improvement over using gasoline.

Since most of the solar energy captured by a plant goes into the production of its cell wall material – polymer sugars consisting of six- and five ring sugars (cellulose and hemicellulose) and lignin – it has long been recognized that the economic conversion of this material into a fuel would be a far better source of biofuel than the conversion of food material. Another enticing possibility would be the use microbes or algae to convert sunlight into hydrocarbons. To give a sense of the ability of plants to convert sunlight in biomass, the best overall quantum efficiency of the sunlight hitting the ground is roughly 1%. The conversion efficiency of single cell organisms can, in principle, be substantially higher since no bio-mass is need to produce root material or other plant structures.

Scientists at Lawrence Berkeley National Laboratory are doing research on many aspects of this problem as part of an effort known dubbed "The Helios Project". Its primary goal is to develop methods to "store" solar energy in the form of renewable transportation fuel. Several approaches under investigation include the generation of

CP1044, *Physics of Sustainable Energy, Using Energy Efficiently and Producing It Renewably*
edited by D. Hafemeister, B. Levi, M. Levine, and P. Schwartz
© 2008 American Institute of Physics 978-0-7354-0572-1/08/$23.00

biofuels from biomass, the generation of biofuels by algae, and the direct conversion of water and carbon dioxide to fuels by biomimetic devices ("artificial photosynthesis").

Researchers from the Helios Project are working in the following organizations:

Energy Biosciences Institute (EBI): The Institute is a collaborative partnership between UC Berkeley, Berkeley Lab, the University of Illinois at Urbana-Champaign and BP, and supported by a grant from the BP Corporation totaling $500 million over 10 years. The EBI is dedicated to the development of the next generation of biofuels and sponsors research on agricultural, scientific, environmental, and socioeconomic aspects of biofuel production. Other areas of research include fossil fuel bio-processing, carbon sequestration and microbially-enhanced hydrocarbon recovery.

Helios Solar Energy Research Center (SERC): Helios SERC scientists are developing solar-driven chemical converters that will create transportation fuels from water and carbon dioxide. Centered at Berkeley Lab, this project also includes researchers from UC Berkeley and several other universities. Research focuses on advanced nanomaterials for use in solar light collectors and electrodes, a new generation of catalysts for energy-efficient chemistry, and specialized soft and hard membranes for integrating the light harvesting, charge separating and fuel forming components.

Joint BioEnergy Institute (JBEI): JBEI, located in Emeryville, CA, is a partnership of researchers from three National labs – Berkeley Lab, Lawrence Livermore, and Sandia, and three academic entities – UC Berkeley, UC Davis and The Carnegie Institution at Stanford – and funded by the US Department of Energy. JBEI's research programs will apply advanced research technologies to develop environmentally friendly biofuels using plant biomass and microbes. In addition to the research partners, JBEI will establish industrial collaborations to bring relevant scientific and market capabilities in areas such as energy, agribusiness, and biotechnology.

Some of the problems being tackled include:

* The improved conversion of sunlight into biomass with more efficient use of water and nutrients. Plants that are drought and pest resistant and can be used on agriculturally marginal land are desirable.

* Engineering organisms to produce fertilizers on-site, especially photosynthetic nitrogen fixation in the form of ammonia. This will improve the energy efficiency for converting corn to fuel by avoiding the large fossil fuel consumption by conventional manufacturing processes and fertilizer transportation. More sustainable sources of phosphorous would have to be found.

* Engineering novel metabolic pathways for the conversion of cellulose and hemicellulose to fuels such as ethanol, methanol, and heavier alcohols, isoprenoid fuels, methane, hydrogen. Possible approaches include the design of synthetic organisms with

several artificial metabolic pathways to incorporate the relevant features of the cellulose degrading machinery and the conversion of simple sugars into the desired hydrocarbon.

* Engineering of green algae and cyanobacteria with improved photosynthesis rates, and on-site coupling of hydrogen production to catalytic conversion to carbon-based fuels.

* Biologically-inspired synthetic catalysts for key chemical bond forming and breaking steps for fuel formation and for interconversion of various forms of fuels. Enhancement of the stability and increase of the rates compared to natural catalysts through synthetic modification and use of nanostructured catalyst supports is an important goal. Development of efficient methods for coupling of catalysts to electron or hole sources, and embedding catalysts in robust nanostructured supports for enhanced stability are crucial tasks.

Why Liquid Fuels?

Energy density plays an important role in the choice of automotive fuels. Extra mass reduces the mileage of the cars, and the extra space increases the drag forces. Figure 1 displays the energy per unit volume as a function of the stored per unit weight. Liquid hydrogen requires 3.5 times the volume of gasoline for a given range. The best commercially available lithium ion batteries have a storage capacity of < 200 Wh/kg = 0.72 MJ/kg. A Prius-scale all-electric vehicle with a modest 40 mile range would need a battery capable of delivering 20kWh of energy (100 kg). In order to be commercially viable, the battery must last ~4,000 deep discharges, which is an order of magnitude longer than the life of existing rechargeable batteries. By comparison, the NiMH batteries in today's hybrid cars seldom are kept close to 50% of full charge capacity and are rarely required to deliver more than 0.4 kWh of energy! For these reasons, it is unlikely that we will have battery powered trucks, trains and airplanes in the near future.

If biofuels are to provide a substantial portion of the world's transportation fuel, a critical question is whether fuel production is scalable so that it can provide some meaningful fraction of the world's transportation fuel. Plant growth requires three basic elements: (1) adequate temperatures, (2) adequate natural water supplies, and (3) adequate sunshine. In Figure 2, black regions are those that are not limited by temperature, natural water supplies or by sunshine. One is immediately struck by the vast areas of our planet that are cold or dry. Even without significant warming of the Earth, much of the world is desert.

World population growth is a concern. Today's population of some 6.5 billion and is predicted to grow to over 9 in this century. The rising population raises the issue of whether there is enough arable land to produce transportation fuels as well as food. That issue has been highlighted by scholars in the past. When I was a college student we were greatly concerned by the words of Biology Professor Paul Ehrlich of Stanford. In his 1968 book, *The Population Bomb*, Ehrlich stated that "The battle to feed all of humanity is over... In the 1970s and 1980s hundreds of millions of people will starve to death in spite of any crash programs embarked upon now."

Why **liquid** fuels?

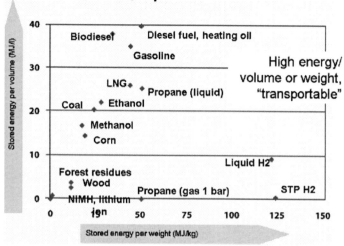

FIGURE 1. Stored energy/volume (MJ/liter) vs. stored energy/weight (MJ/kg).

Limiting factors for plant productivity

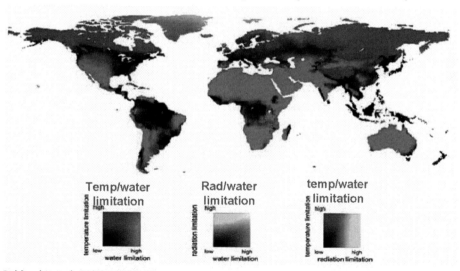

Baldocchi et al. 2004 SCOPE 62

FIGURE 2. Temperature, solar radiation and water limitations for plant productivity.

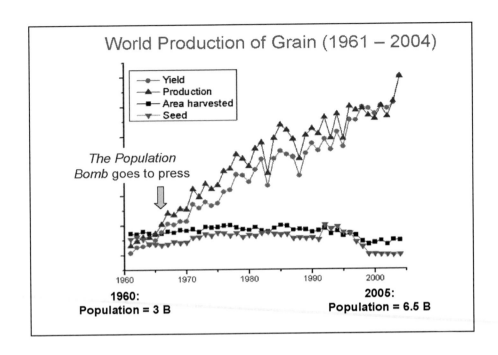

FIGURE 3. World production of grain (1961-2004).

The data of Figures 3 and 4 below paint a different picture. From 1960 to 2005, world population doubled from 3 billion to 6.5 billion. During this period of the so-called *"Green Revolution"*, grain production per acre increased by about a factor of five in Mexico, India, Indonesia and elsewhere. In 1970, Norman Bourlag received the Nobel Peace Prize for his creation of high-yield hybrid wheat that averted mass starvation. The cross-breeding of plants and the production of ammonia for fertilizers has been a big part of this progress. Starvation now is caused not by the lack of global food but by the lack of a completely viable food distribution system. This progress in agriculture has taken place because of enhanced food production rates, allowing a slight reduction in the amount of land cultivated.

Unfortunately, current agricultural practice is not sustainable. In the US mid-West (and in other parts of the world) the underground water tables are dropping and over fertilization have created water pollution problems. Our ability to move water in massive aqueduct systems has led to unwise agricultural land use such as the growing of water intensive crops in arid areas of the world. The predictions of climate change will further strain the world's ability to produce food. Also, the cost of food is rapidly increasing and causing hardship among poor people. A natural question is "Won't growing crops for energy make this problem worse?"

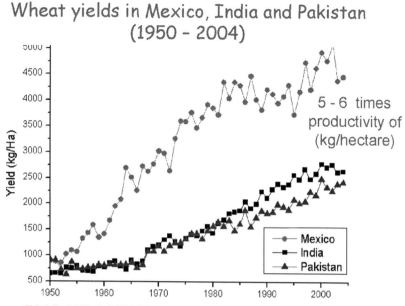

Figure title (in image): Wheat yields in Mexico, India and Pakistan (1950 – 2004)

Text in image: 5 - 6 times productivity of (kg/hectare)

Legend: Mexico, India, Pakistan

Source: FOOD AND AGRICULTURE ORG. (FAO) OF THE U.N. NATIONS

FIGURE 4. Wheat yields in Mexico, India and Pakistan (1950-2004).

The rapid rise in price of foods has been caused by a number of factors that combine in a nonlinear manor. These factors include (1) The rising cost of petroleum and natural gas, since the food industry is energy intensive, (2) the rising wealth of developing countries, which increases the consumption of meat that takes ~10 times more land than plants for the same nutritional value, (3) the impacts of weather (perhaps early signs of climate change?) over the last decade, and well as (4) competition of land use for biofuels. Those that lay the full cause of rise in food prices on biofuels production ignore the fact that only a few percent of the world agricultural output has been diverted to biofuels.

I believe that certain parts of the world (e.g. Brazil and the United States) have tremendous agricultural capacity and can afford to use land for both food and energy. There are a number of ways of producing biofuels would do far more environmental harm than good; for example, as the destruction of tropical rain forests in order to grow palm oil. The type of plant one grows for energy and where it is grown is vital. Sugar cane is a viable energy crop. Corn is not. The University of Illinois has carried out significant research on feedstock grasses (Miscanthus). In their test field, the researchers grew tall grasses (over ten feet high) in fields that were not fertilized or irrigated. The ethanol production potential is 15 times the amount produced per acre compared from corn on identical land. There are 50 million acres of US land that are substandard for agriculture. With the energy crops grown on this 50 million acres and agricultural wastes (wheat straw, corn stover, wood residues, urban waste, animal manure, etc), the US can produce half of all of current US consumption of gasoline.

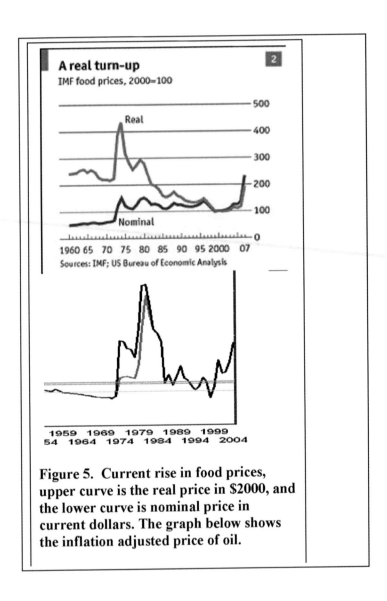

Figure 5. Current rise in food prices, upper curve is the real price in $2000, and the lower curve is nominal price in current dollars. The graph below shows the inflation adjusted price of oil.

FIGURE 6. Feedstock grasses (Miscanthus) is a largely unimproved crop. The woman is 5'4'' tall. The 10–foot stalks are grown with out irrigation and without fertilizers. The grasses are harvested in the fall after growing 3-6 months.

Feedstock grasses are a perennial plant, all of which offer the following advantages:

- They require no tillage for 10-15 years after first planting.
- Their long-lived roots can establish symbiotic interactions with bacteria and other micro-organisms to acquire atmospheric nitrogen and mineral nutrients.
- Some perennials withdraw a substantial fraction (about 50%) of mineral nutrients from above-ground portions of the plant before harvest.
- Perennials have lower fertilizer runoff than annuals. (Compared to corn, switchgrass produces only about 1/8 the nitrogen runoff, which pollutes water supplies, and 1/100 the soil erosion, which depletes top soil.)

It takes energy to make bio-energy. Cellulosic grasses are projected to produce about 10.3 times more bio-energy than is consumed from fossil energy to make it. This is considerably better than by other commercial techniques (see Figure 7). Because gasoline needs fossil energy for refining (cracking and distillation) and for distribution, the net energy content available is reduced to 81%. Electricity from coal is generously labeled at 45% efficient, primarily because of the thermodynamic inefficiency of the fossil fuel power plants . Corn ethanol barely breaks even: It has a modest gain, producing 36% more energy than is consumed by its growth and manufacture. In addition to a much improved energy ratio, ethanol from cellulose reduces carbon dioxide emissions by a factor of 5 (perhaps 10), compared to ethanol produced from corn.

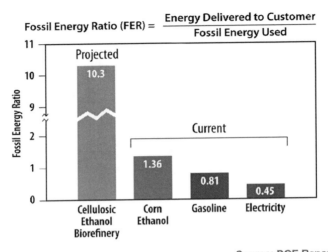

Source: DOE Report
Breaking the Biological Barriers to Cellulosic Ethanol   ÓJune, 2006

FIGURE 7. Fossil Energy Ratio. Cellulosic ethanol is projected to produce ten times the energy need to make it, far in excess of other approaches.

If these grasses are so much better than corn, why aren't they being grown today? The primary answer is the present economic competitiveness. The data in Figure 8 shows that Brazilian ethanol derived from sugar cane is about $40 per barrel, only 40% the cost of crude oil in April 2008. American-made ethanol is about $50 per barrel, but higher now because of the rising price of corn. Ethanol produced from cellulose is about 2.5 to 3 times more costly than sugar cane ethanol at this time. It is projected to cost about the same as corn-based ethanol in the future, but this depends on further research. There are reasons for optimism. The current fossil fuel energy ratio of sugar cane, a perennial crop, is ~5. Some of the cellulose residue (begasse) after the sugar is squeezed from the plant is use to power the bio-fuel refinery, but if would be better utilized if the cellulose were converted into a biofuel. By comparison, half of the corn stover residue has to be plowed back into to soil to partially replenish the soil of depleted nutrients.

Improving the Production of Biofuels

In biomass, most of the mass in the plant cell wall is in the form of polysaccharides (celluloses and hemicelluloses), as summarized in Figure 9. The next most abundant polymer is lignin, which is composed predominantly of phenylpropane building blocks. Lignins perform an important role in strengthening cell walls by cross-linking polysaccharides, thus providing support to structural elements in the overall plant body. This also helps the plant resist moisture and biological attack. These properties of lignin, however, interfere with enzymatic conversion of polysaccharide components. Despite the fact that lignin is not converted readily to ethanol with known technology, the molecules have both high energy content and we should search for other uses in the process if we are to maximize energy yield from biomass.

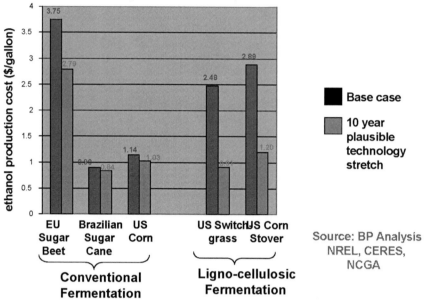

Current and projected product costs of ethanol

Courtesy Steve Koonin, BP Chief Scientist

FIGURE 8. BP estimates of ethanol production costs made in 2005.

The flow diagram for the production of ethanol is shown in Figure 10. In the pre-treatment processes (in blue), woody biomass material is chopped and shredded, and then typically treated with pretreated with steam, caustic high acid or strong bases, to separate the lignin from the long chain polymer sugars. The long chains are then changed into simple sugars with the use of cellulase enzymes extracted from microbes that degrade cellulose. Fig. 11 shows the molecular structure of a cellulase enzyme separating one polymer sugar strand form the original bundle and converting it to simple sugars. The harsh pretreatment conditions usually kill the biological activity of the enzymes, so detoxification steps are necessary. Finally, the simple sugars like glucose are fermented into alcohol using a 5,000 year old technology. After fermentation, the ethanol is separated form the solution by distillation.

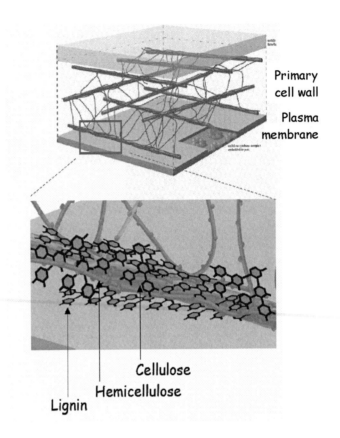

Primary
cell wall

Plasma
membrane

Cellulose

Hemicellulose

Lignin

FIGURE 9. A schematic of the primary cell wall of a plant. The enlarged picture shows how the lignin wraps around the polymer sugars to protect and make the structures more rigid.

There are many potential ways to reduce the cost of the conversion process. We are exploring new, less energy intensive pretreatment processes that are also less toxic to the suite of enzymes being used. Through molecular biology (recombinant) methods, the lignin can be altered so that easily hydrolyzed bonds are introduced into lignin to make those molecules easier to breakdown into smaller molecular units, shown in Figure 12.

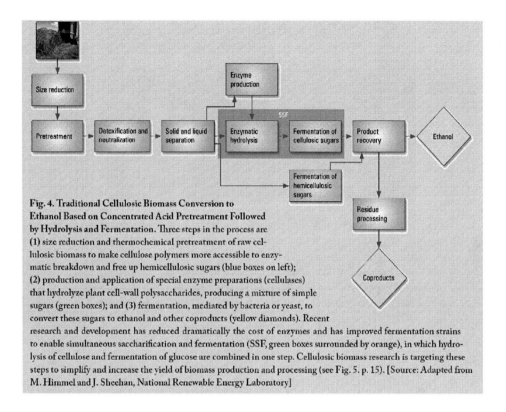

Fig. 4. Traditional Cellulosic Biomass Conversion to Ethanol Based on Concentrated Acid Pretreatment Followed by Hydrolysis and Fermentation. Three steps in the process are (1) size reduction and thermochemical pretreatment of raw cellulosic biomass to make cellulose polymers more accessible to enzymatic breakdown and free up hemicellulosic sugars (blue boxes on left); (2) production and application of special enzyme preparations (cellulases) that hydrolyze plant cell-wall polysaccharides, producing a mixture of simple sugars (green boxes); and (3) fermentation, mediated by bacteria or yeast, to convert these sugars to ethanol and other coproducts (yellow diamonds). Recent research and development has reduced dramatically the cost of enzymes and has improved fermentation strains to enable simultaneous saccharification and fermentation (SSF, green boxes surrounded by orange), in which hydrolysis of cellulose and fermentation of glucose are combined in one step. Cellulosic biomass research is targeting these steps to simplify and increase the yield of biomass production and processing (see Fig. 5. p. 15). [Source: Adapted from M. Himmel and J. Sheehan, National Renewable Energy Laboratory]

FIGURE 10. Taken from "Breaking the Biological Barriers to Cellulosic Ethanol," a research roadmap resulting from the *Biomass to DOE Biofuels Workshop*, December 7–9, 2005. U.S. Department of Energy, DOE/SC–0095 (June 2006).

Yet another avenue is to develop improved catalysts. Figures 13 illustrates work being done to create an artificial cellulosomes containing a series of enzymatic subunits that can work in conjunction more effectively than isolated enzymes. Already, cellulosomes have been shown to increase the rate of polymer breakdown by 2.7 to 4.7. Ideally, a self-reproducing, artificial microbe would be the end goal. As a guide to improved catalysis, researchers are exploring the microbial enzyme systems used by termites and ruminants to digest lignocellulosic material.

Figure 11. An electron density map of a cellulase enzyme converting a strand of polymer sugar into monomer sugars. Over the past decade, two commercial companies, Novozymes and Genencor, found that a cocktail of several different enzymes were found to work more efficiently in converting cellulose into simple sugars. The enzymatic breakdown of the cellulose remains the most expensive part of the conversion of cellulose into ethanol.

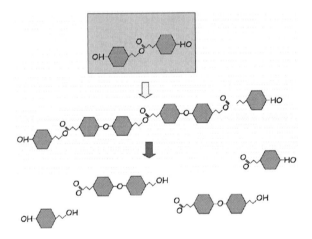

FIGURE 12. Introduction of easily hydrolyzed bonds into lignin.

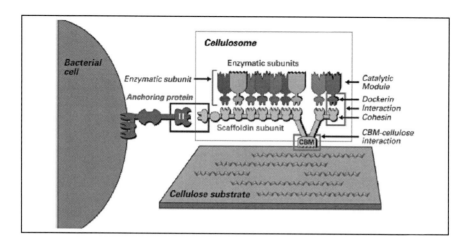

FIGURE 13. A schematic diagram showing how a celluosome can be tailored with many different enzymatic modules.

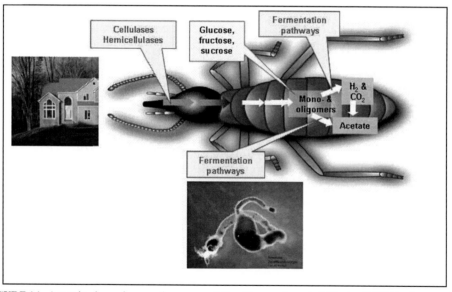

FIGURE 14. A termite devouring your home cannot digest he woody material by itself. Within the several guts of the termite, shown in the inset, as many as 100 different microbes form a community that convert the lingo-cellulose into chemical fuel that the termite can utilize. We are examining the genes of these microbial communities in hope that we can further improve the conversion efficiency of these self-sustaining microbes.

In addition to producing better microbes to convert lingo-cellulose into simple sugars for fermentation, another research goal is to transform convert microbes such as yeast or e-coli into organisms that can convert sugars into a better transportation fuel than ethanol. Ethanol is far from an ideal fuel: it has relatively low energy density (see Figure 1), readily absorbs water so that existing gasoline pipelines can not be used, and can not be blended in arbitrary ratios with gasoline or diesel fuels. The emerging fields of synthetic and systems biology are beginning to allow us to insert not just one gene, but several complete metabolic pathways into a host organism, essentially convert the original microbe into a chemical synthesis factory.

A poster-child of this field was the work of Jay Keasling and his colleagues at Lawrence Berkeley National Lab and UC Berkeley. He bestowed yeast and e-coli with the ability to make artemisinin, a powerful anti-malarial drug that was originally manufactured from a plant extract by introducing over 50 different DNA "parts" into the genome of the microbes. Jay's work got the attention of the Gates Foundation, and through a generous grant, he completed the laboratory scale research and licensed the technology to Amyris Biotechnologies, a company created to commercialize the drug. The drug will soon be ready for distribution at roughly an order of magnitude reduction in cost. The same synthetic biology technology is now currently being used to create organisms that can create gasoline, diesel and airplane-like fuels. There already have been successful "proof of concept" demonstrations of yeast that make gasoline-like fuels that self-separate from water.

Despite this encouraging early progress, we also need to look beyond improving the technologies used to covert bio-mass into transportation. In a future world of 9 – 10 billion people enjoying a middle class standard of living, the world will not have the agricultural and water resources to supply the needed fuel. In the end, we will need to create artificial photosynthetic systems.

Just as we were inspired to learn to fly by studying birds as shown in Figure 15, perhaps we can learn to mimic photosynthetic systems. In the end, there is no law of physics that says we can not do substantially better than the plants and algae we find in nature. Today's 747's do not look anything like a bird, but for our purposes, they work much better. Lifted from the constraints of the wet, warm world of nature, we have access to superior materials. For example, the jet engine turbine blades in today's airplanes are fashioned from single crystal metals. (It is true that 747's don't mate, lay little 747 eggs that hatch into baby airplanes. It is amusing to note that neither birds nor the Wright Brother's airplane had a vertical stabilizer while our largest planes all have large tail fins. While I am not an expert on birds, one can speculate that vertical stabilizers might hinder mating.)

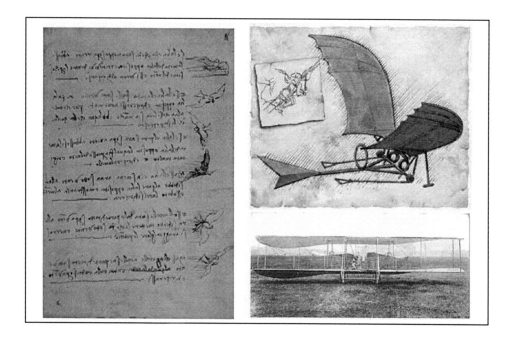

FIGURE 15. The sketch book of Leonardo da Vinci reveals his study of birds flying as the inspiration to one of his flying devices, intended to be powered by the arm and leg muscles of a brave individual. The first powered flight by the Wright Brothers was a hybrid solution. The control of flight was through wings that "warped" like the wings of birds as shown, but the power came from a gasoline engine.

At the Berkeley Lab, we have begun an artificial synthesis project. The first step in this project is the fabrication of a nanostructure imbedded in a membrane that can use the sun's energy to split water into oxygen and protons (H^+) in aqueous solution. The hydrogen ions will then be used to reduce carbon dioxide to carbon monoxide as the next step towards assembling a hydrocarbon fuel. Our dream is to make a cost effective photosynthetic system that will have an overall energy efficiency greater than 1% and will convert a large fraction of the water into a transportation fuel.

We learned to build machines that fly better than birds; it is not preposterous that we can learn make far better energy plants that can turn sunlight into a transportation fuel.

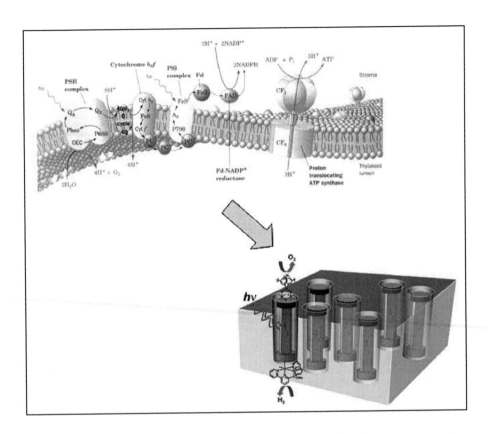

FIGURE 16. A schematic diagram of photosynthesis showing the major components used to convert sunlight energy into chemical fuels. As a first step, can we want to mimic the "water splitting" reaction of Photo-system II.

Batteries for Vehicular Applications

Venkat Srinivasan

Lawrence Berkeley National Lab
1 Cyclotron Road, MS 70R 0108B
Berkeley, CA 94720

Abstract. This paper will describe battery technology as it relates to use in vehicular applications, including hybrid-electric vehicles (HEV), electric vehicles (EV), and plug-in-hybrid-electric vehicles (PHEV). The present status of rechargeable batteries, the requirements for each application, and the scientific stumbling blocks that stop batteries from being commercialized for these applications will be discussed. Focus will be on the class of batteries referred to as lithium batteries and the various chemistries that are the most promising for these applications. While Li-ion is expected in HEVs in the very near future, use in PHEVs are expected to be more gradual and dependent on solving the life, safety, and cost challenges. Finally, batteries for EVs remain problematic because of the range and charging-time issues.

INTRODUCTION TO BATTERIES

Several electrical energy storage and conversion devices have been considered for use in vehicle applications. These are illustrated in Figure 1 in the form of a Ragone plot, wherein the abscissa is specific power (which can be thought of as acceleration in a vehicle) and the ordinate is specific energy (or range in an EV). The graph shows these quantities for various batteries, electrochemical capacitors, and fuel cells. Note that this plot shows specific energy and power on a cell level for batteries made for many different applications, from consumer electronic to vehicles. An additional derating will need to be applied when analysis is done on the pack level. The figure shows that lithium-ion (Li-ion) batteries are superior to nickel metal hydride (Ni-MH) batteries for all three applications from a performance standpoint. The figure also shows that no battery system has the ability to provide energy close to what is possible with gasoline (Internal combustion, or IC-Engine). Finally, the figure suggests that batteries are superior to capacitors for applications where the time of discharge is greater than the order of seconds.

The United States Advanced Battery Consortium (USABC) has set the requirements needed for batteries to be used in EV, PHEV, and HEV application.[1] These requirements cover a wide range of issues and include energy and power. In order to compare the requirements to the performance of these devices illustrated in Figure 1, a few assumptions have to be made, which are done here for the sake of

CP1044, *Physics of Sustainable Energy, Using Energy Efficiently and Producing It Renewably*
edited by D. Hafemeister, B. Levi, M. Levine, and P. Schwartz
© 2008 American Institute of Physics 978-0-7354-0572-1/08/$23.00

simplicity. While the USABC requirements for power represent a peak power for a 10s charge/discharge, the systems in Figure 1 show average power. Similarly, the USABC requirements show energy as an *available* energy, which typically is less than the total energy of the cell. For a HEV, available energy can be as little as 20-30% of the total energy, while for an EV or PHEV, this could be as large as 70-80%.

Despite these differences, a superposition of the USABC requirements onto Figure 1 is illustrative. The figure shows that while Li-ion batteries can easily satisfy HEV requirements, the energy is much smaller than the requirements for EVs. Similarly, the energy needs of a 40 mile PHEV (96 Wh/kg available energy) could be achieved by a high-energy Li-ion cell (similar to the batteries used in the Tesla Roadster).[a]

However, while the plot captures the performance map of various batteries, other criteria need to be considered, including cost, cycle and calendar life, and safety. Moreover, these factors are connected to each other. For example, while PHEVs *appear* possible with Li-ion batteries today from an energy standpoint, this would require the use of batteries that are not made for EV applications and that are cycled over a very wide state of charge (SOC) range, thereby limiting cycle life. One could increase the life by limiting the extent to which these batteries are charged, however, this would limit the energy of the cell, and thereby increase the cost, and the volume and weight of the final battery. Indeed, an analysis of presently-available EV batteries with characteristics that enhance cycle/calendar life suggests that meeting the energy requirements for a 40 mile PHEV is difficult. Finally, the importance of each of these factors changes from application to application. Therefore, while peak power could be an important criterion for a HEV, energy density would be a critical parameter in an EV. Some criteria, like cost and safety, remain challenges in all applications.

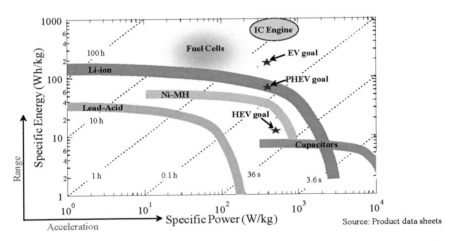

FIGURE 1. Ragone plot (specific power density in W/kg vs. specific energy density in Wh/kg) of various electrochemical energy storage and conversion devices.

---

[a] Assuming a packaging factor of 0.7, a consumer electronics battery with energy of 180 Wh/kg has a pack-level energy density of 126 Wh/kg. A 76% SOC range of cycling results in this battery meeting the requirements. However, note that typical EV batteries have a much lower energy density (pack level of 110 Wh/kg) due to the life requirements.

While complicated, the interplay between these various factors is tractable and suggests that Li-ion batteries remain the most promising candidate for use in vehicular applications. The three main reasons for this conclusion are the higher energy, higher power, and the potential for lower cost of Li-ion batteries when compared to Ni-MH batteries. This conclusion has been the reason why significant research efforts have been directed toward Li-ion batteries. In this paper, we will explore the limitations that Li-ion batteries face when used in each of the three applications (HEV, EV, and PHEV), the underlying technical challenges behind the limitations, and the approach taken by researchers the world over to address these limitations. We begin with a brief tutorial on batteries.

Batteries typically consist of two electrodes, an anode and a cathode with a separator between them to prevent shorting. The cell is filled with electrolyte. Figure 2 illustrates a typical Li-ion cell sandwich consisting of a graphite anode and a lithium cobalt oxide cathode ($LiCoO_2$).

Lithium-ion battery

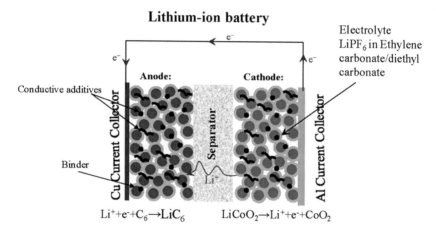

FIGURE 2. Schematic of a typical Li-ion cell.

The electrodes consist of active materials bound together with an electronically insulating binder and conductive additives. Each electrode is pasted onto current collectors. During charge, Li is removed from the cathode (or positive electrode), transferred through the separator *via* the electrolyte and is inserted into the anode. The reverse occurs on discharge. The difference in voltage of the cathode and the voltage of the anode is the cell voltage. The amount of Li that is stored in each of these materials is related to the capacity (often given in mAh/g). The product of the voltage and the capacity is the energy. How quickly the Li is transferred from one electrode to the other (or how quickly the energy is removed) is related to the power. More details on batteries can be found in references 2, 3, 4, and 5.

Figure 3 shows the typical steady-state charge of the anode and cathode of a Li-ion cell with a graphite anode and a $LiCoO_2$ cathode in an organic electrolyte consisting of a Li salt (lithium hexafluoro phosphate, $LiPF_6$) in a solvent (*e.g.*, ethylene carbonate and diethyl carbonate). This is the battery used in laptops and cell phones. The voltage of each electrode is represented with respect to a Li-metal reference electrode. As the Li is removed from the cathode, its potential increases, while the potential of the anode decreases with insertion of Li. The process of Li moving in and out of the electrodes is referred to as intercalation/deintercalation. The voltage of the battery is the difference in voltage of the cathode and the anode, which increases as charge proceeds. The abscissa represents how much Li is stored in the cell, while the ordinate shows at what voltage the Li is inserted/removed from the materials. In order to increase the energy of the battery three avenues can be pursued, namely (i) increase the voltage of the cathode, (ii) decrease the voltage of the anode, and (iii) increase the capacity of the cell. However, the thermodynamics of electrochemical reactions other than the intercalation of Li (referred to as side reactions) limit these quantities.

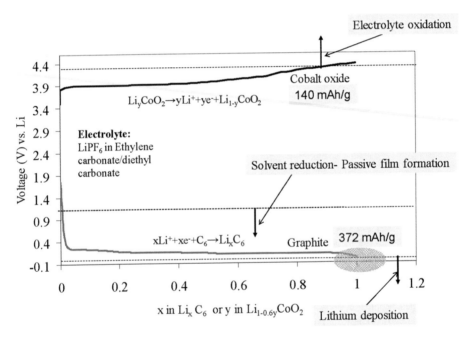

FIGURE 3. Steady state charge curve of a Li-ion cell consisting of a graphite anode and a $LiCoO_2$ cathode. The graph shows the half cell potentials and the thermodynamics potentials for various side reactions.

The three side reactions worth mentioning in this plot are the oxidation of the solvent that occurs above ~ 4.2 V *vs.* Li, Li-metal deposition that occurs below 0.0 V *vs.* Li, and solvent reduction that occurs below ~1 V *vs.* Li. These three reactions not only limit the energy of the cell, they are also implicated in the life and safety problems associated with Li-ion batteries. Staying within the voltage window allows these problems to be minimized, at the loss of energy. The tradeoffs that are needed to balance these various parameters are captured by Figure 3.

Innovation in Li-ion batteries can occur in two ways, (i) *via* engineering advances that reduce, *e.g.*, the thickness of the separator and/or (ii) *via* innovation in the materials used as the active material. For example, the chemistry used in present-day cell phone and laptop batteries (LiCoO$_2$/graphite) has a theoretical energy density of ~360 Wh/kg (this accounts for only the weight of the active material and not the weights of the other components in the cell such as the current collectors, electrolyte, binders, and cell packaging). The practical energy density of a packaged 18650 cell is ~190 Wh/kg. Fifteen years ago, in the early days of Li-ion, this number was ~90 Wh/kg, using the same material sets.[6] In other words, over the last 15 years, engineering advances have resulted in a doubling of the energy density of Li-ion batteries. It is expected that in the future, improvements in performance will occur by moving to new higher energy materials.

Fortunately, it has been observed that Li can intercalate into many different anode and cathode materials. At present, three classes of cathodes, four classes of anodes, and four classes of electrolytes are being considered for use in Li-ion cells. Depending on the combination of the anode, cathode, and electrolyte, one can have a completely new battery with changes to the energy, power, life, safety characteristics, low temperature performance, *etc*. These classes are illustrated in Figure 4 for the three components of the battery.

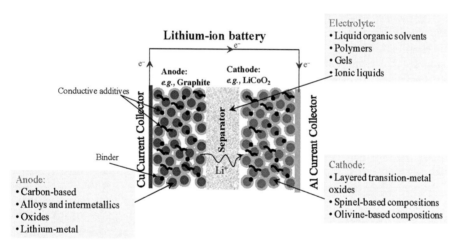

FIGURE 4. Schematic of a Li-ion cell with the various anode, cathode, and electrolytes that are presently being considered. Changing the combination results in changes to the energy, power, safety, life, and cost.

Each of these classes is a collection of numerous specific materials with their own different characteristics. For example, the LiCoO$_2$ electrode used today falls under the class of layered materials. Its capacity is 140 mAh/g. A new cathode emerging in the market is LiNi$_{0.8}$Co$_{0.15}$Al$_{0.05}$O$_2$ (NCA), also a layered material but with a capacity of 185 mAh/g. This increase in capacity means a significant increase in the energy of the cell, reiterating that many different materials combinations can be used to tune the battery to suit the application for which it is used. This behavior of Li batteries makes it very different from other classes of batteries, for example, the lead-acid system where the material set is fixed. It also adds additional degrees of freedom for researchers to find new materials and in effect make a new battery.

One issue worth mentioning is the concept of capacity balancing. Today's Li-ion batteries are typically referred to as cathode limited. In other words, the capacity of the cathode (140 mAh/g for $LiCoO_2$) limits the capacity of the battery, as it is smaller than the capacity of the anode (372 mAh/g for graphite). What this means is that the cathode has a significantly higher impact compared to the anode in increasing the energy of the cell. A quick calculation shows that an order of magnitude increase in the anode capacity increases the capacity of the cell by only 35% (one would also need to see at what voltage this new anode operates in order to estimate the improvement in energy).

While the flexibility of Li batteries provides the means to tune the battery's characteristics, it appears that no ideal combination has been found that satisfies the needs of vehicular applications. We now briefly look at the three applications, the status of batteries in these applications, and the research that is being pursued.

BATTERIES FOR HEVs

Over the last 5 years enormous progress has been made in using Li-ion batteries for HEV applications. The most studied system typically consists of the NCA cathode with a graphite anode. With continuous improvements this system has overcome many of the limitations and has come closer to satisfying a majority of the requirements set out by the USABC. In particular, the calendar life of this chemistry has been projected to be greater than 15 years based on accelerated testing, when the SOC of the battery is controlled carefully. This promising development provides hope for use in vehicles.

Despite these advances, three main barriers remain before commercialization: (i) cost, (ii) low-temperature operation and (iii) safety.[7] It has been shown that a large proportion of the cost of HEV batteries is due to the separator (25%) and the electrolyte (17%) (note that in a HEV cell, where the electrodes are thin, there is more separator area for the same size compared to an EV battery).[8] Therefore, efforts to make low-cost separators would help with the cost reduction in the cell. An ability to make low-cost separators thinner, without compromising safety, would enhance power, and minimize cost simultaneously. In addition, lower cost electrolytes and cathodes would also help in decreasing the cost of the battery. Finally, as the important metric for a HEV is $/kW of power, a new chemistry that promises higher power at the same cost per cell would result in a lower pack cost.

When operating Li-ion batteries at low temperatures (less than 0°C), it has been seen that the power capability decreases significantly because of increased resistance in the cell. It is thought that the increased resistance is caused by the reduction in the kinetic rate constants of the electrochemical charge-transfer reaction.[9] This has implications during regeneration where the anode in the Li-ion cell is accepting Li at high rates. It has been observed that instead of intercalating into the graphite structure, the Li plates on the surface. The Li metal then reacts with the electrolyte and gets consumed, leading to capacity fade. While it is known that the choice of the solvent can have a significant impact on the low-temperature performance, these special solvents come at an additional cost. An alternative technique is to use an anode that allows the lowering of the impedance without causing Li plating. This can be achieved if the voltage of the anode is much higher than the voltage for Li-plating, allowing a bigger window before plating occurs. Lithium titanate ($Li_4Ti_5O_{12}$) is a candidate that allows this feature because its voltage is 1.5 V *vs.* Li. Companies such as Toshiba, Enerdel, and Altair Nano are pursuing this concept to allow low-temperature operation and fast-charge capability (in addition to possible safety and life enhancements). However, the high anode voltage results in a low cell voltage and therefore decreases the energy of the battery.

It also should be noted that if the HEV is designed so that the battery is not used when cold but instead is allowed to warm up using the IC engine, then the problem of low-temperature operation could be circumvented. At present, it appears that the HEVs of the near future will be using this strategy. A pure EV would be prone to have a problem from this low-temperature limitation because no means exist to heat the battery when it is not plugged into an electrical outlet. Here, again, the problem could be solved by not allowing regeneration until the battery warms, and by using the battery to power the car (albeit at lower acceleration) until it self-heats.

The last limitation of Li-ion batteries is the safety of the pack, highlighted by the recent incidents involving fires of consumer-electronics batteries. The main cause of safety limitations in Li-ion cells can be characterized by the three stages in which thermal runaway occurs.[10] In stage 1, an unexpected failure occurs in the cell, *e.g.*, internal shorting (*e.g.*, due to metal particles) or malfunction of the overcharge protection system. This results in the temperature of the cell increasing to around 125°C, where a reaction that involves decomposition of a passive layer on the anode (called the solid electrolyte interphase layer or SEI) occurs; an exothermic reaction that increases the temperature further. As the temperature increases to above 180°C, in some cathodes, the oxygen from the lattice is released, resulting in a rapid increase in heat in a short time. This chain reaction results in the smoke and the fires that have been observed in Li batteries.

One can prevent/retard these reactions in different ways. For example, using a cathode that does not release oxygen could help in suppressing the final exothermic step. Lithium iron phosphate is one such cathode that is being pursued by various companies, notably, A123 systems and Phostech. A second approach is to prevent the cell from approaching these temperatures by preventing the decomposition of the SEI. Lithium titanate is an example of an anode where no SEI forms and so it has been argued that this makes the battery safer. Finally, research efforts are underway to develop overcharge protection mechanisms, such as redox shuttles, electroactive polymers, and high puncture-strength separators to prevent internal short circuits. These ideas are aimed at preventing the cell from reaching the temperatures where the anode reactions occur. Finally, electrolytes that are not flammable and nonvolatile (*e.g.*, ionic liquids) help minimize the impact of these incidents even if the reactions do occur.

To summarize, three problems prevent the widespread commercialization of Li-ion batteries for HEVs today, cost, low-temperature operation, and safety. However, in the short-term, various engineering solutions can be adopted that can prevent the latter two. With cost being a highly socio-economic factor, it is thought that HEVs with Li-ion will start reaching the marketplace within the next few years.

BATTERIES FOR EVs

We will now examine the use of Li-ion batteries for EV applications. As was pointed out previously, Figure 1 suggests that the energy density of Li-ion batteries is not sufficient for use in present-day cars to provide the necessary range to make them commercially viable. A typical rule-of-thumb is that a sedan can go 1 mile on 300 Wh of battery. In other words, a car with 300 mile range requires a 90 kWh pack that can fit in the vehicle with little impact to passenger and luggage space. Presently available EV batteries have a specific energy of ~110 Wh/kg and 160 Wh/l,[11] suggesting that the 90 kWh battery will weigh 1800 lbs and have a volume of 148 gallons; far too large for existing vehicles. These energy numbers are on a pack level and are therefore smaller than those on a cell level. Further, note that these numbers are not meant to be precise, rather they are estimates to illustrate the challenges.

Arguments could be made that one may not need a 300-mile-range vehicle, that one could drive more than 1 mile with 300 Wh (for example, *via* the use of lightweight materials and better body design), and that one could enhance the specific energy on the pack level using lightweight battery packaging.[b] All these would help make EVs a reality. However, the cost of such a vehicle would still remain large, and, arguably, prohibitive. For example, the battery pack in the Tesla Roadster is reported to be $25,000. The fact that the Tesla batteries are made for consumer electronic applications, where the cost of batteries is considerably lower than that for vehicles, highlights the magnitude of the problem. Similar to HEVs, the important quantity is the $/kWh of the battery (a more comprehensive metric would be $/kWh/year or a lifetime cost on an energy basis[c]). Therefore, a higher energy chemistry could be a route to decreasing the cost of these batteries. Three ideas that are being pursued are discussed here.

As pointed out before, the capacity of cathodes used in consumer electronic batteries are limited by the solvent oxidation reaction that occurs at ~4.2 V *vs*. Li. At these limits, only ~50% of the lithium is removed from the lattice of the cathode. Tremendous improvements could be made if the extra lithium can be accessed. Efforts are underway to find new electrolytes that can be used in high-voltage cathodes. However, as of today, there appears to be no electrolyte that shows promise in increasing the voltage significantly. An interesting recent advancement has been in altering the surface of the cathode in order to modify the interface where reaction occurs. For example, it has been shown that coating the cathode with a layer of aluminum phosphate $(AlPO_4)$[12] allows the potential to be pushed to higher values, thereby increasing the capacity and the energy, with little loss in capacity. This

---

[b] One can perform an optimistic calculation using a high-energy Li-ion cell of say 180 Wh/kg specific energy with a packaging factor of 0.8 resulting in a pack level specific energy of 144 Wh/kg. For a 200 mile vehicle, this would require 900 lbs of battery; a more manageable number.
[c] For a Tesla battery, which is a 56 kWh battery which would last, as reported, 5 years, the $/kWh/year is 90. The USABC long-term goal is $10/kWh/year (and 15 for the short-term) for EVs.

modification of the interface allows us to think of new ways to enhance the energy of the cell while maintaining cycle life and safety.

The second possibility is to change the anode to enhance the energy of the cell. Two options are being pursued in this area, (i) the use of anodes that alloy with Li, such as silicon and (ii) the use of Li metal. In area one, anodes, such have silicon, are attractive because they have high capacity (~3700 mAh/g) when compared to graphite (372 mAh/g).[13] Despite having a higher voltage compared to graphite (0.5 V vs. 0.1 V), the use of Si results in an increase in energy density by ~25-35%. However, several problems prevent alloys from being commercialized including a large volume change (~270% for silicon) with cycling that results in particle cracking/isolation, a large 1st cycle loss in capacity that results in Li being consumed to make a passive layer and not being active to provide capacity, and a continuous consumption of Li from cycle to cycle that limits the cycle life.[14] Several approaches are being pursued to alleviate these limitations.

Two that are worth noting are the use of what is termed "active-inactive matrix" where the active component (Si or Sn) is embedded in an inactive matrix (e.g., carbon),[15] and the use of novel structures to accommodate the volume change. In the first approach, the carbon allows the expansion of the alloy without particle cracking. However, the added weight and volume of the inactive component limits the capacity of these anodes to ~1200 mAh/g. This concept is used in the Nexilion cell introduced by Sony Corporation. In the second concept approach, micron-sized silicon pillars that reversibly cycle with high capacity are used.[16] A nano-sized analog of this concept was recently reported by a group at Stanford University.[17] While interesting, these structures require expensive and slow processing.

While these two approaches show promise in accommodating the volume change, efforts need to be made to reduce the 1st cycle capacity loss (which scales with surface area) and the cycle-to-cycle loss of Li. Use of "in situ SEI-formers" like the SLMP powder marketed by FMC Corporation could help with the former.[18] The use of additives could help with the latter.

In area two, using Li metal as the anode brings the promise of a large capacity (3800 mAh/g) at a voltage of 0.0 V, thereby increasing the energy of the cell. Lithium metal has been used in primary Li batteries for decades, however, its use in secondary batteries have not been successful because of problems during charging. During charge, Li plates on the anode and, similar to plating of other metals like zinc, does not plate uniformly. Instead, dendrites of Li grow on the surface and, with time, penetrate the separator and short the cathode. In addition, dendrites break and isolate Li resulting in capacity fade.

There have been numerous efforts over the decade to stop the dendrites from growing by placing the anode against a hard surface, such as a solid polymer electrolyte. It has been hypothesized that the hard surface of the polymer would prevent the dendrites from penetrating. However, it has been observed that as the stiffness of the polymer increases, its conductivity decreases, thereby limiting the power capability of the cell. This interplay between conductivity and stiffness has resulted in the inability to effectively prevent dendrites. The use of block copolymers, where one block is made of a stiff material like polystyrene, and the second block is made of a conducting phase, like polyethylene oxide, has allowed the decoupling of conductivity and stiffness.[19] However, as of today, these polymers operate at 80 C and have significantly lower conductivity compared to liquid electrolytes. The use of polymers for Li batteries is being pursued by companies such as Sion Power and Seeo Inc.

Another concept is the use of single-ion conducting glasses to isolate the Li. This concept has been explored by Polyplus Battery Company where it has been shown that the Li can be completely isolated to the point where it can be dipped in water. This concept is intriguing in that it enables the development of very high energy cells like Li-air and Li-sulfur. These glasses are not very conductive and are prone to defect formation. Advances in the behavior of solid polymers and ionic glasses could provide huge benefits in batteries.

While these approaches provide a pathway for enabling a higher energy cell that could allow the development of EVs, they fail to solve one other problem, namely, charging time. With typical battery charge times ranging from 3-8 hours, a change in lifestyle is needed when compared to using a gasoline car. While fast-charge batteries do exist (*e.g.*, Altair Nano and Toshiba), these come at the expense of energy. Further, infrastructure questions arise concerning the ability of the grid to handle a pure electricity-based transportation economy with fast-charge batteries. The issues of range and charging are circumvented with the use of a PHEV.

BATTERIES FOR PHEVs

The PHEV is an ideal compromise between the minimal fuel-saving advantage of the HEV and the range/charging-time issue of the EV. With the addition of the internal combustion engine for longer trips and a large battery pack to sustain 10-40 miles of equivalent electric range, the PHEV captures the best of both concepts. As suggested previously, the energy of the Li-ion cell is connected with the life and enhancing one typically lessens the other. While high energy Li-ion cells, similar to ones used in consumer electronic applications, could probably satisfy the range requirements for a 40 mile PHEV, the ability of these batteries to last 15 years is highly questionable. Typical EV batteries today do not appear to posses the energy needed for a 40 mile range. However, recent reports suggest that battery manufacturers are increasing moving to lightweight packaging that should allow the specific energy to increase. In addition, presently-available batteries should be sufficient to satisfy the energy requirements of PHEVs that have a lesser EV range (*e.g.*, 20 miles). However, two problems still need addressing, namely, cost, and cycle/calendar life.

Present cost estimates suggest that a 40-mile PHEV battery can cost upwards of $10,000.[d] While battery companies have been reporting cycle life of Li-ion cells in

[d] USABC goal for available energy of a PHEV-40 is 11.6 kWh. Assuming a 70% range, this means that the total battery size would be 16.5 kWh. Assuming a cost of $1000/kWh for a PHEV cell, this would mean a total cost of $16,500 for the battery pack. While not well understood, the life of this battery *could* be 10-15 years. One could perform an alternate calculation: Assuming that one uses a consumer electronic battery with the same characteristics as the battery for a Tesla Roadster, a 40 mile PHEV would cost $5,000 (1/5th the cost of the Tesla battery of total energy 11.2 kWh). Assuming, optimistically, that this battery lasts 5 years, this means a yearly cost of $1000 for the battery pack. Assuming that the battery is cycled 300 times in a year (for a total of 12,000 miles pure EV driving), the cost of electricity would be $270 at $0.08 per kWh. In other words, one would need $1270 per year to pay for the battery and the electricity per year. Assuming a 3 year battery life (arguably more realistic) increases this cost to $1935 per year. To drive the same miles in a gasoline car with 30 mpg mileage would require $1800 per

the 3000-5000 cycle range (approaching the USABC requirements), the more important question is the calendar life of these cells, particularly because these batteries may spend a considerable part of the time at high SOCs, the state where Li-ion batteries are most prone to exhibit capacity fade. Strategies are being pursued that involve decreasing the maximum SOC of the battery in order to enhance the life; however, these come at the expense of energy and therefore increase the cost of the pack. As of today, no clear data exists that shows the ability of Li-ion batteries to sustain 15 year life under PHEV conditions. This is expected to become available in the near future.

Concepts that involve increasing the energy of the battery *via* new materials, described in the EV section, will prove to be very effective in decreasing the cost (by decreasing the $/kWh). In addition, careful use of additives/coatings to modify the electrode/electrolyte interface in order to retard side reactions in the cell can help to enhance the life of the battery, without sacrificing energy. Finally, new materials that show inherent stability could also be useful in improving the life of these batteries.

We end the discussion of PHEVs by reminding the reader that safety remains a concern for all three applications and that the ideas to enhance safety suggested in the HEV section are applicable for PHEVs (and EVs).

FUTURE OF BATTERIES

The energy density of batteries has been increasing at the rate of ~5% per year over the last 15 years, well below the improvements that have been made in semiconductor devices (*e.g.*, Moore's law). The question arises as to how batteries will improve over the next 15 years. While difficult to predict, one can look at advances in the literature and project the impact these advances would have on energy. We choose to look at specific energy, although energy density is also important. This is summarized in Figure 5 where the specific energy is captured at the cell level. We split the plot into "low risk" where two advances are expected. One involves the move to a NCA-like cathode from $LiCoO_2$ (which is already underway) and the second involves the move to an active-inactive alloy (similar to the Nexilion cells, but with higher capacity). We then describe higher risk systems that involve moving to high-voltage-transition-metal-oxide (TMO) cathodes and to alloy anodes that have much larger capacity. These are considered possible, if the concept of coatings work to enhance the capacity of cathodes without compromising safety and life and if alloys, like Si, prove to meet the cycle life requirements. While both these are far from certain, the literature results are encouraging.[12,20] Because of the uncertainty, we denote this as higher risk. These advances, if successful, would result in the doubling of the energy of present day batteries; a significant increase.

Next, we address the issue of the theoretical limit of the energy of batteries. While some use a periodic table to evaluate this number, we choose to highlight systems that are very far from reality, but have been alluded to in the literature. Three of these systems are listed in Figure 6 including zinc-air, lithium-sulfur and lithium-air. Note

---

year, assuming gasoline costs at $4.5 per gallon (costs in the San Francisco Bay Area as of June 2008). Both calculations allude to the need for creative financing options (*e.g.*, battery leasing) to offset the initial investment needed to buy a PHEV.

that the numbers in this figure are theoretical energies (that do not include any weights except that of the active material). The figure shows that batteries today are very far from achieving the theoretically maximum-possible energy. Efforts are underway both in the research community and in industry to examine these systems. For example, the Zinc-air technology is being pursued by ReVolt Technologies in Sweden and the Li-S cell is the focus of Sion Power. However, note that while the figure shows the theoretical specific energy, practical values are far below these quantities. For example, the target for Li-S cells is in the range of 350 Wh/kg; less than a doubling of presently-available systems. This shows the difficulty involved with these chemistries and the long-term nature of the research for them to be commercial.

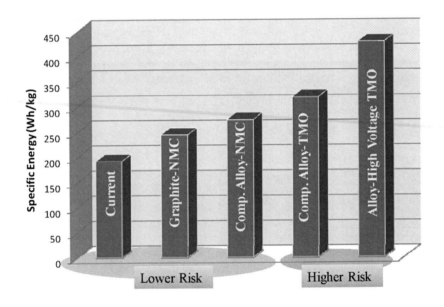

FIGURE 5. Projected increase in energy density (Wh/kg) of Li-ion cells on a cell level; NMC - $LiNi_{1/3}Co_{1/3}Mn_{1/3}O_2$, TMO - transition metal oxide. Based on Draft of FreedomCAR PHEV R&D Plan: http://www1.eere.energy.gov/vehiclesandfuels/features/phev_plan.html

Finally, we examine the need for non-Li-based systems for energy storage for vehicles. While systems are being researched (like magnesium and sodium), these appear further away from commercialization. Considering the wide variety of materials that Li can intercalate into, it appears that Li-based batteries (whether Li-ion or Li-metal) offer many opportunities for improvements that are worth pursuing. While it can be argued that being dependent on a single cathode or anode material (like being dependent on platinum in a fuel cell) would lead to resource limitations, one should note that with the wide material set for the anode and cathode for use in a Li-ion cell, one has opportunities to spread the risk among many different metals. The one common metal that is needed for all Li-based cells is Li. Back of the envelope calculations show that Li-metal limitations should not be an issue for at least the short to intermediate term. However, a comprehensive study is probably needed to ensure that this resource is not limited if widespread penetration of PHEVs and EV do occur.

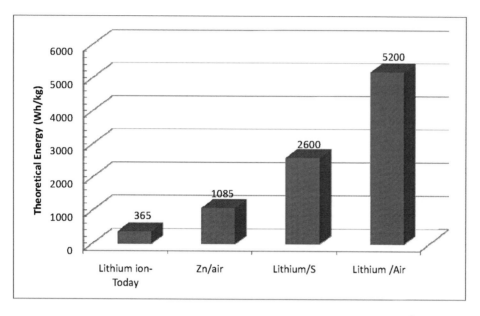

FIGURE 6. Theoretical specific energy (Wh/kg) of various electrochemical couples. All numbers are theoretical energy densities and accounts **only** for the weight of the active material.

CONCLUSIONS

Lithium-ion batteries offer many advantages that make them suitable for use in vehicular applications, including higher energy and power capability and, possibly, lower cost compared to Ni-MH cells. One significant advantage of the Li battery is that the chemistry (*i.e.,* materials for the anode, cathode, and electrolyte) are not fixed and a wide variety can be used, each of which can change the characteristics of the system. This provides battery researchers with additional degrees of freedom. However, with each choice, it has been observed that while some characteristics are improved, others prove lacking. No silver bullet has been found that can be considered an ideal Li chemistry.

While Li-ion is expected in HEVs in the very near future, use in PHEVs is expected to be more gradual and dependent on solving the life and cost challenges. Finally, batteries for pure EVs remain problematic because of the range and charging-time issue. A focus on higher energy systems would help decrease the $/kWh. Similarly, the use of coatings and additives will play a large role in enhancing life. Safety remains a big concern with Li batteries, highlighted by the recent incidents plaguing the consumer electronics market. Both engineering and materials approaches are being pursued to address this challenge. Safety will remain in the forefront of any discussion of battery-powered vehicles and much care and attention is needed to ensure that no incidents occur that undermine this concept.

The future of batteries remains very strong and one can be optimistic as to the advances that will occur in this class of energy storage devices. Considerable improvements can still be made to the concept of a Li-ion battery, both in terms of finding new materials and in alleviating the limitations of existing systems. Batteries exist that promise tremendous improvements compared to the ones presently available; however there are significant challenges in commercializing these systems. The key to enabling these systems to operate with high energy, long life, and good safety characteristics lies at the interface. Modification of the interface will play an increasing important role in batteries in the near future.

ACKNOWLEDGMENT

This work was supported by the Assistant Secretary for Energy Efficiency and Renewable Energy, Office of Vehicle Technologies of the U.S. Department of Energy under Contract No.DE-AC02-05CH11231.

REFERENCES

1. http://www.uscar.org/guest/article_view.php?articles_id=85
2. *Chem. Reviews* 104(10), (October 2004).
3. D.A. Scherson and A. Palsencsar, *The Electrochem. Society Interface* 17 (Spring 2006).
4. R.J. Brodd, *et al*, *J. Electrochem Soc.* 151, K1 (2004).
5. W. van Schalkwijk and B. Scrosati (ed.), *Advances in Lithium-Ion Batteries*, New York, Kluwer Academic/Plenum Publishers (2002).
6. V. Srinivasan and L. Lipp, *J. Electrochem. Soc.* 150, K15 (2003).
7. www1.eere.energy.gov/vehiclesandfuels/pdfs/merit_review_2008/energy_storage/merit08_habib.pdf
8. http://www.transportation.anl.gov/pdfs/TA/149.pdf
9. D.P. Abraham, *et al*, *J. Electrochem. Soc.* 153, A1610 (2006).
10. http://www.prod.sandia.gov/cgi-bin/techlib/access-control.pl/2004/040584.pdf
11. http://www.saftbatteries.com/120-Techno/20-10_produit.asp?sSegment=&sSegmentLien=&sSecteurLien=&secteur=&Intitule_Produit=VLEmodule&page=2
12. J. Cho, *et al*, *Angewandte Chemie* 42, 1618 (2003).
13. J. Li and J.R. Dahn, *J. Electrochem. Soc.* 154, A156 (2007).
14. S.D. Beattie, *et al*, *J. Electrochem. Soc.* 155, A158 (2008).
15. I. Kim, *et al*, *Elec. Solid State Lett.* 7, A44 (2004).
16. US patent 20060097691.
17. C.K. Chan, *et al. Nature Nano.* 3, 31 (2008).
18. http://www.fmclithium.com/products/pds/FMC003_EMTsht_3_06.pdf
19. M. Singh, *et al. Macromolecules* 40, 4578 (2007).
20. M.N. Obravac and L. J. Kruase, *J. Electrochem. Soc.* 154, A103 (2007).

Hydrogen Storage for Automotive Vehicles

J. F. Herbst

Materials and Processes Laboratory
General Motors R&D Center
MC 480-106-224
30500 Mound Road
Warren, MI 48090-9055 USA

Abstract. Hydrogen fuel cells represent a strong potential alternative to internal combustion engines relying on fossil fuels. The benefits and hurdles to hydrogen mobility are discussed, with particular emphasis on methods and materials for storing hydrogen in automotive vehicles.

WHY HYDROGEN FUEL CELL VEHICLES?

Proton exchange membrane fuel cells operating on hydrogen are an attractive alternative to internal combustion engines relying on hydrocarbon fuels for a variety of reasons. Hydrogen is the most abundant element in the universe and occurs in enormous amounts terrestrially as a constituent of water and other compounds. Its use as a fuel addresses several societal issues, including the prospect of eliminating automotive vehicles from the environmental equation. With sufficient fuel cell powered cars and trucks in the vehicle fleet, use of hydrogen from domestic energy sources could diminish our dependence on foreign petroleum and reduce the US overseas trade deficit. Generation of hydrogen from renewable sources would help to bring carbon dioxide emissions back into balance with natural phenomena and ameliorate global climate change. Furthermore, since hydrogen fuel cells generate electricity while producing only water and heat as by-products, local air quality in urban areas would be improved.

CHALLENGES TO HYDROGEN MOBILITY

If the vision of a hydrogen fuel cell transportation future is to be realized, three major hurdles must be overcome. First, light, compact, robust, and affordable fuel cell propulsion systems must be developed. Very substantial progress has been made on fuel cell stacks in recent years, and there is confidence within the automotive industry that this goal can be met.

CP1044, *Physics of Sustainable Energy, Using Energy Efficiently and Producing It Renewably*
edited by D. Hafemeister, B. Levi, M. Levine, and P. Schwartz
© 2008 American Institute of Physics 978-0-7354-0572-1/08/$23.00

Second, a hydrogen production and distribution infrastructure must be established. This will require the leadership of national, state, and local governments to enable construction of refueling facilities. In this regard it is important to note that early analysis of the number of refueling stations needed to support the introduction of fuel cell vehicles is only a fraction of that for the current petroleum refueling infrastructure. Meeting this goal is certainly possible but will require a concerted and coordinated societal effort.

At present H_2 is produced mostly by reforming natural gas, but in the future we can look to splitting water, our greatest source of hydrogen, by means of non-carbon energy sources. Possibilities include electrolysis with electricity from solar, wind, hydroelectric, and nuclear resources; direct H_2 generation via photocatalysis using sunlight and semiconductors; nuclear and solar thermochemical cycles; and biological and bio-inspired techniques. The Department of Energy sponsors considerable research in many of these areas.[1]

The distribution infrastructure for making hydrogen available is dependent to a large extent on the methods for production and storage. A centralized system will require pipelines and/or delivery trucks. There is also the possibility, however, of decentralized sourcing, with electrolyzers, photocatalyzers, or reformers at fueling stations and homes.

The third challenge is the principal subject of this article. A light, compact, durable, affordable, and responsive system for storing hydrogen on a vehicle needs to be developed. This challenge is receiving a great deal of international research attention.

SYSTEM TARGETS

A practical hydrogen storage system must satisfy a number of criteria. The two most important are the gravimetric energy density, the system energy per unit mass, and the volumetric energy density, the system energy per unit volume. Figure 1 shows these two *system* parameters (i. e., not only material properties) for several known methods of storing hydrogen. These methods include H_2 gas compressed to 10,000 psi (700 bar; 70 MPa), for example; liquid hydrogen; a hydride such as MgH_2 having a high hydrogen release temperature at ~1 bar pressure; a medium temperature hydride such as sodium alanate, $NaAlH_4$; and low temperature hydrides such as $LaNi_5H_6$ and $FeTiH_2$. The corresponding parameters for gasoline (~40 MJ/kg, ~30 MJ/liter) are well beyond the upper right corner of Fig. 1. While all the filled circles are in the lower left corner of Fig. 1, Pinkerton and Wicke have estimated that to achieve about 300 miles (500 km) range on an advanced, light-weight fuel cell vehicle the system densities should reside in the "minimum performance goal" area.[2] The even more stringent requirement to provide the same range for any vehicle architecture would demand a hydrogen storage system in the "ultimate technology goal" area.

There are other parameter goals in addition to the gravimetric and volumetric densities. Table 1 includes several of them. It is desirable to consume as little of the energy available in the hydrogen as possible in order to release it; the release temperature should be around 80°C, near the operating temperature of the fuel cell stack; the refueling time should be minimal; hydrogen release needs to be possible over the ±45°C temperature interval to which vehicles can be exposed; and durability of the hydrogen storage system is a *sine qua non*.

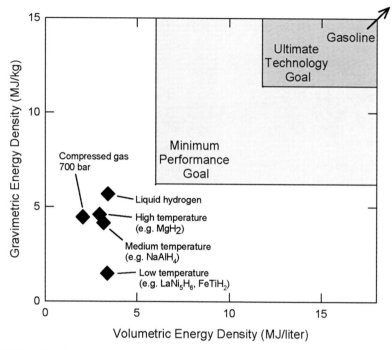

FIGURE 1. Gravimetric vs. volumetric energy density for various hydrogen storage technologies (Reference 2).

TABLE 1. Hydrogen storage parameter goals.	
Metric	**Goal**
System energy per unit mass for conventional vehicles with 300 mile (500 km) range	> 6 MJ/kg
System energy per unit volume for conventional vehicles with 300 mile (500 km) range	> 6 MJ/ℓ
Usable energy consumed in releasing H_2	< 5%
H_2 release temperature	~80°C
Refueling time	< 5 min
H_2 ambient release temperature range	±45°C
Durability (to maintain 80% capacity)	150,000 miles (240,000 km)

CURRENT OPTIONS FOR STORING HYDROGEN

Present methods for storing hydrogen can be grouped roughly into two broad categories, physical storage and chemical storage, as indicated in Fig. 2.

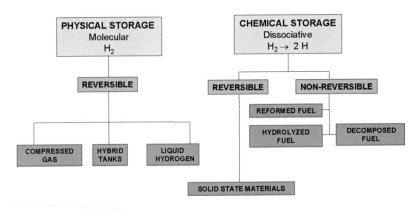

FIGURE 2. Options for storing hydrogen.

Physical Storage

Methods in which hydrogen is present in the form of H_2 molecules are classified here as physical storage. Hydrogen can be contained as a compressed gas under high pressure or as a liquid at cryogenic temperatures (~20 K). A third, less familiar, option is a hybrid of compressed gas and low temperature (~77 K) tanks. As Fig. 2 indicates, all forms of physical storage are reversible - the tank can be refilled on-board the vehicle.

Early commercial prototypes of compressed gas tanks have been developed [Fig. 3(a)]. If this technology is to be used in large-scale automotive applications, however, efficient, high-volume manufacturing processes, as well as less expensive materials, carbon fiber and binder in particular, will be necessary. Furthermore, engineering factors relevant to tank safety requirements and possible failure mechanisms will need to be understood.

(a)

(b)

FIGURE 3. Examples of prototype tanks for storing (a) compressed hydrogen gas (Ref. 3) and (b) liquid hydrogen (courtesy M. Herrmann).

Prototype liquid hydrogen tanks have also been constructed [Fig. 3(b)]. Prospects for commercial-scale implementation would benefit from reduced tank mass and especially reduced system volume, in addition to cost-effective high-throughput manufacturing techniques. The most significant drawback to liquid hydrogen tanks, however, is the increase in pressure and attendant boil off of hydrogen due to heat transfer from the ambient atmosphere. Technology to extend this "dormancy" period is essential. More efficient methods for liquefying hydrogen are also desirable.

Hybrid physical storage involves compressed H_2 at liquid nitrogen temperature. Hybrid storage tanks generally consist of a moderate pressure container mounted within a cryogenic outer shell. The combination of elevated pressure and low temperature results in a greater increase in gas density than by raising pressure at ambient temperature. Further density increase is possible through the use of adsorbents within the pressure vessel, presenting an attractive opportunity for new materials research. The question is whether a hybrid tank represents the best of both compressed gas and cryogenic tank characteristics, or the worst. Several hybrid tank concepts are currently under development by different organizations.

Chemical Storage

Opportunities for storing hydrogen increase on moving from physical to chemical storage, which usually involves the dissociation of H_2 into two hydrogen atoms. As Fig. 2 indicates, chemical storage options can be divided into two broad classes. One class comprises non-reversible materials that release or generate hydrogen on-board and leave dehydrided products that must be physically removed from the vehicle and recharged off-board. In general heat must be removed during hydrogen production on the vehicle, and energy is required to reconstitute the hydrided form.

Of the non-reversible options in Fig. 2, use of an on-board reformer to produce hydrogen from either alcohols or hydrocarbons has been deemed impractical by most automobile manufacturers in view of packaging and cost considerations. Reforming at fueling stations, however, does remain a viable possibility. Hydrolysis hydrides such as NaH and $NaBH_4$ have the disadvantage of producing large amounts of heat during hydrogen generation and requiring substantial energy for rehydriding the products off-board. Decomposition of specialty fuels, liquid hydrocarbons in particular, continues to be an option, but an efficient recycling process engineered for minimum cost and ease of use is imperative.

Reversible solid-state hydrogen storage on-board the vehicle can be considered the ultimate goal for automotive applications. This is a very rich area with enormous potential, driven in part by the fact that solid hydrides often feature volumetric hydrogen density exceeding that of liquid hydrogen. Many research groups around the world are seeking to identify and develop material systems that meet the technology targets. Figure 4 illustrates the improvement in storage capacity that has been achieved over the past few years. Prior to the late 1990s most of the focus on reversible storage was directed toward traditional metal hydrides such as $LaNi_5H_6$, $FeTiH_2$, and Mg_2NiH_4. Thereafter significant interest turned to complex hydrides, those containing hydrogen-bearing chemical complexes such as $(AlH_4)^{-1}$, $(BH_4)^{-1}$, $(NH_2)^{-1}$, and $(NH)^{-2}$. Sodium alanate, $NaAlH_4$, received considerable attention after

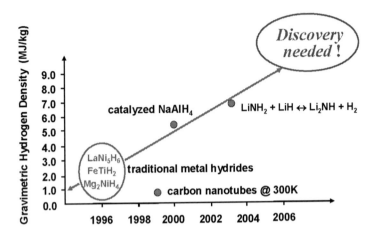

FIGURE 4. Chronology of gravimetric hydrogen capacity for materials considered as candidates for reversible on-board hydrogen storage.

the seminal work of Bogdanović and Schwickardi demonstrating that its dehydriding reactions can be reversed under improved conditions of temperature and pressure by means of suitable catalysts.[4] Subsequently, the reversible reaction of lithium amide ($LiNH_2$) with lithium hydride (LiH) to yield lithium imide (Li_2NH) and H_2 has been studied (see, e. g., Refs. 5-9 and references therein). It features a higher storage capacity than $NaAlH_4$ but releases H_2 at a higher temperature. Carbon nanotubes do not sorb appreciable amounts of hydrogen at room temperature, as Fig. 4 indicates, despite initial claims to the contrary.

RECENT DEVELOPMENTS IN HYDROGEN STORAGE MATERIALS

While no materials system has yet satisfied all the requirements given in Fig. 1 and Table 1, progress continues to be made. This section touches on a few specific aspects of that progress.

Novel Hydrides

Although binary hydrides (AH_x) are exceedingly well known and a prodigious body of literature on ternary hydrides (AB_nH_x) exists, more hydrogen-containing materials certainly await discovery. There is every reason for optimism that new materials may offer improved hydrogen storage properties. In this respect the area can be likened to superconductivity before the advent of $YBa_2Cu_3O_7$ and to permanent magnetism prior to the discovery of $Nd_2Fe_{14}B$.

Figure 5 displays the cubic unit cell of $Li_4BN_3H_{10}$, a novel quaternary hydride containing both $(BH_4)^{-1}$ and $(NH_2)^{-1}$ complexes.[10,11] It releases about 11 mass% hydrogen, but attempts to reverse it with catalysts and additives have so far been

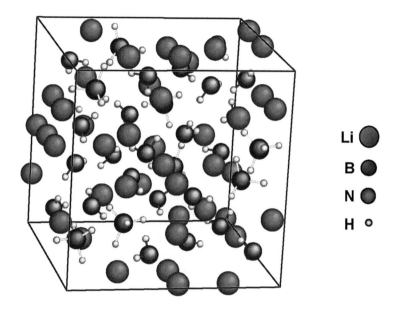

FIGURE 5. Unit cell of body-centered cubic crystal structure of $Li_4BN_3H_{10}$ [= $(LiBH_4)(LiNH_2)_3$].

unsuccessful. Other compounds have been tentatively identified in the rich Li-N-B-H phase diagram.[12]

Destabilized Systems

The equilibrium pressure P and operating temperature T of a hydride system are linked by the enthalpy of formation (or reaction) ΔH by the van't Hoff relation:

$$\ell n\ P(bar) \sim \Delta H/RT - \Delta S/R \quad , \tag{1}$$

where ΔS is the corresponding entropy change and R the gas constant. Light metal hydrides such as LiH, $LiBH_4$, and MgH_2 tend to have large values of ΔH and, hence, high operating temperature at a pressure of 1 bar. Vajo and coworkers,[13] building on early work of Reilly and Wiswall,[14] have emphasized that ΔH can be moderated by reacting a hydride with another compound to produce another phase (or phases) that reduces the overall enthalpy. This has become a very active area of inquiry (see, e. g., Refs. 15-17 and references therein). The following are some examples of destabilized reactions:

$$LiNH_2 + LiH \leftrightarrow Li_2NH + H_2 \tag{2}$$
$$[6.5\ mass\%H;\ T(1\ bar) \sim 275°C] \quad ;$$

303

$$2LiBH_4 + MgH_2 \leftrightarrow 2LiH + MgB_2 + 4H_2 \qquad (3)$$
[11.5 mass%H; T(1 bar) ~ 225°C]   ;

$$6LiBH_4 + CaH_2 \leftrightarrow 6LiH + CaB_6 + 10H_2 \qquad (4)$$
[11.5 mass%H; T(1 bar) ~ 420°C]   .

Reaction 2, included in Fig. 4, can be viewed as destabilizing either $LiNH_2$ or LiH. It features a capacity of 6.5 mass%H and an operating temperature at P = 1 bar, T(1 bar), of about 275°C, much lower than for either of the reactants alone. MgH_2 and CaH_2 are destabilized via reaction with $LiBH_4$ in reaction 3 and reaction 4, respectively. Both reactions produce at least 11 mass%H. The destabilized systems identified thus far, however, are characterized by undesirably high T(1 bar) values, and the reverse (hydrogenation) reactions are too sluggish.

Cryogenic Materials for Hybrid Tanks

Hydrogen molecules can bind to surfaces at low temperatures, so materials with large surface area might enable a hybrid tank with sufficiently improved capacity to offset the penalty for cooling. Considerable research is underway on such materials. Activated carbons can have an effective surface area as large as 2500 m^2/g (1 lb $\leftrightarrow$ 280 acres!) and sorb 5 mass%H at 77K. Metal organic frameworks (MOFs) constitute a new class of porous materials that can feature even larger surface areas and H_2 capacity. Pioneered by Yaghi and coworkers, MOFs consist of inorganic clusters such as ZnO_4 held by organic linkers such as benzenedicarboxylate in a crystalline matrix.[18] More than 10,000 MOFs have been identified, but the hydrogen sorption properties of comparatively few have been measured. Effective surface areas and hydrogen contents at 77K as large as ~5000 m^2/g and 5–7 mass%, respectively, have been observed thus far.[19,20] Figure 6 displays a representative MOF.

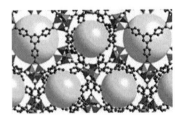

FIGURE 6. Schematic diagram of MOF-177, comprised of zinc clusters and benzenetribenzoate units (Reference 21). The spheres delineate the pores.

Modeling New Materials with Density Functional Theory

Implementation of density functional theory (DFT)[22] with progressively more powerful computers and computational techniques has enabled the efficient calculation of electronic structure and other material properties. DFT has become a valuable and important tool in hydrogen storage research. It has significantly enhanced our understanding of known materials, and it is being imaginatively employed to guide the discovery and development of new hydrides. Recently proposed materials include (i) organometallic buckyballs - C_{60} fullerene molecules decorated with transition metals (TMs) such as scandium;[23] (ii) TM-decorated polymers such as polyacetylene;[24] (iii) TM-ethylene complexes (Fig. 7);[25] and (iv) activated boron nitride nanotubes.[26]

$$C_2H_4(Ti\text{-}5H_2)_2$$

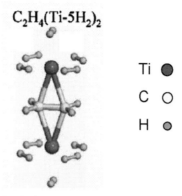

Ti ●

C ○

H ●

FIGURE 7. Ethylene molecule functionalized by two titanium atoms binding ten H_2 molecules (Ref. 25).

SUMMARY

While considerable progress on vehicular hydrogen storage has been made, system properties need to be enhanced and cost structures improved if a broad applications spectrum is to be realized. Liquid and compressed gas storage are technically feasible and are being employed on prototype vehicles. Although both fall short of volumetric and gravimetric goals, innovative architectures could afford efficient packaging and extended vehicle range. Solid state hydrogen storage continues to offer attractive possibilities at both ambient and cryogenic temperatures. The area is opulent with research ideas and effort, encouraging justifiable optimism that further advances are on the horizon.

ACKNOWLEDGMENTS

Valuable discussions with and input from A. Dailly, M. Herrmann, F. E. Pinkerton, J. A. Spearot, B. G. Wicke , and J. Yang are greatly appreciated.

REFERENCES

1. http://www1.eere.energy.gov/hydrogenandfuelcells/
2. F. E. Pinkerton and B. G. Wicke, *Ind. Phys.* 10, 20 (2004).
3. www.nrel.gov/hydrogen/photos.html.
4. B. Bogdanović and M. Schwickardi, *J. Alloys Compd.* 253-254, 1 (1997).
5. P. Chen, Z. Xiong, J. Luo, J. Lin, and K. L. Tan, *Nature* 420, 302 (2002).
6. P. Chen, Z. Xiong, J. Luo, J. Lin, and K. L. Tan, *J. Phys. Chem. B* 107, 10967 (2003).
7. T. Ichikawa, S. Isobe, N. Hanada, and H. Fujii, *J. Alloys Compd.* 365, 271 (2004).

8. G. P. Meisner, F. E. Pinkerton, M. S. Meyer, M. P. Balogh, and M. D. Kundrat, *J. Alloys Compd.* 404-406, 24 (2005).
9. J. F. Herbst and L. G. Hector, Jr., *Phys. Rev.* B 72, 125120 (2005).
10. F. E. Pinkerton, G. P. Meisner, M. S. Meyer, M. P. Balogh, and M. D. Kundrat, *J. Phys. Chem. B Lett.* 109, 6 (2005).
11. Y. E. Filinchuk, K. Yvon, G. P. Meisner, F. E. Pinkerton, and M. P. Balogh, *Inorg. Chem.* 45, 1433 (2006).
12. G. P Meisner, M. L. Scullin, M. P. Balogh, F. E. Pinkerton, and M. S. Meyer, *J. Phys. Chem. B.* 110, 4186 (2006).
13. J. J. Vajo, F. Mertens, C. C. Ahn, R. C. Bowman, Jr., and B. Fultz, *J. Phys. Chem. B* 108, 13977 (2004).
14. J. J. Reilly and R. H. Wiswall, *Inorg. Chem.* 6, 2220 (1967).
15. S. V. Alapati, J. K. Johnson, and D. S. Sholl, *Phys. Chem. Chem. Phys.* 9, 1438 (2007).
16. D. J. Siegel, C. Wolverton, and V. Ozoliņš, Phys. Rev. B 76, 4102 (2007).
17. F. E. Pinkerton, M. S. Meyer, G. P. Meisner, M. P. Balogh, and J. J. Vajo, *J. Phys. Chem C* 111, 12881 (2007).
18. O. M. Yaghi, M. O'Keeffe, N. W. Ockwig, H. K. Chae, M. Eddaoudi, and J. Kim, *Nature* 423, 705 (2003).
19. A. G. Wong-Foy, A. J. Matzger, and O. M. Yaghi, *J. Am. Chem. Soc.* 128, 3494 (2006).
20. M. Latroche, S. Surblé, C. Serre, C. Mellot-Draznieks, P. L. Llewellyn, J-H. Lee, J-S. Chang, S. H. Jhung, and G. Férey, Angew. *Chem. Int. Ed.* 45, 8227 (2006).
21. *Chem. & Eng. News* 83, 11 (2005).
22. W. Kohn and L. Sham, *Phys. Rev.* 140, A1133 (1965).
23. Y. Zhao, Y-H. Kim, A. C. Dillon, M. J. Heben, and S. B. Zhang, *Phys. Rev. Lett.* 94, 155504 (2005).
24. H. Lee, W. I. Choi, and J. Ihm, *Phys. Rev. Lett.* 97, 056104 (2006).
25. E. Durgun, S. Ciraci, W. Zhou, and T. Yildirim, *Phys. Rev. Lett.* 97, 226102 (2006).
26. S-H. Jhi, *Phys. Rev. B* 74, 155424 (2006).

SESSION D

ELECTRICITY FROM RENEWABLE ENERGY

Solar Energy Conversion[*]

George W. Crabtree[a] and Nathan S. Lewis[b]

[a]*Materials Science Division*
Argonne National Laboratory
Argonne, Illinois 60439
[b]*Chemistry Department and Molecular Materials Research Center*
California Institute of Technology
Pasadena, California 91125

Abstract. If solar energy is to become a practical alternative to fossil fuels, we must have efficient ways to convert photons into electricity, fuel, and heat. The need for better conversion technologies is a driving force behind many recent developments in biology, materials, and especially nanoscience.

The Sun provides Earth with a staggering amount of energy—enough to power the great oceanic and atmospheric currents, the cycle of evaporation and condensation that brings fresh water inland and drives river flow, and the typhoons, hurricanes, and tornadoes that so easily destroy the natural and built landscape. The San Francisco earthquake of 1906, with magnitude 7.8, released an estimated 10^{17} joules of energy, the amount the Sun delivers to Earth in one second. Earth's ultimate recoverable resource of oil, estimated at 3 trillion barrels, contains 1.7×10^{22} joules of energy, which the Sun supplies to Earth in 1.5 days. The amount of energy humans use annually, about 4.6×10^{20} joules, is delivered to Earth by the Sun in one hour. The enormous power that the Sun continuously delivers to Earth, 1.2×10^{5} terawatts, dwarfs every other energy source, renewable or nonrenewable. It dramatically exceeds the rate at which human civilization produces and uses energy, currently about 13 TW.

The impressive supply of solar energy is complemented by its versatility, as illustrated in Figure 1. Sunlight can be converted into electricity by exciting electrons in a solar cell. It can yield chemical fuel via natural photosynthesis in green plants or artificial photosynthesis in human-engineered systems. Concentrated or unconcentrated sunlight can produce heat for direct use or further conversion to electricity.[1]

Despite the abundance and versatility of solar energy, we use very little of it to directly power human activities. Solar electricity accounts for a minuscule 0.015% of

---

[*] Reprinted with permission from George W. Crabtree and Nathan S. Lewis, *Physics Today*, Vol. 60, March 2007, pages 37–42. Copyright 2007, American Institute of Physics.

world electricity production, and solar heat for 0.3% of global heating of space and water. Biomass produced by natural photosynthesis is by far the largest use of solar energy; its combustion or gasification accounts for about 11% of human energy needs. However, more than two-thirds of that is gathered unsustainably—that is, with no replacement plan—and burned in small, inefficient stoves where combustion is incomplete and the resulting pollutants are uncontrolled.

Between 80% and 85% of our energy comes from fossil fuels, a product of ancient biomass stored beneath Earth's surface for up to 200 million years. Fossil-fuel resources are of finite extent and are distributed unevenly beneath Earth's surface. When fossil fuels are turned into useful energy though combustion, they produce greenhouse gases and other harmful environmental pollutants. In contrast, solar photons are effectively inexhaustible and unrestricted by geopolitical boundaries. Their direct use for energy production does not threaten health or climate. The solar resource's magnitude, wide availability, versatility, and benign effect on the environment and climate make it an appealing energy source.

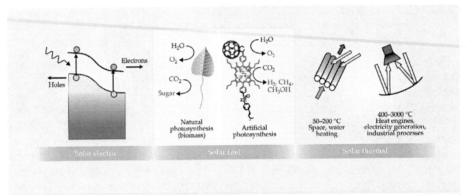

FIGURE 1. Solar photons convert naturally into three forms of energy—electricity, chemical fuel, and heat—that link seamlessly with existing energy chains. Despite the enormous energy flux supplied by the Sun, the three conversion routes supply only a tiny fraction of our current and future energy needs. Solar electricity, at between 5 and 10 times the cost of electricity from fossil fuels, supplies just 0.015% of the world's electricity demand. Solar fuel, in the form of biomass, accounts for approximately 11% of world fuel use, but the majority of that is harvested unsustainably. Solar heat provides 0.3% of the energy used for heating space and water. It is anticipated that by the year 2030 the world demand for electricity will double and the demands for fuel and heat will increase by 60%. The utilization gap between solar energy's potential and our use of it can be overcome by raising the efficiency of the conversion processes, which are all well below their theoretical limits.

RAISING EFFICIENCY

The enormous gap between the potential of solar energy and our use of it is due to cost and conversion capacity. Fossil fuels meet our energy demands much more cheaply than solar alternatives, in part because fossil-fuel deposits are concentrated sources of energy, whereas the Sun distributes photons fairly uniformly over Earth at a more modest energy density. The use of biomass as fuel is limited by the production capacity of the available land and water. The cost and capacity limitations on solar energy use are most effectively addressed by a single research objective: cost effectively raising conversion efficiency.

The best commercial solar cells based on single-crystal silicon are about 18% efficient. Laboratory solar cells based on cheaper dye sensitization of oxide semiconductors are typically less than 10% efficient, and those based on even cheaper organic materials are 2–5% efficient. Green plants convert sunlight into biomass with a typical yearly averaged efficiency of less than 0.3%. The cheapest solar electricity comes not from photovoltaics but from conventional induction generators powered by steam engines driven by solar heat, with efficiencies of 20% on average and 30% for the best systems. Those efficiencies are far below their theoretical limits. Increasing efficiency reduces cost and increases capacity, which raises solar energy to a new level of competitiveness.

TABLE 1. Photovoltaic conversion efficiencies. *As verified by the National Renewable Energy Laboratory. Organic cell efficiencies of up to 5% have been reported in the literature.

	Laboratory best*	Thermodynamic limit
Single junction		31%
Silicon (crystalline)	25%	
Silicon (nanocrystalline)	10%	
Gallium arsenide	25%	
Dye sensitized	10%	
Organic	3%	
Multijunction	32%	66%
Concentrated sunlight (single junction)	28%	41%
Carrier multiplication		42%

Dramatic cost-effective increases in the efficiency of solar energy conversion are enabled by our growing ability to understand and control the fundamental nanoscale phenomena that govern the conversion of photons into other forms of energy. Such phenomena have, until recently, been beyond the reach of our best structural and spectroscopic probes. The rise of nanoscience is yielding new fabrication techniques based on self-assembly, incisive new probes of structure and dynamics at ever-smaller length and time scales, and the new theoretical capability to simulate assemblies of thousands of atoms. Those advances promise the capability to understand and control the underlying structures and dynamics of photon conversion processes.

ELECTRICITY

Solar cells capture photons by exciting electrons across the bandgap of a semiconductor, which creates electron–hole pairs that are then charge separated, typically by p–n junctions introduced by doping. The space charge at the p–n junction interface drives electrons in one direction and holes in the other, which creates at the external electrodes a potential difference equal to the bandgap, as sketched in the left panel of Figure 1. The concept and configuration are similar to those of a semiconductor diode, except that electrons and holes are introduced into the junction by photon excitation and are removed at the electrodes.

With their 1961 analysis of thermodynamic efficiency, William Shockley and Hans Queisser established a milestone reference point for the performance of solar cells.[2] The analysis is based on four assumptions: a single p–n junction, one electron–hole pair excited per incoming photon, thermal relaxation of the electron–hole pair energy in excess of the bandgap, and illumination with unconcentrated sunlight. Achieving the efficiency limit of 31% that they established for those conditions remains a research goal. The best single–crystal Si cells have achieved 25% efficiency in the laboratory and about 18% in commercial practice. Cheaper solar cells can be made from other materials,[3] but they operate at significantly lower efficiency, as shown in the table above. Thin-film cells offer advantages beyond cost, including pliability and potential integration with preexisting buildings and infrastructure. A new approach, (Figure 2) used flexible, processable material, consisting of crystalline Si wires embedded in a plastic (polydimethylsiloxane) The wires allow effective capture and conversion of sunlight from inexpensive silicon, while the embedded structure offers the processability of organic materials such as photographic film. Achieving high efficiency from inexpensive materials with so-called third-generation cells, indicated in Figure 3, is the grand research challenge for making solar electricity dramatically more affordable.

The Shockley–Queisser limit can be exceeded by violating one or more of its premises. Concentrating sunlight allows for a greater contribution from multi-photon processes; that contribution increases the theoretical efficiency limit to 41% for a single-junction cell with thermal relaxation. A cell with a single p–n junction captures only a fraction of the solar spectrum: photons with energies less than the bandgap are not captured, and photons with energies greater than the bandgap have their excess energy lost to thermal relaxation. Stacked cells with different bandgaps capture a greater fraction of the solar spectrum; the efficiency limit is 43% for two junctions illuminated with unconcentrated sunlight, 49% for three junctions, and 66% for infinitely many junctions.

The most dramatic and surprising potential increase in efficiency comes from carrier multiplication,[4] a quantum-dot phenomenon that results in multiple electron hole pairs for a single incident photon. Carrier multiplication was discussed by Arthur Nozik in 2002 and observed by Richard Schaller and Victor Klimov two years later. Nanocrystals of lead selenide, lead sulfide, or cadmium selenide generate as many as seven electrons per incoming photon, which suggests that efficient solar cells might be made with such nanocrystals. In bulk-semiconductor solar cells, when an incident photon excites a single electron–hole pair, the electron–hole pair energy in excess of the bandgap is likely to be lost to thermal relaxation, whereas in some nanocrystals most of the excess energy can appear as additional electron–hole pairs. If the nanocrystals can be incorporated into a solar cell, the extra pairs could be tapped off as enhanced photocurrent, which would increase the efficiency of the cell.

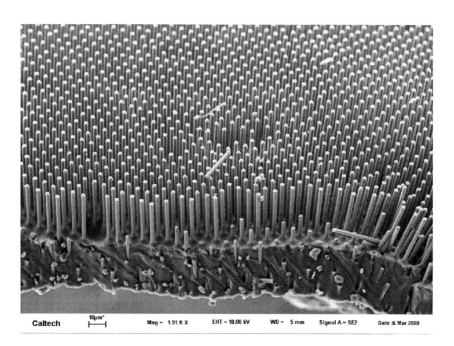

Caltech   10µm*   Mag = 1.51 K X   EHT = 10.00 kV   WD = 5 mm   Signal A = SE2   Date :6 Mar 2008

FIGURE 2. Photograph (upper) and scanning electron microscope (lower) of a flexible, processable material consisting of crystalline Si wires embedded in a plastic (polydimethylsiloxane). The wires allow effective capture and conversion of sunlight from inexpensive silicon, while the embedded structure offers the processability of organic materials such as photographic film. [J. Spurgeon, K. Plass, and N. Lewis, Division of Chemistry and Chemical Engineering, Caltech]

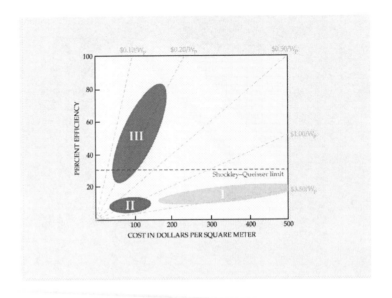

FIGURE 3. The three generations of solar cells. First-generation cells are based on expensive silicon wafers and make up 85% of the current commercial market. Second-generation cells are based on thin films of materials such as amorphous silicon, nanocrystalline silicon, cadmium telluride, or copper indium selenide. The materials are less expensive, but research is needed to raise the cells' efficiency to the levels shown if the cost of delivered power is to be reduced. Third-generation cells are the research goal: a dramatic increase in efficiency that maintains the cost advantage of second-generation materials. Their design may make use of carrier multiplication, hot electron extraction, multiple junctions, sunlight concentration, or new materials. The horizontal axis represents the cost of the solar module only; it must be approximately doubled to include the costs of packaging and mounting. Dotted lines indicate the cost per watt of peak power (W_p). (Adapted from ref. 2, Green.)

Hot-electron extraction provides another way to increase the efficiency of nanocrystal-based solar cells: tapping off energetic electrons and holes before they have time to thermally relax.[5] Hot electrons boost efficiency by increasing the operating voltage above the bandgap, whereas carrier multiplication increases the operating current. Femtosecond laser and x-ray techniques can provide the necessary understanding of the ultrafast decay processes in bulk semiconductors and their modification in nanoscale geometries that will enable the use of hot-electron phenomena in next-generation solar cells.

Although designs have been proposed for quantum-dot solar cells that benefit from hot electrons or carrier multiplication, significant obstacles impede their implementation. We cannot attach wires to nanocrystals the way we do to bulk semiconductors; collecting the electrons from billions of tiny dots and putting them all into one current lead is a problem in nanoscale engineering that no one has solved yet. A second challenge is separating the electrons from the holes, the job normally done by the space charge at the p–n junction in bulk solar cells. Those obstacles must be overcome before practical quantum-dot cells can be constructed.[5]

Dye-sensitized solar cells, introduced by Michael Grätzel and coworkers in 1991, create a new paradigm for photon capture and charge transport in solar conversion.[6] Expensive Si, which does both of those jobs in conventional cells, is replaced by a hybrid of chemical dye and the inexpensive wide-bandgap semiconductor titanium dioxide. The dye, analogous to the light-harvesting chlorophyll in green plants, captures a photon, which elevates one of its electrons to an excited state. The electron is then quickly transferred to the conduction band of a neighboring TiO_2 nanoparticle, and it drifts through an array of similar nanoparticles to the external electrode. The hole left in the dye molecule recombines with an electron carried to it through an electrolyte from the counter electrode by an anion such as I–. In addition to using cheaper materials, the scheme separates the absorption spectrum of the cell from the bandgap of the semiconductor, so the cell sensitivity is more easily tuned to match the solar spectrum. The cell efficiency depends on several kinds of nanoscale charge dynamics, such as the way the electrons move across the dye–TiO_2 and dye–anion interfaces, and the way charges move through the dye, the TiO_2 nanoparticle array, and the electrolyte. The development of new dyes and shuttle ions and the characterization and control of the dynamics through time-resolved spectroscopy are vibrant and promising research areas. An equally important research challenge is the nanoscale fabrication of dye-sensitized cells to minimize the transport distances in the dye and semiconductor and maximize the electron-transfer rate at the interfaces.

FUEL

Over the past 3 billion years, Nature has devised a remarkably diverse set of pathways for converting solar photons into chemical fuel. An estimated 100TW of solar energy go into photosynthesis, the production of sugars and starches from water and carbon dioxide via endothermic reactions facilitated by catalysts. Although plants have covered Earth in green in their quest to capture solar photons, their overall conversion efficiency is too low to readily satisfy the human demand for energy. The early stages of photosynthesis are efficient: Two molecules of water are split to provide four protons and electrons for subsequent reactions, and an oxygen molecule is released into the atmosphere. The inefficiency lies in the later stages, in which carbon dioxide is reduced to form the carbohydrates that plants use to grow roots, leaves, and stalks. The research challenge is to make the overall conversion process between 10 and 100 times more efficient by improving or replacing the inefficient stages of photosynthesis. There are three routes to improving the efficiency of photosynthesis-based solar fuel production: breeding or genetically engineering plants to grow faster and produce more biomass, connecting natural photosynthetic pathways in novel configurations to avoid the inefficient steps, and using artificial bio–inspired nanoscale assemblies to produce fuel from water and CO_2. The first route is the occupation of a thriving industry that has produced remarkable increases in plant yields, and we will not discuss it further. The second and third routes, which involve more direct manipulation of photosynthetic pathways, are still in their early stages of research.

Nature provides many examples of metabolic systems that convert sunlight and chemicals into high-energy fuels. Green plants use an elaborate complex of chlorophyll molecules coupled to a reaction center to split water into protons, electrons, and oxygen. Bacteria use the hydrogenase enzyme to create hydrogen molecules from protons and electrons. More than 60 species of methane-producing archaea, remnants from early Earth when the atmosphere was reducing instead of oxidizing, use H_2 to reduce CO_2 to CH_4. Anaerobic organisms such as yeasts and bacteria use enzymes to ferment sugars into alcohols.

In nature, the metabolic pathways are connected in complicated networks that have evolved for organisms' survival and reproduction, not for fuel production. The efficient steps that are relevant for fuel production might conceivably be isolated and connected directly to one another to produce fuels such as H_2, CH_4, or alcohols. Hybridizing nature in that way takes advantage of the elaborate molecular processes that biology has evolved and that are still beyond human reach, while eliminating the inefficient steps not needed for fuel production. For example, the protons and electrons produced in the early stages of photosynthesis could link to hydrogenase to produce H_2, and a further connection to methanogenic archaea could produce CH_4. The challenges are creating a functional interface between existing metabolic modules, achieving a competitive efficiency for the modified network, and inducing the organism hosting the hybrid system to reproduce. The ambitious vision of hybrids that produce energy efficiently sets a basic research agenda to simultaneously advance the frontiers of biology, materials science, and energy conversion.

Artificial photosynthesis takes the ultimate step of using inanimate components to convert sunlight into chemical fuel.[7,8] Although the components do not come from nature, the energy conversion routes are bio-inspired. Remarkable progress has been made in the field.[8] Light harvesting and charge separation are accomplished by synthetic antennas linked to a porphyrin-based charge donor and a fullerene acceptor, as shown in Figure 4. The assembly is embedded in an artificial membrane, in the presence of quinones that act as proton shuttles, to produce a light-triggered proton gradient across the membrane. The proton gradient can do useful work, such as powering the molecular synthesis of adenosine triphosphate by mechanical rotation of natural ATP synthase inserted into the membrane. Under the right conditions, the required elements self-assemble to produce a membrane-based chemical factory that transforms light into the chemical fuel ATP, molecule by molecule at ambient temperature, in the spirit of natural photosynthesis.

Such remarkable achievements illustrate the promise of producing fuel directly from sunlight without the use of biological components. Many fundamental challenges must be overcome, however. The output of the above energy conversion chain is ATP, not a fuel that links naturally to human–engineered energy chains. The last step relies on the natural catalyst ATP synthase, a highly evolved protein whose function we cannot yet duplicate artificially. Laboratory approximations of biological catalysts have catalytic activities that are often orders of magnitude lower than those of their biological counterparts, which indicates the importance of subtle features that we are not yet able to resolve or to reproduce.

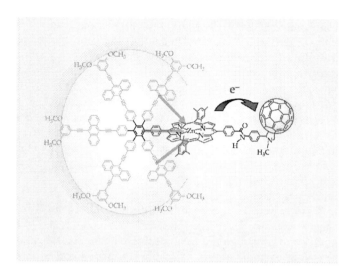

FIGURE 4. An artificial antenna–reaction-center complex that mimics the early stages of photosynthesis. The central hexaphenylbenzene core provides structure and rigidity for the surrounding wheel of five bis(phenylethynyl)anthracene antennas that gather light at 430–475 nanometers. The energy is transferred to a porphyrin complex in 1–10 picoseconds (orange arrows), where it excites an electron that is transferred to the fullerene acceptor in 80 ps; the resulting charge-separated state has a lifetime of 15 nanoseconds. Complexes such as the one shown provide the first steps in artificial photosynthesis. They have the potential to drive further chemical reactions, such as the oxidation of water to produce H_2 or the reduction of CO_2 to CH_4, alcohol, or other fuel. (Adapted from ref. 16.)

Solar fuels can be created in an alternate, fully nonbiological way based on semiconductor solar cells rather than on photosynthesis. In photoelectrochemical conversion, the charge-separated electrons and holes are used locally to split water or reduce CO_2 at the interface with an electrolytic solution, rather than being sent through an external circuit to do electrical work.[9] Hydrogen was produced at the electrode–water interface with greater than 10% efficiency by Adam Heller in 1984 and by Oscar Khaselev and John Turner in 1998, but the fundamental phenomena involved remain mysterious, and the present devices are not practical. A promising way to improve them is by tailoring the nanoscale architecture of the electrode–electrolyte interface to promote the reaction of interest. A better understanding of how individual electrons negotiate the electrode–electrolyte interface is needed before H_2 can be produced with greater efficiency or more complex reactions can be designed for reducing CO_2 to useful fuels.

HEAT

The first step in traditional energy conversion is the combustion of fuel, usually fossil fuel, to produce heat. Heat produced by combustion may be used for heating space and water, cooking, or industrial processes, or it may be further converted into motion or electricity. The premise of solar thermal conversion is that heat from the

Sun replaces heat from combustion; fossil–fuel use and its threat to the environment and climate are thus reduced. Unconcentrated sunlight can bring the temperature of a fluid to about 200 °C, enough to heat space and water in residential and commercial applications. Many regions use solar water heating, though in only a few countries, such as Cyprus and Israel, does it meet a significant fraction of the demand. Concentration of sunlight in parabolic troughs produces temperatures of 400 °C, and parabolic dishes can produce temperatures of 650°C and higher.[10,11] Power towers, in which a farm of mirrors on the ground reflects to a common receiver at the top of a tower, can yield temperatures of 1500 °C or more.[10,12] The high temperatures of solar power towers are attractive for thermo–chemical water splitting and solar–driven reforming of fossil fuels to produce H_2.[11]

The temperatures produced by concentrated sunlight are high enough to power heat engines, whose Carnot efficiencies depend only on the ratio of the inlet and outlet temperatures. Steam engines driven by solar heat and connected to conventional generators currently supply the cheapest solar electricity. Nine solar thermal electricity plants that use tracking parabolic-trough concentrators were installed in California's Mojave Desert between 1984 and 1991. Those plants still operate, supplying 354 MW of peak power to the grid. Their average annual efficiency is approximately 20%, and the most recently installed can achieve 30%.

Although those efficiencies are the highest for any widely implemented form of solar conversion, they are modest compared to the nearly 60% efficiency of the best gas-fired electricity generators. Achieving greater efficiency for solar conversion requires large-scale plants with operating temperatures of 1500°C or more, as might be produced by power towers. Another alternative, still in the exploration stage, is a hybrid of two conversion schemes: A concentrated solar beam is split into its visible portion for efficient photovoltaic conversion and its high-energy portion for conversion to heat that is converted to electricity through a heat engine.[10]

Thermoelectric materials, which require no moving parts to convert thermal gradients directly into electricity, are an attractive possibility for reliable and inexpensive electricity production.[13] Charge carriers in a thermal gradient diffuse from hot to cold, driven by the temperature difference but creating an electric current by virtue of the charge on each carrier. The strength of the effect is measured by the thermopower, the ratio of the voltage produced to the applied temperature difference. Although the thermoelectric effect has been known for nearly 200 years, materials that can potentially convert heat to electricity efficiently enough for widespread use have emerged only since the 1990s.[13] Efficient conversion depends on minimizing the thermal conductivity of a material, so as not to short–circuit the thermal gradient, while maximizing the material's electrical conductivity and thermopower. Achieving such a combination of opposites requires the separate tuning of several material properties: the bandgap, the electronic density of states, and the electron and phonon lifetimes. The most promising materials are nanostructured composites. Quantum-dot or nanowire substructures introduce spikes in the density of states to tune the thermopower (which depends on the derivative of the density of states), and interfaces between the composite materials block thermal transport but allow electrical transport, as discussed by Lyndon Hicks and Mildred Dresselhaus in 1993.[14] Proof of concept for interface control of thermal and electrical conductivity was achieved by 2001 with thin-film superlattices of Bi_2Te_3/Sb_2Te_3 and $PbTe/PbSe$, which performed twice as well as bulk-alloy thermoelectrics of the same materials. The next challenges are to achieve the same performance in nanostructured bulk materials that can handle large amounts of power and to use nanodot or nanowire inclusions to control the thermopower. Figure 5 shows encouraging progress: structurally distinct nanodots in a bulk matrix of the thermoelectric material $Ag_{0.86}Pb_{18}SbTe_{20}$. Controlling the size, density, and distribution of such nanodot inclusions during bulk synthesis could significantly enhance thermoelectric performance.[15]

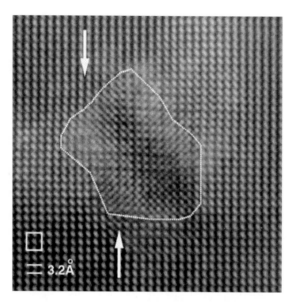

FIGURE 5. A nanodot inclusion in the bulk thermoelectric material $Ag_{0.86}Pb_{18}SbTe_{20}$, imaged with high-resolution transmission electron microscopy. Despite a lattice mismatch of 2–5%, the nanodot (indicated by the dotted line) is almost perfectly coherently embedded in the matrix. The arrows show two dislocations near the interface, and the white box indicates the unit cell. The nanodot is rich in silver and antimony relative to the matrix. (Adapted from ref. 15.)

STORAGE AND DISTRIBUTION

Solar energy presents a scientific challenge beyond the efficient conversion of solar photons to electricity, fuel, and heat. Once conversion on a large scale is achieved, we must find ways to store the large quantities of electricity and heat that we will produce. Access to solar energy is interrupted by natural cycles of day–night, cloudy–sunny, and winter–summer variation that are often out of phase with energy demand. Solar fuel production automatically stores energy in chemical bonds. Electricity and heat, however, are much more difficult to store. Cost effectively storing even a fraction of our peak demand for electricity or heat for 24 hours is a task well beyond present technology.

Storage is such an imposing technical challenge that innovative schemes have been proposed to minimize its need. Baseload solar electricity might be generated on constellations of satellites in geosynchronous orbit and beamed to Earth via microwaves focused onto ground-based receiving antennas. A global superconducting grid might direct electricity generated in sunny locations to cloudy or dark locations where demand exceeds supply. But those schemes, too, are far from being implemented. Without cost-effective storage and distribution, solar electricity can only be a peak–shaving technology for producing power in bright daylight, acting as a fill for some other energy source that can provide reliable power to users on demand.

OUTLOOK

The Sun has the enormous untapped potential to supply our growing energy needs. The barrier to greater use of the solar resource is its high cost relative to the cost of fossil fuels, although the disparity will decrease with the rising prices of fossil fuels and the rising costs of mitigating their impact on the environment and climate. The cost of solar energy is directly related to the low conversion efficiency, the modest energy density of solar radiation, and the costly materials currently required. The development of materials and methods to improve solar energy conversion is primarily a scientific challenge: Breakthroughs in fundamental understanding ought to enable marked progress. There is plenty of room for improvement, since photovoltaic conversion efficiencies for inexpensive organic and dye-sensitized solar cells are currently about 10% or less, the conversion efficiency of photosynthesis is less than 1%, and the best solar thermal efficiency is 30%. The theoretical limits suggest that we can do much better. Solar conversion is a young science. Its major growth began in the 1970s, spurred by the oil crisis that highlighted the pervasive importance of energy to our personal, social, economic, and political lives. In contrast, fossil-fuel science has developed over more than 250 years, stimulated by the Industrial Revolution and the promise of abundant fossil fuels. The science of thermodynamics, for example, is intimately intertwined with the development of the steam engine. The Carnot cycle, the mechanical equivalent of heat, and entropy all played starring roles in the development of thermodynamics and the technology of heat engines. Solar–energy science faces an equally rich future, with nanoscience enabling the discovery of the guiding principles of photonic energy conversion and their use in the development of cost–competitive new technologies.

This article is based on the conclusions contained in the report[1] of the US Department of Energy Basic Energy Sciences Workshop on Solar Energy Utilization, April 18–21, 2005. We served as chair (Lewis) and co-chair (Crabtree) of the workshop and were principal editors of the report. We acknowledge the US Department of Energy for support of both the workshop and preparation of the manuscript.

REFERENCES

1. N.S. Lewis, G.W. Crabtree (ed.), *Basic Research Needs for Solar Energy Utilization: Report of the Basic Energy Sciences Workshop on Solar Energy Utilization*, April 18–21, 2005, US Department of Energy Office of Basic Energy Sciences (2005), available at http://www.sc.doe.gov/bes/reports/abstracts.html#SEU.
2. W. Shockley and H.J. Queisser, *J. Appl. Phys.* 32, 510 (1961); Green, M.A., *Third Generation Photovoltaics: Advanced Solar Energy Conversion*, Springer, New York (2003); Hamakawa, Y. (ed.), *Thin-Film Solar Cells: Next Generation Photovoltaics and Its Applications*, Springer, New York (2006); Würfel, P., *Physics of Solar Cells: From Principles to New Concepts*, Wiley, Hoboken, NJ (2005).
3. J. Liu, et al., *J. Am. Chem. Soc.* 126, 6550 (2004).
4. A.J. Nozik, *Physica E* (Amsterdam)14, 115 (2002); R. D. Schaller, V.I. Klimov, *Phys. Rev. Lett.* 96, 097402 (2006); J.E. Murphy et al., *J. Am. Chem. Soc.* 128, 3241 (2006).
5. A.J. Nozik, *Inorg. Chem.* 44, 6893 (2005).
6. B. O'Regan, M. Grätzel, *Nature* 353, 737 (1991); M. Grätzel, *Nature* 414, 338 (2001).
7. M.R. Wasielewski, *J. Org. Chem.* 71, 5051 (2006).
8. D. Gust, T. Moore, A. Moore, *Acc. Chem. Res.* 34, 40 (2001).
9. N.S. Lewis, Inorg. Chem. 44, 6900 (2005); A. Heller, *Science* 223, 1141 (1984); O. Khaselev, J.A. Turner, *Science* 280, 425 (1998).
10. D. Mills, *Sol. Energy* 76, 19 (2004).
11. A. Steinfeld, *Sol. Energy* 78, 603 (2005).
12. C. Dennis, *Nature* 443, 23 (2006).
13. R. Service, *Science* 306, 806 (2004).
14. L.D. Hicks, M.S. Dresselhaus, *Phys. Rev.* B47, 12727 (1993); 47, 16631 (1993); R. Venkatasubramanian et al., *Nature* 413, 597 (2001).
15. E. Quarez et al., *J. Am. Chem. Soc.* 127, 9177 (2005).
16. G. Kodis et al., *J. Am. Chem. Soc.* 128, 1818 (2006).

Organic Semiconductors for Low–Cost Solar Cells

Michael D. McGehee and Chiatzun Goh

Department of Materials Science and Engineering
Stanford University
Stanford, California 94305

Abstract. The current cost of solar electricity derived from silicon photovoltaics is about 30 to 40 cents per kilowatt–hour. This cost is similar to peak–power charges in California during the height of summer, thus establishing a partial path to economic viability. However, this competitiveness is not viable in other seasons and many other locations. This paper will discuss the basic theory and progress of a new class of photovoltaic semiconductors derived from organic polymer materials. These materials have obtained promising results with 5% conversion efficiency. In addition, these materials can be manufactured relatively easily by using printing technologies and roll-to-roll coating machines, similar to those used to make photographic film or newspapers. Solar cells made this way would not only be cheaper, but could also be incorporated into roofing materials to reduce installation costs. Organic semiconductors can be dissolved in common solvents and sprayed or printed onto substrates, so they are very promising candidates for the solar production of electricity.

INTRODUCTION

Currently the world consumes an average of 13 TW for all types of power. By the year 2050, this amount will likely grow to 30 TW as the world's population increases and the standard of living in developing countries improves. If this power is provided by burning fossil fuels, the concentration of carbon dioxide in the atmosphere will more than double and substantial global warming, along with many undesirable consequences, will likely occur. Therefore, one of the greatest and most important challenges that engineers face is finding a way to provide the world with 30 TW of power without releasing carbon into the atmosphere. While it is possible that this may be done by using carbon sequestration along with fossil fuels or by greatly expanding our use of nuclear power plants, it is clearly desirable to develop renewable sources of energy. The sun deposits 120,000 TW of radiation on the surface of Earth, so there is clearly enough power available if an efficient means of harvesting solar energy can be developed.

Today, only a very small fraction of our power is generated by solar cells, which convert solar energy into electricity, because they are too expensive.[i] More than 95% of the solar cells in use are made of crystalline silicon. The efficiency of the most common panels is approximately 12% (but can be as high as 18%), while the cost is $420/m^2$. Put another way, the cost of the panels is $3.50 per watt of electricity produced in peak sunlight. This results from a cell cost of $2.50/W_{peak}$ and a module manufacturing cost of $1.00/W_{peak}$. The cost of the DC to AC converter is about $0.50/W_{peak}$. The cost of installation, panel support and wiring is about $4–5/W_{peak}$. The total price is approximately $8–9/W_{peak}$. Small units will cost more and large units will cost less. Over the lifetime of the panels, which is approximately 30 years, the average cost of the electricity generated in California is $0.34/kW-hr at 6% interest on the loan to buy and install the complete solar system. In California, the average electricity cost is about $0.13/KW-hr, and we can expect this to rise in the future. Peak rates in summer time can be over $0.30/kw-hr for larger consumers or for those opting to sell electricity to the utility with a time–of–day meter. Thus, silicon photovoltaics begin to show competitiveness only in summer peak hours in California. However, we see that it costs approximately five times more to get electricity from silicon solar cells as compared to the average national cost of electricity. Thus, if the cost of producing solar cells could reduced by a factor of five to ten, it would not only be environmentally favorable to use solar cells, but it would also be economically favorable. If there were a reduction of a factor of three in the total cost of solar cells, there would be considerable market growth. In 2007, capital investments in solar photovoltaics totaled $750 million.

Although crystalline silicon solar cells will get cheaper as economies of scale are realized, it is clear that dicing and polishing wafers will always be somewhat expensive and that it is desirable to find a cheaper way to make the cells. The ultimate method of manufacturing would be depositing patterned electrodes and semiconductors on rolls of plastic or metal in roll-to-roll coating machines, similar to those used to make photographic film or newspapers. Solar cells made this way would not only be cheaper, but could also be directly incorporated into roofing materials to reduce installation costs. Organic semiconductors can be dissolved in common solvents and sprayed or printed onto substrates, so they are very promising candidates for this application.

Organic semiconductors operate in a fundamentally different way from conventional inorganic semiconductors due to a difference in bonding systems. Inorganic semiconductors are held together by strong covalent bonds extending three-dimensionally, resulting in electronic bands that give rise to its semiconducting properties. Organic materials have similar intramolecular covalent bonds but are held together only by weak intermolecular van der Waals interactions. The electronic wave function is thus strongly localized on individual molecules. The weak intermolecular interactions instigate a narrow electronic bandwidth formed in molecular solids. The semiconducting nature of organic semiconductors arises from the π electron bonds that exist when molecules are fully conjugated (i.e. have alternating single and double bonds). The weakly held π electrons are responsible for all interesting optical and electronic transitions in organic semiconductors. The π to π^* excited state transitions in organic semiconductors typically fall in the range of $1.4 - 2.5$ eV, which overlaps well with the solar spectrum, rendering them very promising candidates as the active light absorber in solar cells. A few examples of organic semiconductors used in solar cells are shown in Figure 1.

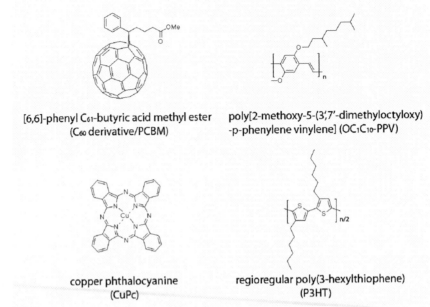

[6,6]-phenyl C₆₁-butyric acid methyl ester
(C₆₀ derivative/PCBM)

poly[2-methoxy-5-(3′,7′-dimethyloctyloxy)
-p-phenylene vinylene] (OC₁C₁₀-PPV)

copper phthalocyanine
(CuPc)

regioregular poly(3-hexylthiophene)
(P3HT)

FIGURE 1. The chemical structures of four different organic semiconductors used in organic solar cells.

ORGANIC VS. INORGANIC SEMICONDUCTORS

The main difference between organic semiconductors and inorganic semiconductors as photovoltaic materials is that optical excitations of organic semiconductors create bound electron-hole pairs called excitons that are not effectively split by the electric field.[ii] In order to separate the bound electrons and holes, there must be a driving force to overcome the exciton binding energy of typically 0.1- 0.4 eV. Excitons in organic semiconductors that are not split eventually recombine either radiatively or nonradiatively, and this recombination reduces the quantum efficiency of a solar cell. In inorganic semiconductors, however, the attraction between an electron-hole pair is less than the thermal energy kT and thus no additional driving force is needed to generate separated carriers. Research has shown that excitons in organic semiconductors can be efficiently split at a heterojunction of two materials with dissimilar electron affinity or ionization potential, as described in later paragraphs.

The narrow electronic bandwidth in organic semiconductors has a few consequences. First, the absorption spectrum bandwidth is accordingly narrower compared to conventional inorganic semiconductors. Consequently a single organic material can be potentially photoactive in only a narrower optical wavelength range of the solar spectrum (Figure 2). Although this is a disadvantage in terms of harvesting solar flux, the use of multiple absorbers in stacks of solar cells connected in series can be engineered to expand the absorption range. Because the valence band and conduction band are concentrated in narrower energy regions, the absorption coefficient resulting from the excitation of electrons from the valence band to the conduction band is very strong. The absorption coefficient of typical organic semiconductors at peak absorption is $> 10^{-5}$ cm^{-1}. The high absorption coefficient of organic semiconductors means only a thin (100 – 200 nm) film is sufficient to absorb most incident light. This is attractive for solar cells because less material is needed to make devices. Second, the charge carriers do not exhibit band-like transport as in inorganic semiconductors, but instead move around by a hopping mechanism between localized states. The charge carrier mobilites in organic semiconductors are therefore inherently low, with typical values of $< 10^{-2}$ cm^2/Vs. The low charge carrier mobility puts a constraint on the thickness of organic materials that can be used in a solar cell since recombinative loss increases with increasing thickness. Fortunately, this drawback is offset by the need to only use a very thin layer of organic semiconductors because it is highly absorptive. Organic solar cells may potentially perform better at higher temperature, because hopping is a thermally activated process. Inorganic solar cells, on the other hand, typically suffer from reduced performance at increasing operating temperature. One other key difference between organic and inorganic semiconductors is that organic materials do not possess dangling bonds at surfaces. Organic-organic junction or organic-metal junctions employed in organic solar cells therefore do not exhibit interface states that potentially act as charge carrier recombination sites.

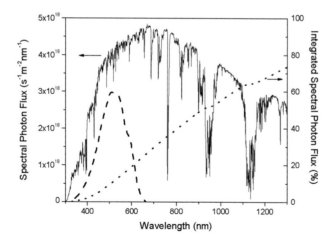

FIGURE 2. Overlap of the absorption spectrum of P3HT (dashed line) with the AM 1.5 solar spectrum (solid line). Integrated spectral photon flux below each photon wavelength is also shown (dotted line).

The simplest organic solar cells can be made by sandwiching thin films of organic semiconductors between two electrodes with different work functions. The work function is the amount of energy needed to pull an electron from a material. When such a diode is made, electrons from the low work function metal flow to the high work function metal until the Fermi levels are equalized throughout the structure. This sets up a built-in electric field in the semiconductor. When the organic semiconductor absorbs light, electrons are created in the conduction band and holes (positive charge carriers) are created in the valence band. In principle, the built-in electric field can pull the photogenerated electrons to the low-work function electrode and holes to the high-work function electrode, thereby generating a current and voltage (Figure 3a). In practice, however, these cells have very low power conversion efficiency (< 0.1 %) because the electric field is insufficient to separate the strongly bound excitons, the excited state species formed in organic semiconductors that has been described earlier.

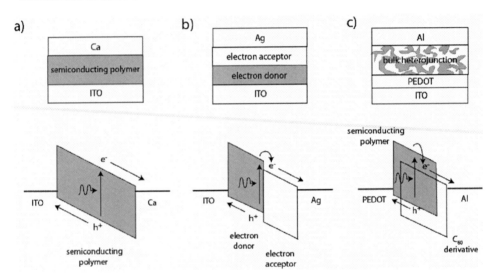

FIGURE 3. A schematic energy band diagram of: a) a single layer semiconductor polymer solar cell, in which indium-tin oxide (ITO) serves as a transparent high work function electrode and Ca serves as the low work function electrode, b) photoinduced electron transfer from the lowest unoccupied molecular orbital (LUMO) of an electron donor to the LUMO of an electron acceptor in a planar heterojunction cell, c) a bulk heterojunction solar cell based on semiconducting polymer and C_{60} derivative.

One significant improvement in the performance of organic solar cells was achieved by Tang.[iii] The device consisted of a heterojunction formed between donor and acceptor semiconductors, resembling a p-n junction in conventional solar cells (Figure 3b). The benefit of the device derived from the employment of two organic materials with offset electron affinity (lowest unoccupied molecular orbital, LUMO) or ionization potential (highest occupied molecular orbital, HOMO). Excitons that diffuse to the interface undergo efficient charge transfer, as this offset in the energy levels provides for a sufficient chemical potential energy to overcome the intrinsic exciton binding energy. Upon charge transfer, the electrons are transported in the acceptor material, and the holes in the donor material, to their respective electrodes. However, the efficiency of this type of planar heterojunction devices is limited by the exciton diffusion length, which is the distance over which excitons travel before undergoing recombination. The exciton diffusion length in most organic semiconductors are found to be approximately 5-10 nm, which means excitons formed at a location larger than this distance away from the heterojunction cannot be harvested. The active area of this type of solar cells is thus limited to only a very thin region close to the interface, which is not enough to adsorb most of the solar radiation flux.

In the mid 1990's it was shown that excitons can be rapidly split by electron transfer before the electron and hole recombine if C_{60} derivatives are blended into the polymer (Figure 3c).[iv] Solar cells were made simply by blending the C_{60} derivative, which acts as an electron acceptor, into the polymer at concentrations in the range of 18-80 wt.%. At these concentrations, the polymer and the C_{60} derivatives form a connected network to each electrode. The key to making efficient blend solar cells is to ensure that the two materials are intermixed very closely at a length scale less than the exciton diffusion length so that every exciton formed in the polymer can reach an interface with C_{60} to undergo charge transfer. At the same time, the film morphology has to enable charge carrier transport in those two different phases to minimize recombination. The concentration of materials, film-casting solvent, annealing time and temperature are a few of the parameters that determine the film morphology (i.e. phase separation between the two materials) and ultimately the efficiency of the devices. Solar cells made by this method have continuously improved to over 2% power efficiency under solar AM 1.5 conditions over the last few years (Shaheen *et al*[v] and Padinger *et al*[vi]), and recently, an efficiency of 5% was reported by Ma *et al*.[vii] The work on polymer/C_{60} derivative blend cells has created a new paradigm in the field of organic-based solar cells, which is the notion of bulk heterojunction devices, wherein two semiconductors with offset energy levels are interpenetrated at a very small length scale to create high interface area for achieving high efficiency devices. Since then, similar bulk heterojunction devices employing other electron acceptors such as CdSe nanorods by Huynh *et al*,[viii] a second semiconducting polymer by Granstrom *et al*[ix], or titania nanocrystals by Arango *et al*[x] instead of the C_{60} derivative have been demonstrated, albeit with slightly lower efficiency.

To understand what limits the performance of bulk heterojunction devices and make plans for improving them, it is important to consider all of the processes that must occur inside the cells for electricity to be generated. These processes, shown in Figure 4, are light absorption (1), exciton transport to the interface between the two semiconductors (2), forward electron transfer (3) and charge transport (4). One must also consider undesirable recombination processes that can limit the performance of the cell, such as geminate recombination of electrons and holes in the polymer (5) and back electron transfer from the electron acceptor to the polymer (6).

Consideration of the need to absorb most of the solar spectrum (process 1) sets two requirements. The first is that the band gap should be small enough to enable the polymer to absorb most of the light in the solar spectrum. Calculations by Coakley and McGehee to determine the band gap that optimizes the amount of the light that can be absorbed and the voltage that can be generated show that the ideal band gap is approximately 1.5 eV, depending on the combination of semiconducting polymers and electron acceptors.[xi] The second requirement is that the film should be thick enough to absorb most of the light. For most organic semiconductors, this means that the films need to be 150-300 nm thick, depending on how much of the films consists of a non-absorbing electron acceptor. As elucidated earlier, the optimum film thickness is one that absorbs much incident light without being so thick that recombination losses become significant.

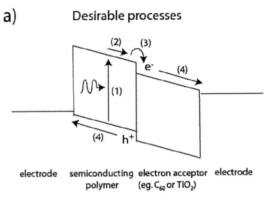

a)   Desirable processes

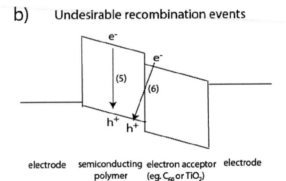

b)   Undesirable recombination events

FIGURE 4. Schematic energy diagrams of the semiconductors and energy levels in a bulk heterojunction solar cell showing the (a) desirable and (b) undesirable processes that can occur. These processes are described in the main text.

Once an exciton is created in the polymer, it must diffuse or travel by resonance energy transfer (process 2) to the interface with the other semiconductor and be split by electron transfer before it recombines (process 5). Experiments have shown that an exciton can diffuse approximately 5-10 nm in most semiconducting polymers before recombination. It is therefore, important to make sure that there are no regions in the polymer that are farther than 5-10 nm from an interface. Templating or nanostructuring of the donor and acceptor phases to fabricate ordered bulk heterojuction with controlled dimensions is an attractive approach, suggested by Coakley and McGehee, to achieving full exciton harvesting (Figure 5).[xii] There are some small molecule semiconductors with larger exciton diffusion lengths.[xiii] Research is under way to improve exciton transport in organic semiconductors, for example by the use of resonance energy transfer to funnel excitons directly to an absorber located at the charge splitting interface or with the incorporation of phosphorescent semiconductors, which exhibit longer excited state lifetimes (Liu *et al*[xiv] and Shao and Yang[xv]).

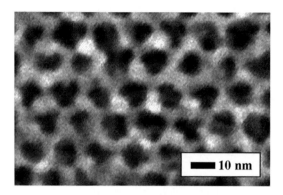

FIGURE 5. SEM image of mesoporous titania film used in fabricating ordered bulk heterojunction solar cell as described in Coakley et. al. (Coakley and McGehee, 2003).

The actual process of charge transfer (process 3) requires that the offset in LUMO levels of the donor and acceptor to be sufficient to overcome the exciton binding energy. However, this drop in energy should not be in excess either, because the maximum voltage attainable from this type of bulk heterojunction solar cell is determined by the gap between the HOMO of the electron donor and the LUMO of the acceptor. This gap becomes smaller as the LUMO of electron acceptors is moved farther away from the LUMO of polymer, which corresponds to a larger driving force for charge transfer. As can be seen from the processes of 1, 2, and 3, the design of an efficient organic solar cells involves optimizing the various energy levels to achieve the optimum level of extracted current with respectable voltage, as the power supplied by a solar cell is the product of current and voltage. Fortunately, the wealth of chemical synthetic knowledge and the dependence of electronic properties of organic molecules on its molecular structures allow for flexible tuning of the bandgap and energy levels of organic semiconductors by chemical synthesis. Significant research is currently targeted towards band engineering of this type and should yield very promising results in the near future.

After forward electron transfer, the holes in the polymer and the electrons in the electron acceptor must reach the electrodes (process 4) before the electrons in the acceptor undergo back electron transfer to the polymer (process 6). In the best bulk heterojunction cells, this competition limits the efficiency of the cells. The problem is usually avoided to some extent by making cells that are only 100-nm thick so that the carriers do not have to travel very far. Unfortunately, most of the light is not absorbed by films this thin. If the films are made thick enough to absorb most of the light, then only a small fraction of the carriers escape the device. Many researchers are now striving to optimize the interface between the two semiconductors and improve charge transport in the films so that the charge can be extracted from 300-nm-thick films before recombination occurs.

The outlook for organic solar cells is very bright. Efficiency greater than 5% has been achieved (Xue *et al*[xvi] and Ma *et al*[xvii]) and many are optimistic that 20% can be achieved by optimizing the processes described above. Once this goal is achieved, a primary research challenge will be making cells that are stable in sunlight and that can handle the wide temperature swings that solar cells must endure. The facts that many organic pigments in car paints are able to survive in sunlight and that organic light-emitting diodes with operational lifetimes greater than 50,000 hours are being made provide good hope that the required stability can be achieved. The final challenge will be scaling up the process and manufacturing the cells at a cost of approximately $30/m^2. Several approaches to making organic solar cells are reviewed in more detail in the January 2005 issue of the Materials Research Society Bulletin.

REFERENCES

[i] N.S. Lewis, and G. Crabtree, *Basic research needs for solar energy utilization* (2005). Available online at: *<http://www.sc.doe.gov/bes/reports/files/SEU_rpt.pdf>*.

[ii] B. Gregg, "Excitonic solar cells," *Journal of Physical Chemistry B* 107(20), 4688-4698 (2003).

[iii] C.W. Tang, "2-layer organic photovoltaic cell," *Applied Physics Letters* 48(2), 183-185 (1986).

[iv] G. Yu, J. Gao, J.C. Hummelen, F. Wudl and A.J. Heeger, "Polymer photovoltaic cells: Enhanced efficiencies via a network of internal donor-acceptor heterojunctions," *Science* 270(5243), 1789-1791 (1995).

[v] S.E. Shaheen, C.J. Brabec, N.S. Sariciftci, F. Padinger, T. Fromherz and J.C. Hummelen, "2.5% efficient organic plastic solar cells," *Applied Physics Letters* 78(6): 841-843 (2001).

[vi] F. Padinger, R.S. Rittberger and N.S. Sariciftci, "Effects of postproduction treatment on plastic solar cells," *Advanced Functional Materials* 13(1), 85-88 (2003).

[vii] W. Ma, W., C. Yang, X. Gong, K. Lee and A.J. Heeger, "Thermally stable, efficient polymer solar cells with nanoscale control of the interpenetrating network morphology," *Advanced Functional Materials* 15(10), 1617-1622 (2005).

[viii] W.U. Huynh, J.J. Dittmer and A.P. Alivisatos, "Hybrid nanorod-polymer solar cells," *Science* 295(5564), 2425-2427 (2002).

[ix] M. Granstrom, K. Petritsch, A. Arias, A. Lux, M. Andersson and R. Friend, "Laminated fabrication of polymeric photovoltaic diodes," *Nature* 395(6699), 257-260 (1998).

[x] A.C. Arango, S.A. Carter and P.J. Brock, "Charge transfer in photovoltaics consisting of polymer and TiO$_2$ nanoparticles," *Applied Physics Letters* 74(12), 1698-1700 (1999).

[xi] K.M. Coakley and M.D. McGehee, "Conjugated polymer photovoltaic cells," *Chemistry of Materials* 16(23), 4533-4542 (2004).

[xii] K.M. Coakley, and M.D. McGehee, "Photovoltaic cells made from conjugated polymers infiltrated into mesoporous titania," *Applied Physics Letters* 83(16), 3380-3382 (2003).

[xiii] P. Peumans, A. Yakimov and S.R. Forrest, "Small molecular weight organic thin-film photodetectors and solar cells.," *Journal of Applied Physics* 93(7), 3693-3723 (2003).

[xiv] Y. Liu, M.A. Summers, C. Edder, J.M.J. Fréchet and M.D. McGehee, "Using resonance energy transfer to improve exciton harvesting in organic-inorganic hybrid photovoltaic cells," *Advanced Materials* (2005, in press).

[xv] Y. Shao and Y. Yang, "Efficient organic heterojunction photovoltaic cells based on triplet materials," *Advanced Materials* (2005, in press).

[xvi] J.G. Xue, S. Uchida, B.P. Rand and S.R. Forrest, "Asymmetric tandem organic photovoltaic cells with hybrid planar-mixed molecular heterojunctions," *Applied Physics Letters* 85(23), 5757-5759 (2004).

[xvii] W. Ma, C. Yang, X. Gong, K. Lee and A.J. Heeger, "Thermally stable, efficient polymer solar cells with nanoscale control of the interpenetrating network morphology," *Advanced Functional Materials* 15(10): 1617-1622 (2005).

Concentrating Solar Power

Mark Mehos

National Renewable Energy Laboratory
Golden, CO 80401

Abstract. Concentrating Solar Power (CSP) has the potential to contribute significantly to the generation of electricity by renewable energy resources in the U.S.. Thermal storage can extend the duty cycle of CSP beyond daytime hours to early evening where the value of electricity is often the highest. The potential solar resource for the southwest U.S. is identified, along with the need to add power lines to bring the power to consumers. CSP plants in the U.S. and abroad are described. The CSP cost of electricity at the busbar is discussed. With current incentives, CSP is approaching competiveness with conventional gas-fired systems during peak-demand hours when the price of electricity is the highest. It is projected that a mature CSP industry of over 4 GWe will be able to reduce the energy cost by about 50%, and that U.S capacity could be 120 GW by 2050.

TECHNOLOGY OVERVIEW

Mirrors can concentrate solar flux to obtain temperatures sufficient to melt iron. This approach can be expanded to produce electricity from steam cycles. Solar heat can raise the temperature of water to create steam directly, or indirectly using a heat transfer fluid which vaporizes water through a steam generator. Steam drives a turbine and generates electricity in the conventional manner used for over a century. A concentrating solar power (CSP) system of mirrors replaces the fuel source used in conventional power plants. This increases the capital cost of the system, but minimizes or eliminates the fuel costs associated with running the plant. A tension thus exists between the higher capital costs for the solar field and the savings associated with eliminating fuel costs.

CSP technologies are classed into two groups: (1) CSP systems *with thermal storage* dispatch electrical power to high value periods (*dispatchable CSP*). These systems generally include parabolic troughs, and molten salt central receivers, Storage systems are currently designed to work for 6-7 hours although longer periods are conceivable as the cost of storage is reduced over time. (2) CSP systems *without thermal storage* lack the ability to dispatch on demand (*nondispatchable*). These systems include steam-based trough and tower systems, parabolic dish/engine engine systems and concentrating photovoltaic systems.

Dispatchable 250-MWe CSP plants (or multiple plants in power parks) are being considered built for intermediate and baseload power markets. They have moderate

CP1044, *Physics of Sustainable Energy, Using Energy Efficiently and Producing It Renewably*
edited by D. Hafemeister, B. Levi, M. Levine, and P. Schwartz
© 2008 American Institute of Physics 978-0-7354-0572-1/08/$23.00

solar-to-electric efficiencies on an annualized basis of about 14-16%. Thermal storage offers the ability to operate in the near-term at capacity factors approaching 40% and in the long-term at capacity factors up to 70%.

In the U.S. Southwest, the demand curve for electrical power peaks during the daytime and continues into the early evening hours (Fig. 1). The dashed line corresponds to the hourly load for electricity consumption. The solar curve peaks at noon is the solar resource. The box in the foreground of Fig. 1 denotes the time width of the electricity developed from a parabolic trough systems with molten-salt storage. Storage units provide higher value because power production can better match the needs of the utility. Storage lowers costs because storage is cheaper than the incremental costs for additional turbines needed when operating CSP in the nondispatchable mode.

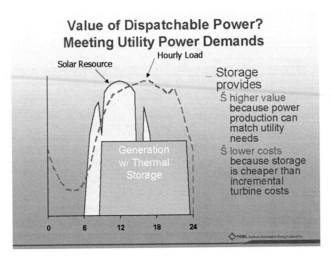

FIGURE 1. The daily power curve for the solar resource vs. the hourly load. Thermal storage extends the solar resource into the off-peak hours.

Parabolic troughs give directional reflections from mirrored surface that are deployed along the east–west direction. The absorber tube is placed near the approximate focal axis of the parabolic trough. The focal axis shifts as the sun moves in an arc from east to west. By rotating the parabolic trough during the day, the solar flux remains focused on the tube. The parabolic mirrors are rotated in the north–south direction to maintain focus on the tube. The working fluid in the tube is typically a synthetic oil, which can heat molten salt for storage. The Solar Energy Generating System (SEGS) plant at Kramer Junction, California, consists of five 30-MWe hybrid plants, for a total of 150 MWe, which was commissioned in 1986–1988. Four additional SEGS plants are at Daggett and Harper Lake, California, for a total peak power of 354 MW.

Figure 2 describes the thermal cycle for a parabolic trough power plant with thermal storage. It is similar to a coal-fired plant in that steam is delivered to a turbine that operates in a Rankine cycle. Coal plants can attain 40% efficiency with super-heated steam. A parabolic trough plant produces steam at 390°C has and therefore has a slightly lower thermal efficiency. When considering optical and thermal losses associated with the solar field, the annual solar-to-electric efficiency of the systems is about 14%. Higher temperatures yield higher efficiencies and reduce storage volume, both resulting in a drop in the cost of delivered energy.

Parabolic Trough Power Plant with Thermal Storage

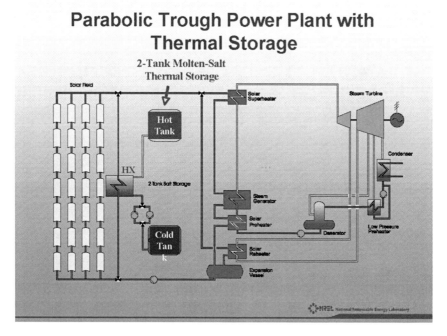

FIGURE 2. Parabolic trough plant with thermal storage. Current systems use a synthetic oil as the working fluid (HTF). Molten salt is heated by the HTF during the peak solar hours. During off-peak hours, molten-salt storage is used to make steam for the turbines.

Solar Resource Potential in the Southwest U.S.

The Southwest region is ideally suited for CSP due to the high solar resource availability. The states of Arizona, California, Colorado, Nevada, New Mexico, Texas and Utah are especially attractive A screening analysis is used to identify regions in the Southwest that are most favorable economically to construct large-scale CSP systems. This analysis is used in conjunction with transmission and market analysis. Figure 3 displays the unfiltered data for the Southwest region. Figure 4 displays the filtered data at a slope of less than 3%. The criteria to determine which locations would be acceptable for suitable development are as follows:

1. Begin with direct-normal solar resource estimates derived from 10–km-resolution satellite data that take into account cloudiness.
2. Eliminate locations with less than 6.0 kWh/m^2-day.

3. Exclude environmentally sensitive lands, major urban areas, and water features.
4. Remove land areas with greater than 1% average slope for parabolic trough systems (or 3% for central-receiver and dish systems).
5. Eliminate areas with contiguous areas of less than 1 square kilometer.

Southwest Solar Resources - Unfiltered Data

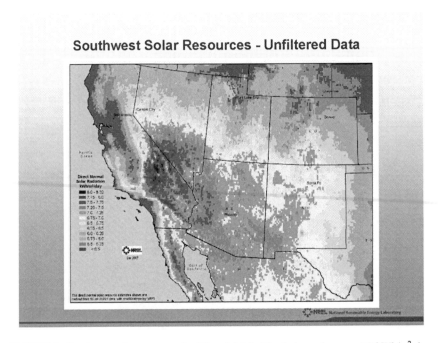

FIGURE 3. Southwest solar resources (unfiltered data). The darker colors are at 8 kWh/m²-day, whereas the lighter colors are at 6 kWh/m²-day.

Figure 4 displays the solar resource on lands with slopes less than 1% for parabolic trough CSP. The table displays the appropriate land area available, solar power capacity in MW, and solar energy capacity in GWh. The Southwest region is capable of generating some 26 million GWh/year at the most-favorable CSP sites. For comparison, a 1–GWe conventional power plant with a load factor of 0.8 generates 7,000 GWh/year. The prime solar area of the Southwest has the potential to generate the power of 4,000 1–GWe baseload plants. The U.S. has about one-fourth this capacity at 1,000–GWe capacity, generating 4 million GWh/year. CSP power is intended for customers in the Southwest; but in the future, one could envision direct-current (DC) power lines carrying power at greater distances. Figure 5 indicates locations where the assigned cost has been determined.

Resulting CSP Resource Potential

State	Land Area (mi²)	Solar Capacity (MW)	Solar Generation Capacity GWh
AZ	13,613	1,742,461	4,121,268
CA	6,278	803,647	1,900,786
CO	6,232	797,758	1,886,858
NV	11,090	1,419,480	3,357,355
NM	20,356	2,605,585	6,162,729
TX	6,374	815,880	1,929,719
UT	23,288	2,980,823	7,050,242
Total	87,232	11,165,633	26,408,956

The table and map represent land that has no primary use today, exclude land with slope > 1%, and do not count sensitive lands.
Solar Energy Resource 6.0
Capacity assumes 5 acres/MW
Generation assumes 27% annual capacity factor

Current total nameplate capacity in the U.S. is 1,000GW w/ resulting annual generation of 4,000,000 GWh

FIGURE 4. Southwest lands that satisfy the five selection criteria, with slopes of <1%. The calculations assume a 27% annual capacity factor (no storage), 5 acres/MW, and solar resource of >6 kWh/m²-day.

Optimal CSP Sites
from CSP Capacity Supply Curves

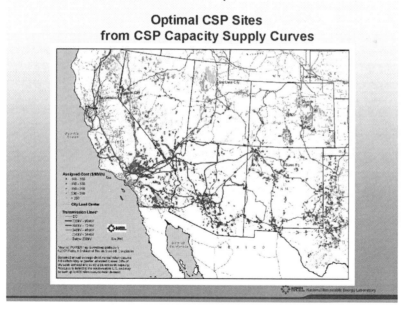

FIGURE 5. Optimal CSP locations. It is projected that the best sites could produce 300 GWe with present technology at $140–$160/MWh (14–16 cents/kWh).

U.S. and International Project Development

A federal investment tax credit (ITC) of 30% has spurred the development of CSP plants in the U.S., listed in Table 1 below. In addition, the following state governments are requiring utilities in their states to have a renewable portfolio standard (RPS) in terms of a minimum percentage for the state: Arizona (15% by 2025), California (20% by 2010), Colorado (20% by 2020), Nevada (20% by 2015, 5% solar), New Mexico (20% by 2015), and Texas (about 4.2% by 2015). The ten U.S. CSP projects in Table 1 (early 2008) have a projected total power of 2.5–3.0 GWe. The 12 international CSP projects in Table 2 total 1.3 GWe. CSP projects are very active in Spain, which has similar solar characteristics as the Southwest U.S.

TABLE 1. U.S. CSP projects (early 2008) total 2.5 to 3.0 GW.
To the first order, it takes about 3 kW of CSP to sustain a household.

CSP Projects – early 2008

U.S. projects: enabled by 30% investment tax credit and State renewable portfolio standards

State	RPS Requirement
Arizona	15% by 2025
California	20% by 2010
Colorado	20% by 2020
Nevada	20% by 2015, 5% Solar
New Mexico	20% by 2015
Texas	5,880MW (~4.2%) by 2015

Utility/State	Capacity (MW)	Technology -Status
Arizona Public Service (APS)	1	Trough – completed and in operation 2006 (Acciona)
Nevada Power	64	Trough – completed and in operation June 2007 (Acciona)
Southern Cal Edison and San Diego Gas and Electric	500/300	Dish – signed power purchase agreement (SES)
Pacific Gas & Electric	550	Trough – signed power purchase agreement for four plants (Solel)
Pacific Gas & Electric	170	CLFR – signed power purchase agreement (Ausra)
Pacific Gas & Electric	500	Tower – MOU signed (Bright Source)
Florida Power and Light	300	CLFR or Trough
Arizona Public Service	280	Trough – signed power purchase agreement (Abengoa)
SW Utility joint venture (APS)	Est. 250	TBD – multiple expressions of interest submitted
New Mexico Utility Joint Venture	50-500	TBD – initial stages

TABLE 2. International CSP projects total 1.3 GW.

CSP Projects – International

Country/Company	Capacity (MW)	Technology -Status
Spain: Solar Millenium	4 x 50MW with storage	Trough – Andosol 1 &2 under construction.
Spain: Abengoa/Solucar	5 x 50MW	Trough – 1st plant under construction
Spain: Abengoa/Solucar	11MW &20MW	Power Tower (saturated steam) – PS10 operational. PS20 under construction
Spain: SENER	17MW	Power Tower (molten salt) – contract terms under discussion
Spain: various	TBD	Projects under various stages of development due to tariff for 500MWs of CSP capacity. Cap likely to be raised to 1000MWs.
Algeria: Abener	150MW	Integrated Solar Combined Cycle System (ISCCS) – 25MW Solar Capacity
Egypt: TBD	140MW	ISCCS – 25MW Solar Capacity, negotiations in progress
Mexico: TBD	TBD	ISCCS – RFP issued
Morocco: TBD	230MW	ISCCS – 35 MW Solar Capacity
Israel: Solel	2 x 125MW	Trough – Northern Negev. Waiting approval from Interior Ministry
Australia: SHP	15MW,th	Linear Fresnel – under construction for integration into feed water heaters in existing coal plant
Greece: TBD	TBD	Tariff for CSP recently enacted. Similar in design to Spanish feed-in tariff

Cost Targets and Market Penetration

Cost targets were obtained using the California Energy Commission Market Price Referent (MPR) as a proxy. The methodology is based on capacity and energy costs associated with "conventional" baseload combined-cycle generation plant and utility time-of-delivery (TOD) values. We focused on California MPR because its RPS currently calls for 20% of the state's generation to come from renewables by 2010. The 2007 baseload MPR for plants built in 2011 is \$0.10/kWh. The allowable price for CSP was based on TOD factors. We assumed dispatchable parabolic trough systems with thermal storage and used TOD values for the three California utilities (San Diego Gas and Electric, Pacific Gas and Electric, and Southern California Edison). The result is a cost of \$0.12–\$0.14 per kWh for the initial penetration into California's intermediate-load markets.

Figure 6 displays a projected drop in electric cost from trough CSP systems. Assuming a 10% investment tax credit, the cost for the initial plant in the Mojave desert is expected to be \$0.16/kWh (nominal LCOE at busbar). Cost reductions are expected to bridge the gap depending on deployment, plant size, financing, and R&D. This analysis does not include the 30% ITC which would reduce the cost by and additional 15-20%. The 2015 DOE goal is \$0.10/kWh (nominal) although this goal will be revised given the current escalation in energy costs from conventional plants. The long-term conclusion is that the cost of CSP electricity will be reduced by 50% when the first 4 GWe have been built. The ITC is necessary in the near term to initiate deployment by creating conditions to lower per-unit costs.

Bridging the Cost Gap

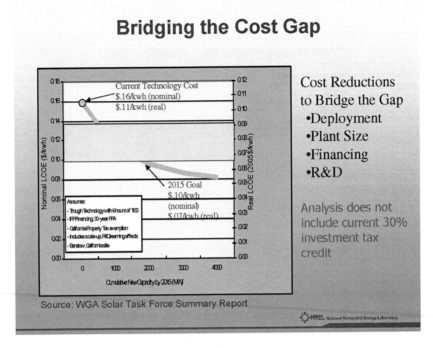

FIGURE 6. Projected drop in cost of electric from a trough CSP system located in California by 2015 after 2–4 GW have been installed.

A Southwest market analysis was performed using the NREL developed Regional Electricty Deployment System (REEDS) model. REEDS is a multi-regional, multi-time-period model of capacity expansion in the electric sector of the U.S. that is focused on competing renewables (currently wind and CSP) with current and future fossil fuel (with and w/o carbon sequestration) and nuclear plants. It was designed to estimate market potential of wind and solar energy in the U.S. for the next 20–50 years under different technology development and policy scenarios. The estimates for cumulative CSP capacity were considered for the case of no extension of the ITC and for the case of an 8–year extension of a declining tax credit. The extended tax credit encourages a faster rise in CSP power plants to 80 GW in 2040, compared to the non–extension case with 50 GW by 2040. Both approaches result in 115 GW in 2050—with 15% on the existing grid, 85% on new transmission lines—with 10% used in the regions where the electricity is generated.

SUMMARY

- CSP technologies, especially those that incorporate near-term thermal storage, offer a combination of low cost and high value to utility-scale markets.

- The solar resource in the Southwest is immense, resulting in a generation potential of CSP greater than six times the current U.S. demand.

- Capacity supply curves based on the screening analysis demonstrate that suitable lands are located close to existing transmission, minimizing the cost required to access high-value solar resources.

- Near-term U.S. market penetration is a challenge, but largely based on continuation of current investment tax credit and Southwest state policies attractive to large-scale solar.

- Preliminary market-penetration analysis indicates up to 30 GW of CSP capacity could be achieved by 2030 in the U.S., with 120 GW by 2050.

The Status and Future of
Wind Energy Technology

Robert Thresher[a], Michael Robinson[a] and Paul Veers[b]

[a]National Wind Technology Center[1]
National Renewable Energy Laboratory
Golden, CO 80401–3393
[b]Wind Energy Technology Department[2]
Sandia National Laboratories
Albuquerque, NM 87185

INTRODUCTION TO WIND ENERGY AND MODERN WIND TECHNOLOGIES IN THE UNITED STATES

From the birth of modern electricity-generating wind turbines in the late 1970s to now, wind energy technology has dramatically improved. Capital costs have plummeted, reliability has improved, and efficiency has increased. High-quality turbine manufacturers exist around the world, and wind plants of 300 megawatts (MW) and larger are being integrated into the electrical grid to exacting utility specifications. These modern wind plants are now routinely produced by multi-national manufacturing companies at a cost of energy approaching – and in some cases below – that of fossil fuel generating plants.

At the end of 2006, the total U.S. wind energy capacity had grown to 11,603 MW, or enough to provide the electrical energy needs of more than 2.9 million American homes. Wind capacity in the United States and in Europe has grown at a rate of 20% to 30% per year over the past decade (Figure 1). Despite this rapid growth, wind currently provides less than 1% of total electricity consumption in the United States.

---

[1] This work has been authored by an employee or employees of the Midwest Research Institute under Contract No. DE-AC36-99GO10337 with the U.S. Department of Energy. The United States Government retains and the publisher, by accepting the article for publication, acknowledges that the United States Government retains a non-exclusive, paid-up, irrevocable, worldwide license to publish or reproduce the published form of this work, or allow others to do so, for United States Government purposes.

[2] Sandia National Laboratories is a multiprogram laboratory operated by Sandia Corporation, a Lockheed Martin Company, for the United States Department of Energy under Contract DE-AC04-94AL85000.

CP1044, Physics of Sustainable Energy, Using Energy Efficiently and Producing It Renewably
edited by D. Hafemeister, B. Levi, M. Levine, and P. Schwartz
© 2008 American Institute of Physics 978-0-7354-0572-1/08/$23.00

The vision of the wind industry in the United States and in Europe is to increase wind's fraction of the electrical energy mix to more than 20% within the next two decades.

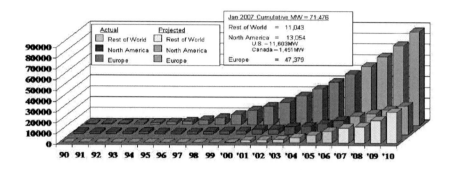

FIGURE 1. Worldwide growth of wind energy installed capacity.

Wind energy is one of the fastest-growing electrical energy sources in the United States, second only to natural gas in the past two years. The United States installed 2,400 MW in 2006, and experts are forecasting an additional 3,000 MW in 2007, with the majority of the new additions expected in the West. The state distribution of wind capacity is illustrated in Figure 2.

The United States is blessed with an abundance of wind energy potential. The land-based and offshore wind resource has been estimated to be sufficient to supply the electrical energy needs of the entire country several times over. The Midwest region, from Texas to North Dakota, is particularly rich in wind energy resources (as illustrated in Figure 3a). It is important to note that the wind energy potential is a function of height above the ground, and many states that show limited or negligible wind resource in the Figure 3a wind resource map drawn for 50 m above ground may have much greater wind potential at the 100-m elevation, which is the operating hub height for many modern turbines. There is an ongoing effort by the Federal Wind Energy Program, in partnership with the states, to model the wind flow using mesoscale models (continental numerical weather prediction models) at higher elevations above ground to generate improved resource estimates at the 100-m level. Indiana, for example, does not appear to have much resource at 50 m, but the 100-m resource map (Figure 3b) depicts a strong, widespread resource. It is almost certain that these improved resource estimates will reveal more U.S. land areas that could potentially be used for wind deployment.

FIGURE 2. Installed wind capacity in the United States at December 31, 2007.

Today's Commercial Wind Technology

Modern wind turbines deployed throughout the world today have three-bladed rotors with diameters of 70 to 80 meters mounted atop 60- to 80-m towers (Figure 4). The typical turbine installed in the United States in 2006 can produce about 1.5 MW of electrical power. The turbine power output is controlled by rotating the blades about their long axis to change the angle of attack with respect to the relative wind as the blades spin about the rotor hub, which is referred to as "controlling the blade pitch." The turbine is pointed into the wind by rotating the nacelle about the tower, which is called "yaw control." Almost, all modern turbines operate with the rotor positioned on the windward side of the tower, which is referred to as an "upwind rotor." Wind sensors on the nacelle tell the yaw controller where to point the turbine, and when combined with sensors on the generator and drive train, tell the blade pitch controller to regulate the power output and rotor speed and to prevent overloading structural components. A turbine will generally start producing power in winds of about 12 mph and reach maximum power output at about 28 to 30 mph. The turbine will "feather the blades" (pitch them to stop power production and rotation) at about 50 mph.

342

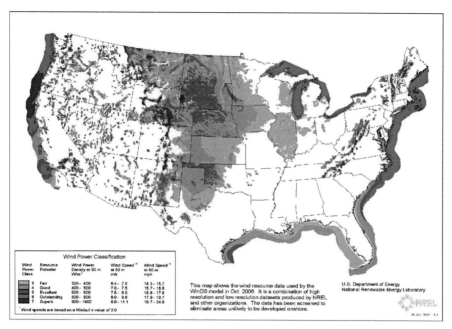

FIGURE 3a. The wind resource potential at 50m on land and offshore.

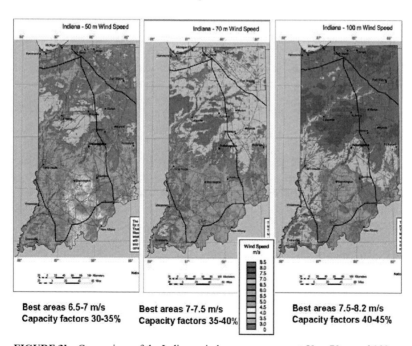

Best areas 6.5-7 m/s
Capacity factors 30-35%

Best areas 7-7.5 m/s
Capacity factors 35-40%

Best areas 7.5-8.2 m/s
Capacity factors 40-45%

FIGURE 3b. Comparison of the Indiana wind energy resource at 50m, 70m, and 100m.

Rotor Blades 37m:
- Shown Feathered
- 37-m length

Rotor Hub

Nacelle Enclosing:
- Low-speed Shaft
- Gearbox
- Generator 1.5 MW
- Electrical Controls

Tower 80m

Minivan

FIGURE 4. A modern 1.5-MW wind turbine Installed in a wind farm
(photo by Mark Rumsey, Sandia National Laboratories).

The amount of energy in the wind available for extraction by the turbine increases with the cube of wind speed; thus a 10% increase in wind speed means a 33% increase in available energy. However, a turbine can only capture a portion of this cubic increase in energy because power above the level for which the electrical system has been designed (referred to as the "rated power") is allowed to pass through the rotor (as will be described later).

The height and the size of wind turbines have increased to capture the more energetic winds at higher elevations. For land-based turbines, size is not expected to grow as dramatically in the future as it has in the past. Many turbine designers don't expect land-based turbines to become much larger than about 100 meters in diameter, with corresponding power outputs of about 3 to 5 MW. Larger sizes are physically possible; however, the logistical constraints of transporting the components over the highway and obtaining cranes large enough to lift the components are potential barriers.

Turbine Performance and Price

The performance of commercial turbines has increased with time and, as a result, capacity factors have slowly increased. Figure 5 presents capacity factor versus the commercial operation date. The data show that turbines in the Lawrence Berkeley National Laboratory database that began commercial operation prior to 1998 have an average capacity factor of about 22%, whereas the turbines that began commercial operation after that show an increasing capacity factor trend, reaching 36% in 2004-05. This increasing capacity factor is expected to continue over time (for reasons that will be discussed later).

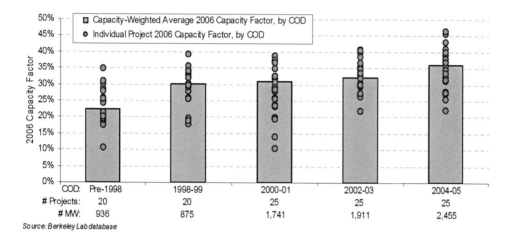

FIGURE 5. Turbine Capacity Factor by Commercial Operation Date Using 2006 Data (reference 1).

The cost of wind-generated electricity has dropped dramatically since 1980, when the first commercial wind farms began operation in California. Figure 6 depicts price data for some more recent wind energy projects from public records. This chart shows that in 2006, the price paid for electricity generated in large wind farms was between 3 and 6.5 cents per kilowatt-hour with an average near 5 cents per kilowatt-hour (1cent/kWh = 10$/MWh). These figures represent the electricity price as sold by a wind farm owner to the utility. The price includes the benefit of the federal production tax credit, and any state incentives, as well as revenue from the sale of any renewable energy credits. Thus the true cost of the delivered electricity would be higher by approximately 1.9 cents per kilowatt-hour, which is the value of the federal tax credit. Accounting for the tax credit then indicates that the unsubsidized cost for wind-generated electricity for projects completed in 2006 ranges from about 5 to 8½ cents per kilowatt-hour.

The reasons generally offered for the increasing price of wind-generated electricity after the long downward price trend of the past 25 years include:

- Turbine and component shortages due to the dramatic recent growth of the wind industry in the United States and Europe.
- The weakening U.S. dollar relative to the Euro (because many major turbine components are imported from Europe). There are relatively few wind turbine component manufacturers in the United States.

- A significant rise in material costs such as steel and copper, as well as transportation fuels, over the past three years.
- The on-again and off-again cycle of the wind energy production tax credit. This uncertainty hinders investment in new turbine production facilities and encourages hurried and expensive production, transportation, and installation of projects when the tax credit is available.

To decrease wind energy costs below the 2003 level will require further research and development effort and will be considered next.

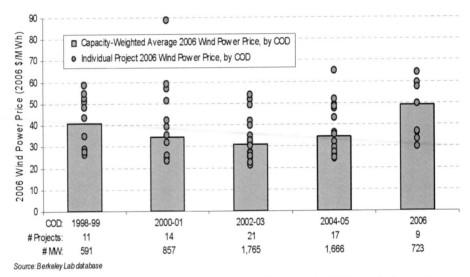

FIGURE 6. Wind energy price by commercial operation date using 2006 data (reference 1).

The History of Wind Technology Development

Until the early 1970s, wind energy filled a small niche market, providing mechanical power for grinding grain and pumping water. With the exception of a small number of battery chargers and the rare experiments with larger electricity-producing machines, the windmill of 1850, or even 1950, differed little from the primitive devices from which they were derived. But the latter half of the 20th century saw spectacular changes in the technology. Blades that had once been made of sail or sheet metal progressed through wood to advanced fiberglass composites. The DC alternator gave way to the induction generator that was grid synchronized. From mechanical cams and linkages that feathered or furled a machine, designs moved to high-speed digital controls. Airfoils are tested in wind tunnels and are designed for insensitivity to surface roughness and dirt. Our knowledge of aeroelastic loads and the ability to incorporate this knowledge into finite element models and structural dynamics codes make the machine of today more robust but much less expensive than those of a decade ago.

346

Turbine Size

Over the past 20 years, average wind turbine ratings have grown almost linearly (Figure 7), with current commercial machines rated at 1.5 MW. Each group of wind turbine designers has predicted that their machines are as large as they will ever be. However, with each new generation of wind turbines, the size has increased along the linear curve and has achieved reductions in life-cycle cost of energy.

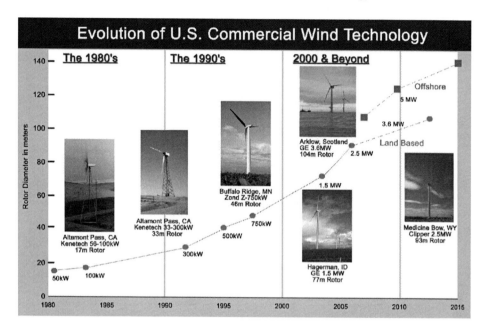

FIGURE 7. The Development Path and Size Growth of Wind Turbines.

The long-term drive to develop larger turbines stems from a desire to take advantage of wind shear by placing rotors in the higher, much more energetic winds at a greater elevation above ground (wind speed increases with height above the ground). This is a major reason for the increase in capacity factor over time (Figure 5). However, there are constraints to this continued growth to larger sizes; in general, it costs more to build a larger turbine.

The primary argument for a size limit for wind turbines is based on the "square-cube law." Roughly stated, it says that "as a wind turbine rotor increases in size, its energy output increases as the rotor-swept area (the diameter squared), while the volume of material, and therefore its mass and cost, increases as the cube of the diameter." In other words, at some size the cost for a larger turbine will grow faster than the resulting energy output revenue, making scaling a losing economic game. Engineers have successfully skirted this law by changing the design rules with increasing size and removing material or by using material more efficiently to trim

347

weight and cost. Clearly, turbine performance has improved and cost per unit of output has been reduced (Figures 5-7).

Studies have shown that in recent years, blade mass has been scaling at roughly an exponent of 2.3 versus the expected 3 [reference 2]. This WindPACT study shows how successive generations of blade design have moved off the cubic weight growth curve to keep weight down (Figure 8). If advanced research and development were to provide even better design methods, as well as new materials and manufacturing methods that allowed the entire turbine to scale as the diameter squared, then it would be possible to continue to innovate around this limit to size.

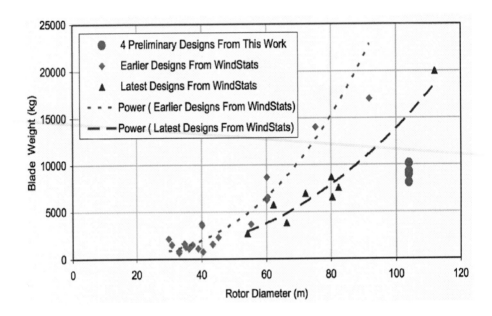

FIGURE 8. WindPACT Study Results indicating the lowering of growth in blade weight due to the introduction of new technology (reference 2).

Land transportation constraints can also pose limiting factors to wind turbine growth for turbines installed on land. Cost-effective transportation can only be achieved by remaining within standard over-the-road trailer dimensions of 4.1 m high by 2.6 m wide. Rail transportation is even more dimensionally limited, eliminating that option.

Unfortunately, other constraints limit the practical size of wind turbines. Crane requirements are quite stringent because of the large nacelle mass in combination with the height of the lift and the required boom extension. As the height of the lift to install the rotor and nacelle on the tower increases, the number of available cranes with the capability to make this lift is fairly limited. Other limiting factors are that cranes with large lifting capacities are difficult to transport, require large crews, and therefore have high operation, mobilization, and de-mobilization costs.

The Rotor

As wind turbines grow in size, so do their blades – from about 8m in 1980 to more than 40m for many land-based commercial systems. Improved blade designs have enabled the weight growth to be kept to a much lower rate than simple geometric scaling, as already described. Today's blade designs are subjected to rigorous evaluation using the latest computer analysis tools so that excess weight can be removed. Designers are also starting to work with lighter and stronger carbon fiber in highly stressed locations to stiffen the blade and improve fatigue resistance while reducing blade weight. However, carbon fiber must be used judiciously because its cost is about 10 times the cost of fiberglass.

Figure 9 shows the power curve for a typical modern turbine and illustrates the different control regions for the turbine. Typically, a turbine will cut-in and begin to produce power at a wind speed of about 12 mph. It will reach its rated power at about 28 to 30 mph, where the pitch control system begins to limit power output and prevent overloading the generator and drive train. At around 50 mph, the control system pitches the blades to stop rotation (which is referred to as feathering the blades) to prevent overloads and damage to the turbine's components.

All of the energy present in a stream of moving air cannot be extracted; some air must remain in motion after extraction or no new, more energetic air can enter the device. Building a brick wall would stop the air at the wall, but the free stream of energetic air would just flow around the wall. On the other end of the spectrum, a device that does not slow the air is not extracting any energy either. The solution for the optimal blockage is generally attributed to a man named Betz and is called the Betz limit. At best, a device can extract a theoretically maximum 59% of the energy in a stream with the same area as the working area of the device.

The aerodynamic performance of a modern wind turbine has improved dramatically over the past 20 years. The rotor system can be expected to capture about 80% of the theoretically possible energy in the flow stream. This has been made possible through the design of custom airfoils for wind turbines. In fact, it is now commonplace for turbine manufacturers to have special airfoil designs for each individual turbine design. These special airfoils attempt to optimize low-speed wind aerodynamic efficiency and limit aerodynamic loads in high winds. These new airfoil designs also attempt to minimize sensitivity to blade fouling, due to dirt and bugs that accumulate on the leading edge and can greatly reduce efficiency. Although rotor design methods have improved significantly, there is still much room for improvement.

Controls

Today's controllers integrate the signals from dozens of sensors to control rotor speed, blade pitch angle, generator torque, and power conversion voltage and phase. The controller is also responsible for critical safety decisions, such as shutting down the turbine when extreme conditions are realized. Today most turbines operate variable-speed, and the control system regulates the rotor speed to obtain peak efficiency in fluctuating winds by continuously updating the rotor speed and generator loading to maximize power and reduce drive train transient torque loads. Operating variable speed requires the use of power converters, which also enables turbines to deliver fault ride through protection, voltage control, and dynamic reactive power support to the grid.

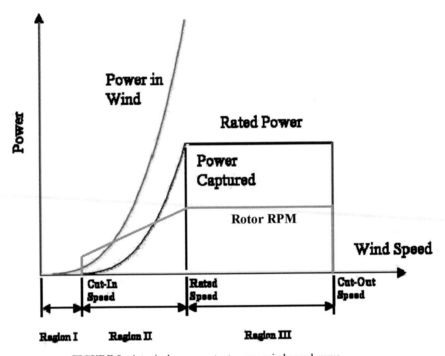

FIGURE 9. A typical power output versus wind speed curve.

The Drive Train (Gearbox, Generator, and Power Converter)

Wind generation of electricity places an unusual set of requirements on electrical systems. Most applications for electrical drives are aimed at using electricity to produce torque, rather than using torque to produce electricity. The applications that generate electricity from torque usually operate at a constant rated power. Wind turbines, on the other hand, must generate at all power levels and spend a substantial amount of time at low power levels. Unlike most electrical machines, wind generators must operate at the highest possible aerodynamic and electrical efficiencies in the low-power/low-wind region to squeeze every kilowatt-hour out of the available energy. Traditional electrical machines and power electronics disappoint because in most motor applications, there is power to spare and efficiency is less important in this low-power region. For wind systems, it is not critical for the generation system to be efficient in above-rated winds where the rotor is letting energy flow through to keep the power down to the rated level. Therefore, wind systems can afford inefficiencies at high power while they require maximum efficiency at low power – just the opposite of almost all other electrical applications in existence.

Converting torque to electrical power has historically been achieved using a speed-increasing gearbox and an induction generator. Many current megawatt-scale turbines use a three-stage gearbox consisting of varying arrangements of planetary gears and parallel shafts. Generators are either squirrel cage induction or wound-rotor induction, with some newer machines using the doubly fed induction design for variable speed, in which the rotor's variable frequency electrical output is fed into the collection system through a solid state power converter. Full power conversion and synchronous machines are drawing interest due to their fault-ride-through and other grid support capacities.

Due to fleet-wide gearbox maintenance issues and related failures with some past designs, it has become standard practice to perform extensive dynamometer testing of new gearbox configurations to prove durability and reliability prior to introducing them into serial production. The long-term reliability of the current generation of megawatt-scale drive trains has not yet been fully verified with long-term real world operating experience. There is a broad consensus that wind turbine drive train technology will evolve significantly in the next several years.

The Tower

The tower configuration used almost exclusively is a steel monopole tower on a concrete foundation that is custom designed depending on the local site conditions. The major tower variable is the height. Depending on the site's wind characteristics, the tower height is selected to optimize energy capture with respect to the tower's cost. Generally, a turbine will be placed on a tower of 60 to 80 m, but 100-m towers are being used more frequently. There are ongoing efforts to develop advanced tower configurations that are less costly and more easily transported and installed.

Balance of Station

The balance of the wind farm station consists of turbine foundations, the wind farm electrical collection system, wind farm power conditioning equipment, supervisory control and data acquisition systems, access and service roads, maintenance buildings, service equipment, and engineering permits. The combination contributes about 20% to the installed cost of a wind farm.

Operations, Reliability, and Availability

Operations and maintenance (O&M) costs have also dropped significantly since the 1980s due to improved designs and increased quality. Reference 1 presents data that show O&M expenses are a significant portion of total system cost of energy. O&M costs are reported to be as high as 3-5 cents/kWh for wind farms with 1980s technology, whereas the latest generation of turbines has reported O&M costs below 1 cent/kWh. Availability, defined as the fraction of time during which the equipment is ready to operate, is now over 95% and often reported to exceed 98%.

WIND TECHNOLOGIES OF THE FUTURE

There is no "big breakthrough" on the horizon for wind technology. However, many evolutionary steps executed with technical skill can cumulatively bring about a 30% to 40% improvement in the cost effectiveness of wind technology over the next decade.

Advanced Rotors

As turbines grow larger and larger, rotors must improve their ability to handle large dynamic loads with increased structural efficiency to avoid the costly cubic weight growth described previously. Several approaches are being developed and tested to help alleviate these load levels or create load-resistant designs. High strength-to-weight ratio carbon fibers are now being incorporated into the high-stress areas of wind turbine blades, which reduces overall blade weight.

One approach to reducing cost involves developing new blade airfoil shapes that are much thicker where the blade needs it most, producing inherently better structural properties. In general, thin flat structures like airfoils are very inefficient at carrying structural loads in the flat direction. The trick is to make a thick and structurally efficient blade airfoil shape that doesn't give up much in aerodynamic performance.

Another approach to increasing blade length while restraining the weight and cost growth is to reduce the fatigue loading on the blade. There can be a big payoff in this approach because the approximate rule of thumb for fiberglass blades is that a 10% reduction in cyclic stress can provide about an order of magnitude increase in fatigue life. Blade fatigue loads can be reduced by controlling the blade's aerodynamic response to turbulent wind inputs by actively flying the blade using the pitch control system of the turbine. This approach is being explored using modern state space control strategies so that future turbines can take advantage of this innovation.

An elegant concept is to build passive means of reducing loads directly into the blade structure. By carefully tailoring the structural properties of the blade using the unique attributes of composite materials, the blade can be built in a way that couples the bending deformation of the blade to twisting deformation. This is referred to as flap-pitch, or bend-twist, coupling and allows the outer portion of the blade to twist as it bends (Figure 10). This is accomplished by designing the internal structure of the blade, or orienting the fiberglass and carbon plies within the composite layups, in such a way as to make the blade twist as it is bent. This twisting changes the angle of attack over much of the blade. If properly designed, this change in angle of attack will reduce the lift as wind gusts begin to load the blade and therefore passively reduce the fatigue loads. Another approach to achieving pitch-flap coupling is to build the blade in a curved shape so that the aerodynamic load fluctuations apply a twisting moment to the blade, which will vary the angle of attack. These new blade designs are complex and must be developed, tested, and optimized so as not to adversely impact energy production.

Concepts such as on-site manufacturing and segmented blades are also being explored to help reduce transportation costs. It may be possible to segment molds and

352

move them into temporary buildings close to the site of a major wind installation so that the blades can be made close to or at the wind farm site.

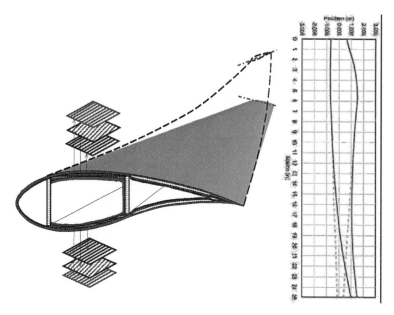

FIGURE 10. A Twist – Flap Coupled Blade design to alleviate fatigue loads (on the left with material coupling and on the right with a curved blade).

Advanced Drive Trains

Several unique designs are under development to reduce drive train weight and cost while improving reliability. These have been explored in design studies under the WindPACT project described in the reference 3 report. One approach for improving reliability is to build a direct drive generator that eliminates the complexity of the gearbox. The tradeoff is that the slowly rotating generator must have a high pole count and is large in diameter. Depending on the design, the generator can be in the range of 4 to 10 m in diameter and can be quite heavy.

The decrease in cost and increase in availability of rare earth permanent magnets is expected to significantly affect the size and cost of future permanent- magnet generator designs. Permanent-magnet designs tend to be quite compact and lightweight and reduce electrical losses in the windings. A 1.5-MW direct-drive generator using rare earth permanent magnets has been studied and a prototype has been built. This design uses 56 poles and is only 4 meters in diameter, versus the 10 meters for a wound rotor design [reference 4]. This machine is undergoing testing at NREL's National Wind Technology Center.

A hybrid of the direct-drive approach that offers promise for future large-scale designs is the single-stage drive using a low-speed generator. This allows the use of a low-speed generator that is significantly smaller than a comparable direct-drive design. The WindPACT drive train project has developed a prototype for such a drive train. This design uses a single-stage planetary drive operating at a gearbox ratio of 9.16:1. This gearbox drives a 190 RPM, 72-pole, permanent-magnet generator. This approach, which reduces the diameter of a 1.5-MW generator to 2 m [reference 3], was fabricated and is also under test on the dynamometer at NREL's National Wind Technology Center.

FIGURE 11. Clipper Windpower's Distributed Drive Train.

Another approach that offers promise for reduced size, weight, and cost is the distributed drive train. This concept is based on splitting the drive path from the rotor to drive several parallel generators. Studies have shown that by distributing the rotor torque on the bull gear over a number of parallel secondary pinions, a significant size and weight reduction is achieved. In 2006, Clipper Windpower developed a 2.5-MW prototype (Figure 11), which incorporates this approach and is currently in the new 2.5-MW Liberty turbine.

Here again the development of new technology for incorporation into a production turbine takes significant R&D resources and a number of years to insure a reliable production product.

Innovative Towers

The cost impact of extremely large cranes and the transport premiums for large tower sections and blades is driving the exploration of novel tower design approaches. Several concepts are under development or being proposed that would eliminate the need for cranes for very high, heavy lifts. One concept is the telescoping or self-erecting tower. Other self-erecting designs include lifting dollies or tower-climbing cranes that use tower-mounted tracks to lift the nacelle and rotor to the top of the tower. Further information on innovative towers can be found in reference 5.

Summary of Potential Future Turbine Technology Improvements

The DOE Wind Program has conducted cost studies under the WindPACT Project that identified a number of areas where technology advances would result in changes to the capital cost, annual energy production, reliability, operations and maintenance, and balance of station. Many of these potential improvements, summarized in Table 1, would have significant impacts on annual energy production and capital cost.

Table 1 also includes the manufacturing learning-curve effect generated by several doublings of turbine manufacturing output over the coming years. The learning-curve effect on capital cost reduction is assumed to range from zero in a worst case to the historic level in a best-case scenario, with the most likely outcome halfway in between. The most likely scenario is a sizeable increase in capacity factor (over the 2006 level) with a modest drop in capital cost.

OFFSHORE WIND ENERGY

U.S. Offshore wind energy resources are abundant, indigenous, and broadly dispersed among the most expensive and highly constrained electric load centers. The U.S. Department of Energy's Energy Information Agency shows that the 28 states in the contiguous 48 states with a coastal boundary use 78% of the nation's electricity.

Nineteen offshore wind projects now operate in Europe with an installed capacity of 900.6 MW. All installations have been in water depths less than 22 m. Although some projects have been hampered by construction overruns and higher-than-expected maintenance, projections show strong growth in many EU markets. In the United States, approximately 10 offshore projects are being considered. Proposed locations span both state and federal waters and total more than 2,400 MW.

355

		Cost Increments (Best/Expected/Least, Percent)	
Technical Area	**Potential Advances**	**Annual Energy Production**	**Turbine Capital Cost**
Advanced Tower Concepts	* Taller towers in difficult locations * New materials and/or processes * Advanced structures/foundations * Self-erecting, initial or for service	+11/+11/+11	+8/+12/+20
Advanced (Enlarged) Rotors	* Advanced materials * Improved structural-aero design * Active controls * Passive controls * Higher Tip Speed/lower acoustics	+35/+25/+10	-6/-3/+3
Reduced Energy Losses and Improved Availability	* Reduced blade soiling losses * Damage tolerant sensors * Robust control systems * Prognostic maintenance	+7/+5/0	0/0/0
Drivetrain (Gearboxes and Generators and Power Electronics)	* Fewer gear stages or direct drive * Medium/low speed generators * Distributed gearbox topologies * Permanent-magnet generators * Medium voltage equipment * Advanced gear tooth profiles * New circuit topologies * New semiconductor devices * New materials (GaAs, SiC)	+8/+4/0	-11/-6/+1
Manufacturing and Learning Curve*	* Sustained, incremental design and process improvements * Large-scale manufacturing * Reduced design loads	0/0/0	-27/-13/-3
Totals		+61/+45/+21	-36/-10/+21

TABLE 1. Areas of potential technology improvement.

* The learning curve results from Reference [TIO, 2007] are adjusted from 3 doublings in the reference to the 4.7 doublings in the 20% scenario.

The shallow-water baseline offshore wind turbine is basically an upgraded version of the standard land-based turbine with some system redesigns to account for ocean conditions. These modifications include structural upgrades to the tower to address the added loading from waves, pressurized nacelles, and environmental controls to prevent corrosive sea air from degrading critical drive train and electrical components, and personnel access platforms to facilitate maintenance and provide emergency shelter. To minimize expensive servicing, offshore turbines may be equipped with enhanced condition monitoring systems, automatic bearing lubrication systems, on-board service cranes, and oil temperature regulation systems, all of which exceed the standard for land-based designs.

Today's offshore turbines range from 2 to 5 MW in size and are typically represented by architectures that comprise a three-bladed, horizontal-axis upwind rotor, nominally 80 m to 126 m in diameter. Tower heights offshore are lower than land-based turbines because wind shear profiles are less steep, tempering the energy capture gains sought with increased elevation. The offshore foundations differ substantially from land-based turbines.

The baseline offshore technology is deployed in arrays using monopiles at water depths of about 20 m. Monopiles are large steel tubes with wall thickness of up to 60 mm and diameters of 6 m. The embedment depth will vary with soil type, but a typical North Sea installation will require a pile that is embedded 25 m to 30 m below the mud line and that extends above the surface to a transition piece with a leveled and grouted flange on which the tower is fastened. Mobilization of the infrastructure and logistical support for a large offshore wind farm is a significant portion of the system cost.

Current estimates indicate that the cost of energy from offshore wind plants is above 10 cents/kWh and that the O & M costs are also higher than for land-based turbines due to the difficulty of accessing turbines during storm conditions.

There are three logical pathways (Figure 12) representing progressive levels of complexity and development that will lead to cost reductions and greater offshore deployment potential. The first path is to lower costs and remove deployment barriers for **shallow water technology** in water depths of 0 to 30 meters. The second path is **transitional depth technology,** which is needed for depths where current technology

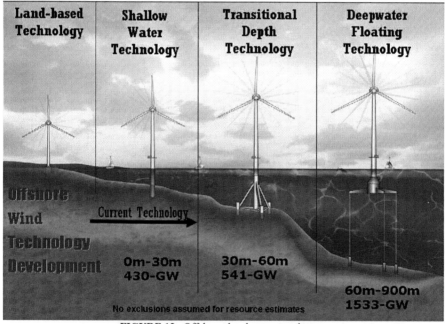

FIGURE 12. Offshore development pathways.

no longer works. This technology deals mostly with substructures that are adapted from existing offshore oil and gas practices. Transitional depths are defined to be 30 to 60 meters. The third path is to develop technology for **deep water**, defined by depths between 60 and 900 meters. This technology will probably use floating systems, which require more R&D to design turbines that are lighter and can survive the added tower motion on anchored, buoyant platforms.   The ultimate vision for offshore wind energy is that it would open up major areas of the outer continental shelf to wind energy development, where the turbines would not be visible. This would require the use of deep water floating platforms that could be mass produced and assembled in dry docks and then floated out and anchored without extensive assembly at sea. Deep water technology also avoids the need for long-distance transmission because the wind farms can be located much closer to load centers.

New offshore technologies will be required to grow wind turbines into 5- to 10-MW sizes or greater. These technologies may include lightweight composite materials and composite manufacturing, lightweight drive train, modular pole direct drive generators, hybrid space frame towers, and large gearbox and bearing designs that are tolerant of slower speeds and large scale. The cost of control systems and sensors that monitor and diagnose turbine status and health will not grow substantially as turbine size increases, and high reliability will be essential due to the limited access during severe storm conditions, which can persist for extended periods.

It is expected that over the next five years, one or more offshore wind farms will be deployed in the United States. They will be installed in shallow water and supply electricity to nearby onshore utilities serving large population centers. If they are successful, the technology will develop more rapidly and the move to deep water systems will progress at a more rapid rate. However, the path toward floating systems must be supported by an extensive R&D program over a decade or more. Further information on the viability of offshore wind energy is provided in reference [reference 6].

SUMMARY

Power production from wind technology has evolved very rapidly over the past decade. Capital costs have plummeted, reliability has improved, and efficiency has dramatically increased, resulting in robust commercial market product that is competitive with conventional power generation. Investments in R&D as well as the development of robust standard design criteria have helped to mitigate technology risk and attract market capital for development and deployment of large commercial wind plants. High-quality products are provided by every major turbine manufacturer, and complete wind generation plants are now being engineered to seamlessly interconnect with the grid infrastructure to provide utilities with dependable energy supply, free of the risks of future fuel price escalation inherent in conventional generation.

The cost-of-energy metric remains the principal technology indicator, incorporating the key elements of capital cost, efficiency, reliability, and durability. The unsubsidized cost of wind-generated electricity ranges from about 5 to 8.5 cents/kWh for projects completed in 2006.

No major technical breakthroughs in land-based technology are needed for a broad geographic penetration of wind power on the electric grid. Advancement requires a systems development and integration approach, reflecting the high level of engineering already incorporated into modern machines. No single component improvement in cost or efficiency can achieve significant cost reductions or dramatically improved performance. Capacity factor can be increased over time using enlarged rotors on taller towers. Market incentives will remain necessary to sustain the industry growth in the near term, but in the longer term subsidies can probably be eliminated. In addition, with continued R&D, offshore wind energy has great potential to allow the United States to greatly expand its electrical energy supply.

IMPORTANT REFERENCES FOR FURTHER READING

1. R. Wiser and M. Bolinger, *Annual Report on U.S. Wind Power Installation, Cost, and Performance Trends: 2006* (May 2007). Available at www.osti.gov/bridge.
2. D.A. Griffin, (2001). *WindPACT Turbine Design Scaling Studies Technical Area 1 -- Composite Blades for 80- to 120-Meter Rotor; 21 March 2000 - 15 March 2001*, 44 pp. (2001), NREL Report No. SR-500-29492. Available at www.nrel.gov/publications/
3. R. Poore and T. Lettenmaier, *Alternative Design Study Report: WindPACT Advanced Wind Turbine Drive Train Designs Study; November 1, 2000 -- February 28, 2002*, 556 pp. (2003), NREL Report No. SR-500-33196. Available at www.nrel.gov/publications/
4. G. Bywaters, V. John, J. Lynch, P. Mattila, G. Norton, J. Stowell, M. Salata, O. Labath, A. Chertok, and D. Hablanian, *Northern Power Systems WindPACT Drive Train Alternative Design Study Report; Period of Performance: April 12, 2001 to January 31, 2005.* 404 pp., NREL Report No. SR-500-35524. Available at www.nrel.gov/publications/
5. M.W. LaNier, *LWST Phase I Project Conceptual Design Study: Evaluation of Design and Construction Approaches for Economical Hybrid Steel/Concrete Wind Turbine Towers; June 28, 2002 - July 31, 2004.* 698 pp. (2005), NREL Report No. SR-500-36777. Available at www.nrel.gov/publications/
6. W. Musial, "A Viable Energy Option for Coastal USA Offshore Wind Electricity," to be published in the *Marine Technology Society Journal* (2008).

Future Roles for Nuclear Energy

Lance Kim[a,b] and Per F. Peterson[a]

[a]Department of Nuclear Engineering
[b]Goldman School of Public Policy
University of California, Berkeley
Berkeley, CA 94720-1730

Abstract. An introduction to the operation of nuclear fission reactors, current status of the commercial nuclear sector, expansion opportunities, and research and development priorities.

INTRODUCTION

Following the discovery of nuclear fission in the late 1930s, the implications of a neutron-induced fission chain reaction were quickly recognized both as source of peaceful energy and for its destructive potential in nuclear weapons. Rapid progress on both fronts soon followed driven by the exigencies of global conflict and growing appetites for energy. On the civilian side, the expansion of nuclear energy stalled following the energy crisis of the 1970s when expectations of growing demand for energy failed to materialize in general and construction cost overruns and highly varied operating performance stymied the further expansion of nuclear power.

Today, the greatly improved performance of existing nuclear power plants and nuclear energy's role as the largest source of non-carbon electricity has rekindled interest in constructing new reactors. With 34 nuclear power plants under construction around the globe (IAEA 2008) and several reactors in the early stages of the U.S. Nuclear Regulatory Commission's streamlined licensing process (Nuclear Energy Institute 2008), a resurgence of nuclear energy appears imminent.

NUCLEAR IN A NUTSHELL

Nuclear power plants harness the energy generated by the neutron-induced fission of heavy isotopes such as uranium and plutonium. Designers of reactor cores arrange material in a critical configuration to sustain a fission chain reaction. By moving up the curve of binding energy, the fission of heavy elements liberates an enormous amount of energy per unit mass in comparison to typical chemical reactions. Per day of operation, a typical nuclear power plant with a rated capacity of 1000 megawatts of electric power (MW_e) (enough power for roughly one million average U.S. homes) consumes only 3.2 kilograms of uranium fuel. By comparison, a 1000 MW_e coal-fired

CP1044, *Physics of Sustainable Energy, Using Energy Efficiently and Producing It Renewably*
edited by D. Hafemeister, B. Levi, M. Levine, and P. Schwartz
© 2008 American Institute of Physics 978-0-7354-0572-1/08/$23.00

power plant demands over seven million kilograms of coal per day and releases comparable quantities of CO_2, NO_x, SO_x and ash to the environment.

The fate of the neutrons, fission products, and neutron activation products produced in a nuclear fission reaction present a number of challenges to system designers and policy-makers. The production of enriched uranium to fuel reactors and the breeding of plutonium 239 raise proliferation concerns given the utility of these materials in nuclear weapons. Radioactive byproducts of fission continue to produce decay heat following the shutdown of the reactor, necessitating highly reliable safety systems to prevent core melt accidents. Additionally, long-lived radionuclides in spent nuclear fuel require long-term sequestration to limit impacts on the environment and public health.

Managing these broad range of issues are high priorities in the design of nuclear reactors and nuclear energy policy. The system of safeguards administered by the International Atomic Energy Agency has been effective in preventing the diversion of declared nuclear material. National measures to provide physical protection of nuclear materials from theft have been effective, but merit further strengthening in some countries. The safety of nuclear reactors is achieved with redundant and diverse systems performing safety functions with high reliability, a multi-barrier approach to containing radionuclides, and inherent negative feedback effects. For new reactors, a shift in design philosophy towards passive safety systems offers a more robust and less expensive approach to reactor safety.

Nuclear Waste Management

The management of nuclear waste continues to be one of the dominant issues influencing the acceptability of nuclear energy. In contrast to the challenge of sequestering the byproducts of the combustion of fossil fuels and accommodating the health and environmental impacts of the pollutants discharged into the biosphere, the high energy density of nuclear fuel makes managing the correspondingly small quantities of spent nuclear fuel a far more manageable task. To this end, international research and development efforts have improved the understanding of the management of nuclear waste to minimize impacts on the environment and public health. Broad scientific consensus exists that geologic isolation can manage nuclear waste safely and reversibly over long periods of time.

The U.S. will have an approved geologic repository for the disposal of spent nuclear fuel should the U.S. Nuclear Regulatory Commission (NRC) grant a construction license. Over two decades of scientific and technical study led to a positive site suitability decision for Yucca Mountain in 2002. The Department of Energy (DOE) has completed its license application in June 2008. The NRC is scheduled to complete its scientific and technical review of this application by 2011.

In granting the license, the DOE and NRC will have demonstrated compliance with Environmental Protection Agency (EPA) standards far more stringent than those used to regulate chemical hazards. The EPA's draft one million year safety standard for Yucca Mountain limits the maximum impact on an individual using ground water to less than 15 mrem/year for the first 10,000 years followed by a 300mrem/year (equivalent to average background radiation exposure) for the next million years. The limited groundwater contamination that could result represents a marginal impact on the environment in comparison to the background of existing contaminated water supplies - a problem that current public health systems already understand how to manage.

More importantly, the politically sensitive question of a second repository will reemerge when Yucca Mountain's legislated limit of 63,000 MT of spent nuclear fuel is surpassed (around 2014 at the expected rate of nuclear power production). The legislated limit is, however, far below the technical limit of 120,000 – 300,000 MT for the 2000 acre footprint of the repository. Furthermore, advanced fuel cycles that recycle the heavy elements in spent fuel would increase the capacity of Yucca Mountain by a factor of roughly 50. Nevertheless, resolving this question is not particularly urgent as above ground dry cask storage of spent nuclear fuel provides a short-term alternative.

RENEWED INTEREST IN NUCLEAR ENERGY

The performance of existing nuclear power plants is one of the drivers for the renewed interest in nuclear energy. Nuclear production costs (fuel plus operations and maintenance) are competitive with coal and less volatile in price in comparison to natural gas. As a consequence, existing nuclear power plants have quietly achieved unprecedented levels of performance through innovations in operations and maintenance processes (including advances in human performance and risk management). On average, U.S. nuclear power plants currently operate roughly 90% of the year, going offline at scheduled times to refuel and perform maintenance. In comparison, the fleet average during the 1980s was a mere 60%. The frequency of forced reactor outages caused by unanticipated events has also declined, resulting in a forced capability loss of less than 2% (median value), an additional indicator of competent management.

Expectations of operating in a carbon-constrained economy are also contributing to the renewed interest in nuclear energy. With life-cycle CO_2 emissions comparable to renewable technologies, nuclear's role as the largest source of low-carbon electricity and the most realistic alternative to coal-generated base load electricity cannot be ignored. With some in the European Union considering a return to coal-fired generation (Rosenthal 2008), we are presented with stark policy alternatives between coal and nuclear for base load generation. The French present one model for energy development. Following a 20-year construction effort, nuclear energy now constitutes the vast majority of domestic primary energy production and the French were able to close their last coal mine in early 2004.

Near-Term Prospects

The major near-term question to address as we consider an expansion of nuclear generation is whether new nuclear power plants can be built on schedule and at a reasonable cost. Capital costs for U.S. reactors increased dramatically for reactors coming online through the 1980s and 1990s due to the high interest rates, public opposition, delays from poor project management, a burdensome licensing process, and safety upgrades following the Three Mile Island accident. For new reactors, the overnight capital cost for new nuclear power plants vary widely from publicized vendor estimates of $1500 per kW of installed capacity to a recent estimate of $2950/kW (Keystone Center 2007).

Where actual costs will fall is debatable. The high degree of optimization for new nuclear infrastructure may offer significant cost savings. Advanced light water reactors featuring passive safety systems provide substantial improvements over earlier designs with reductions in equipment requirements and building volumes contributing to lower cost. Streamlined licensing and modular construction processes call for a high degree of standardization, decreasing costly construction delays and more rapidly moving down the learning curve. On the other hand, the hiatus on nuclear plant construction and uncertain expansion plans have resulted in short-term bottlenecks in the supply chain for nuclear reactors, placing upward pressure on prices for nuclear components. The scarcity of a qualified nuclear workforce compounds the problem. The degree to which these growing pains will persist with the expansion of the nuclear energy sector remains to be seen.

In the long run, resource inputs provide an indication of future costs of energy technologies. The inflation of nuclear construction cost estimates has been attributed in part to the near doubling of construction commodity costs (for example, steel reached prices of some $600 per tonne in March 2008). But, the total cost of commodities (steel, concrete, copper, etc.) used in nuclear construction amounts to only $36/kW. Thus the cost of nuclear construction is not sensitive to the cost of the material inputs, but instead depends upon the costs for manufacturing, construction, and project management and financing. As with reactor operations costs, these costs will depend primarily on the competence of the companies that build nuclear power plants.

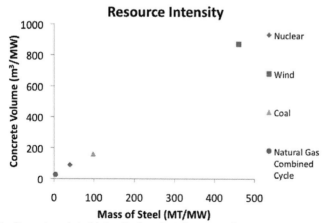

FIGURE 1. Concrete and steel intensity for selected sources of energy per average megawatt of capacity (e.g. corrected for capacity factor)

Comparing resource inputs also provide an indication of the environmental impacts and expected competitiveness of energy technologies. Concrete and steel represent over 95% of the carbon emissions associated with construction inputs and are strongly correlated to capital cost. The relatively low requirements for these two inputs demonstrates that constructions costs for nuclear energy are less sensitive to commodity costs in comparison to wind and, to a lesser extent, coal (see Figure 1) while avoiding the high production costs and volatility of natural gas prices.

Ultimately, capturing the external costs of fossil fuel combustion can be expected to shift economic incentives in favor of nuclear for base load generation capacity. This can be expected to occur in the near term as carbon cap and trade programs are implemented to reduce U.S. carbon emissions. Under potential carbon cap-and-trade programs, the profitability for existing nuclear power plants is expected to increase dramatically. (Smith 2008) Furthermore, new nuclear power plants are estimated to become the least cost source of base load electricity with a carbon tax of roughly $45 (Congressional Budget Office 2008) to $100 per ton of carbon (Sailor, et al. 2000).

Applications of Nuclear Electricity and Process Heat

Additional demand may emerge for nuclear energy beyond the traditional role for nuclear in the market for base load electricity. The cogeneration of electricity and process heat from nuclear reactors broadens opportunities for nuclear energy. The transportation sector may be one area of demand growth. The benefits of adopting alternative vehicle technologies operating entirely or partially on stored electrical energy (e.g. plug-in hybrids), hydrogen, or low-carbon fuel (e.g. ethanol, nuclear-assisted heavy oil extraction) will depend largely upon access to low-carbon sources of electricity and process heat. Reactors such as the Pebble Bed Modular Reactor under development in South Africa can be configured to produce efficiently produce electricity and process heat to flexibly meet demand.

CONCLUSIONS

Activity in the nuclear energy sector has been substantial. Improvements in performance at existing nuclear power plants, demand for low-carbon energy, and the selection of Yucca Mountain has resulted in a dramatic turnaround in the prospects for nuclear energy. Nevertheless, the future remains uncertain and expansion could be derailed. Fully implementing the provisions of the Energy Policy Act of 2005 to overcome the disincentives faced by first-movers will support near term expansion of advanced light water reactors. Key research and development priorities include demonstrating the cogeneration of electricity and process heat to expand demand for nuclear energy. In the longer-term, the development of advanced nuclear reactor and fuel cycle technologies promise to achieve higher levels of economic performance, safety, sustainability, physical security, and proliferation resistance.

REFERENCES

1.Congressional Budget Office. "Nuclear Power's Role in Generating Electricity." 2008.

2.IAEA. *Power Reactor Information System.* 2008. http://www.iaea.org/programmes/a2/ (accessed May 8, 2008).

3.Keystone Center. *Nuclear Power Joint Fact-Finding.* Keystone: Keystone Center, 2007.

4.Nuclear Energy Institute. *New Nuclear Plant Status.* 2008. http://www.nei.org/resourcesandstats/documentlibrary/newplants/graphicsandcharts/newnuclearplantstatus/ (accessed May 8, 2008).

5.Rosenthal, Elisabeth. "Europe Turns Back to Coal, Raising Climate Fears." *New York Times,* April 23, 2008.

6.Sailor, William C., David Bodansky, Chaim Braun, Steve Fetter, and Bob van der Zwaan. "A Nuclear Solution to Climate Chanage." *Science,* May 2000: 1177-1178.

7.Smith, Rebecca. "Carbon Caps May Give Nuclear Power a Lift." *Wall Street Journal,* May 19, 2008.

Carbon Capture and Geologic Storage

Larry R. Myer

Earth Sciences Division
Lawrence Berkeley National Laboratory
Berkeley, CA 94720

Abstract. This paper will briefly discuss carbon capture and storage options, mechanisms and costs. Risks from geologic storage risks will be addressed and the need for monitoring. Some current field studies will be described.

INTRODUCTION

Carbon capture and storage, often referred to as CCS, is a technology for mitigation of CO_2 emissions from large point sources such as fossil fuel power plants, and industrial sources such as cement plants and refineries. The conventional approach to CCS involves four steps. The first step is to capture the CO_2 from flue gas or other process streams. Commercially available technology exists for capture, but it is expensive. The captured CO_2 is then dried and compressed to about 100 bars for transport in pipelines. Pipeline transport of CO_2 is considered to be a commercially available technology. A major CO_2 pipeline currently connects naturally occurring CO_2 deposits in Utah with oil fields in Texas, where the CO_2 is used for enhancing the production of oil from reservoirs in the Permian Basin. Another major pipeline carries CO_2 from a coal gasification plant in North Dakota to oil fields in southern Saskatchewan.

The fourth and final step in the CCS process is injection into the deep subsurface where it can be stored for hundreds to thousands of years (or longer). Characterization of potential storage sites will be required to assure that the site has appropriate geologic characteristics to securely store CO_2 for this period of time. Monitoring will be carried out to confirm that the storage site is performing as expected.

Technologies associated with these steps in the CCS process will be discussed in the following sections.

CP1044, *Physics of Sustainable Energy, Using Energy Efficiently and Producing It Renewably*
edited by D. Hafemeister, B. Levi, M. Levine, and P. Schwartz
© 2008 American Institute of Physics 978-0-7354-0572-1/08/$23.00

Options for CO$_2$ Capture

The largest point sources of CO$_2$ are fossil fueled power generating facilities. There are three primary technological approaches for capture of CO$_2$ from large fossil fuel combustion sources: post-combustion; pre-combustion; and oxy-combustion.

Post-combustion involves removal of CO$_2$ from the flue gas stream. The CO$_2$ is removed by amine based chemical solvents. This technology is considered to be an established, commercial technology, though it has not yet been applied at the scale required by a large coal fired power plant. Another advantage of the technology is that it is readily applied as a retrofit to existing plants. It can also be applied to removal of CO$_2$ from numerous industrial process streams. The major disadvantage of the technology is that there is a large energy penalty, primarily due to the heat required to release the CO$_2$ from the solvent. This penalty can be as high as 30% for a coal fired power plant.

Pre-combustion capture refers to capture from systems which employ gasification of the fuel before combustion. Gasification is applicable to biofuels as well as coal. The result of gasification is a syngas, mostly CO, which is then combined in a shift reaction with H$_2$O to produce CO$_2$ and H$_2$. The CO$_2$ is usually captured using a physical solvent liquid absorber/stripper system. The energy penalty is less than the post-combustion because much of the CO$_2$ flashes out of the physical solvent once the pressure is reduced for stripping. Since the partial flashing of the CO$_2$ occurs at moderate (3 to 6 atmospheres) pressure, compression power requirements and costs are also lower. Gasification systems are much more complex than pos-combustion systems. They have been employed extensively in chemical processing applications but rarely at the scale needed for large power generating plants.

Oxy-combustion refers to power generating systems in which fuel is combusted in the presence of oxygen. Combustion with oxygen in place of air results in a flue gas of mostly CO$_2$ along with water. Since the flue gas stream is about 97% CO$_2$, the capture process could be much simplified, consisting essentially of just dehydration before the stream is compressed for transport/storage. Another advantage of oxyfuel combustion for CO$_2$ capture is its potential ability to retrofit existing combusting systems. The major disadvantages are the large capital costs and power requirements of the oxygen production systems. Oxy-combustion for CO$_2$ capture is the least developed of the three options. It has only been tested at relatively small pilot plant scale. A 49MW oxy-combustion power generating plant combined with geologic storage is planned for operation near Bakersfield California by 2012 in conjunction with the WESTCARB Department of Energy research project.

The cost of carbon capture and storage (CCS) will increase the cost of electrical power. Estimates of costs require a large number of assumptions and are subject to considerable uncertainty. A recent assessment conducted for California [1] concluded that power generation from coal would rise by $35 - 45/ MWh (3.5–4.5 cents/kWh). This amounts to $50 – 60/tonne CO$_2$ avoided. Power generation from natural gas would increase by about $30/MWh, amounting to $80/tonne CO$_2$ avoided. The larger cost per tonne for natural gas is primarily due to the lower concentration of CO$_2$ in the flue gas stream.

These cost estimates[i] took into account the construction environment in California and included a contingency for first-of-a-kind units. Conventional technologies for combustion and capture were assumed. It was also assumed that 90% of the CO_2 would be captured and that the sources were large – such that cost savings associated with "economies of scale" could be assumed. Transport and storage costs of $10/tonne were included. Transport was assumed to be by pipeline, and it is noted that pipeline construction costs can vary widely, depending on many factors.

Industrial plants such as refineries, cement plants, and steel mills are also large point sources of CO_2. Some industrial processes, such as natural gas processing, ethanol production, ammonia production and some types of hydrogen production processes in refineries, produce nearly pure CO_2 streams. The cost of adding CCS to such industrial processes would be about $20 – 30/tonne CO_2 avoided.

Enhanced oil recovery (EOR) using CO_2 would offset the above costs by saving some $20/tonne, depending upon oil prices. Though volumes are not large, it is apparent that CCS, where certain industrial sources can be paired with EOR operations, is economically viable now.

Storage Mechanisms

The primary storage options for CO_2 are:

- oil and gas reservoirs
- deep, unminable coal beds, and
- saline formations.

Depleted oil reservoirs could be used for storage only, or, if conditions are appropriate, storage could be carried out in conjunction with production of additional oil (EOR using CO_2) to help offset CCS costs. Assuming current petroleum industry practice, depleted gas reservoirs would be used for storage only. Research indicates that CO_2 could be used to enhance natural gas recovery (EGR), but EGR using CO_2 is not carried out commercially. Oil/gas reservoirs are ideal CO_2 storage sites because these reservoirs have stored buoyant fluids for millions of years. Many oil/gas reservoirs also contain CO_2 as well. The existence of these reservoirs demonstrates that the rocks have the inherent properties necessary to securely hold CO_2 for geologic time.

Storage in unminable coal beds is advantageous because CO_2 adsorbs to the surface of the coal exposed in cracks and fractures, and is securely held in place. In addition, if methane is present, CO_2 will preferentially adsorb to the coal and displace the methane, which can then be produced to offset CCS costs. This is referred to as storage in conjunction with enhanced coal bed methane (ECBM) recovery. The disadvantages of CO_2 storage with ECBM are that the injectivity of deep coals is often low, and the overall potential storage capacity of coals is small.

Saline formation refers to sedimentary rocks, such as sandstone, which contain only salt water. The amount of salinity is important. In the US only formations with salinity greater than 10,000ppm total dissolved solids are suitable storage sites because the Environmental Protection Agency (EPA) protects all waters that have salinity less than that. Oil and gas reservoirs are also found in sedimentary rocks. In general, oil and gas reservoirs can be though of as localized regions in saline formations where hydrocarbons have accumulated. Saline formations are broadly distributed around the world and have potential storage capacity much greater than other storage options. In the long term, saline formation storage will predominate in CCS.

The physical and chemical mechanisms which trap the CO_2 in the subsurface are:

- physical/structural trapping
- dissolution
- phase trapping
- mineralization, and
- surface adsorption.

Surface adsorption is specific to coal as discussed above. The other mechanisms are applicable in saline formations (and oil/gas reservoirs within the saline formations). In almost all storage applications, CO_2 will be less dense than the fluids into which it is injected in the subsurface. The one exception is storage in a dry methane gas reservoir, in which the CO_2 will be denser than the CH_4.

If the CO_2 is less dense than the other fluids in the pores of the rock, it will be buoyant, so the primary mechanisms for keeping the CO_2 in place are the seal overlying the reservoir and any structural trap which may be present. Figure 1 is a schematic representation of these mechanisms. Suitable sealing rocks (called "roof rock" in figure) are shales, siltstones, or other rocks which have high capillary entry pressures, low permeability and are free of fractures or faults which could conduct fluid across the seal. Figure 1 also shows two types of geologic structures which are ideal traps: one is an anticline where the buoyant CO_2 collects at the highest point in the reservoir; and the other is a fault which has offset the reservoir rock so that it "dead-ends" in the updip direction against seal rock. These structures are also examples of common traps for hydrocarbon accumulations. It is worth emphasizing that the mere existence of a fault would not disqualify a site for storage. A necessary part of site characterization is to determine if faults are present, and if so, to evaluate if they are conductive or not.

Some CO_2 will dissolve into the pore water. The solubility of CO_2 depends on pressure, temperature, and salinity, but typically will be on the order of a couple per cent by mass. The CO_2 will also react with the minerals in the rock. This is the most secure form of trapping, but reaction rates are slow, ie, on the order of thousands of years, so mineralization does not contribute significantly to CCS storage security during time periods of greatest interest to society.

The CO_2 is stored in the rock's pore space, which makes up about $10 - 30\%$ of the volume of a typical reservoir rock like sandstone. The pore space is a dense, complex, interconnected network of passageways, the dimensions of which, in cross section, vary from a micron or less to tens of microns. CO_2 is non-wetting with respect to water, and as the CO_2 plume moves through the system, interference between phases leads to trapping of the CO_2. This trapping mechanism is very important in stabilizing plumes after injection stops, particularly in absence of a structural trap.

369

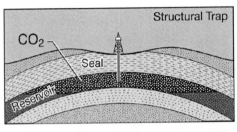

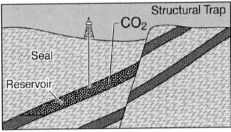

FIGURE 1. Illustration of physical/structural trapping mechanisms (modified after W Gunter, ARC).

Technological Experience

The technology for large scale injection of CO_2 into the subsurface already exists, and derives from decades of experience in the oil and gas industry. The use of CO_2 for enhanced oil recovery is an established commercial technology. In the United States, CO_2 is transported by pipeline from naturally occurring reservoirs in Colorado to oil fields in Texas. These operations have been underway since the 1970s. In 2006 over 48 M tonnes of CO_2 were injected for EOR in the U. S. In these operations, most of the injected CO_2 is produced back along with the oil. This CO_2 is separated and re-injected.

Around the world, injection of CO_2 into a saline formation for purposes of sequestration is currently taking place as part of two commercial natural gas production projects (Figure 2). The first project, developed by Statoil, is the Sleipner project in the North Sea. In this project natural gas containing CO_2 is produced from a reservoir. The CO_2 is separated form the natural gas, and injected into a saline formation located above the gas-producing horizon. About 1M tonnes per year have been injected since 1999. The project is particularly noteworthy for successful application of 3-D seismic technology for monitoring of the CO_2 plume.

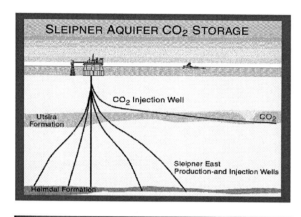

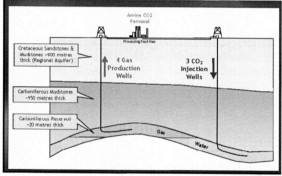

FIGURE 2. Schematic illustration of two current commercial saline storage projects, Sleipner (above) and In Salah (below) (graphics courtesy of Statoil and BP).

The second project, led by BP, is the In Salah project, located in Algeria. This project also involves production of natural gas containing CO_2, which is separated and injected into the saline formation which is contiguous with, but down-dip from, the gas reservoir. Because the rock is relatively low permeability (~10mD), long horizontal wells are used to maximize the area of rock in contact with the injected CO_2.

Storage Capacity

Prospective saline formation storage is broadly distributed around the Earth. The Intergovernmental Panel on Climate Change (IPCC)[ii] concluded in its 2005 Special Report on Carbon Dioxide Capture and Storage that "It is likely that the technical potential for geological storage is sufficient to cover the high end of the economic potential range (2200 $GtCO_2$), but for specific regions, this may not be true." In California regional studies (Figure 3) have identified multiple sedimentary basins containing saline formations which have an enormous potential storage, with initial estimates placing the resource at about 75 to 300Gt of CO_2 storage capacity.[iii] The storage capacity of gas and oil reservoirs found within these saline formations has been estimated at 1.7Gt and 3.6Gt, respectively. These resource estimates do not take into account regulatory or economic constraints, or local geologic conditions which could still preclude specific sites from consideration.

FIGURE 3. Sedimentary basins in California containing saline formations with CO_2 storage potential, indicated as "included for further investigation" (from Myer et al, ref. 3).

Risk

Once CO_2 is injected into the subsurface, the main risks to human health and the environment arise from unintended leakage from the storage reservoir. The primary paths for this leakage, either to the surface or to groundwater resources, would be improperly installed and/or abandoned wells, and undiscovered geologic discontinuities such as faults. Over injection of CO_2 could result in unwanted intrusion of saline fluids, and excessive fluid pressurization due to injection could also induce seismic events. These risks are, however, manageable. On the subject of risk, the 2005 IPCC report [ref. 2] on CCS states: *"With appropriate site selection informed by available subsurface information, a monitoring program to detect problems, a regulatory system, and the appropriate use of remediation methods to stop or control CO_2 releases if they arise, the local health, safety, and environment risks of geological* *storage would be comparable to risks of current activities such as natural gas storage, enhanced oil recovery, and deep underground disposal of acid gas."*

A proper site characterization and monitoring program will actually lead to a decrease in the risk of leakage over the life of a geologic storage project. The risk of leakage is highly dependent upon uncertainty in subsurface conditions and this uncertainty is reduced through data collected during site characterization and from monitoring measurements during operation. The uncertainty arises in large part because the structure and properties of rock masses are inherently heterogeneous at all scales and access to the subsurface is always limited. Wells drilled during operation of a project provide direct access to subsurface information on heterogeneity which was not available during site characterization. Monitoring of the plume movement also provides information on structure and properties. The data obtained as operation continues will lead to improved geologic models and higher confidence in models predicting subsurface behaviour of the CO_2.

After injection ceases, the risk of leakage decreases rapidly as fluid pressures decay and the plume becomes immobilized. The primary mechanisms for plume stabilization are dissolution of CO_2 into the pore water, and residual phase trapping. Results of numerical simulation in Figure 4 provide an example illustration of the relative contribution of these mechanisms and the time frame over which a plume might become immobilized. A key parameter is the residual gas saturation, which controls how much CO_2 is immobilized by residual phase trapping.

Monitoring

As implied above, monitoring will be an essential part of a geologic storage project. Monitoring data will be used to:

- Confirm storage efficiency and processes
- Ensure effective injection controls
- Detect plume location and leakage from storage formation
- Ensure worker and public safety
- Design and evaluate remediation efforts
- Detect and quantify surface leakage
- Provide assurance and accounting for monetary transactions
- Settle legal disputes

The monitoring program will be tailored to geologic conditions at each site. Not all techniques work under all conditions. However, based primarily on decades of work in petroleum exploration and production, a substantial portfolio of monitoring techniques is available:

- Seismic and electrical geophysics
- Well logging
- Hydrologic pressure and tracer measurements
- Geochemical sampling
- Remote sensing

373

•CO_2 sensors
•Surface measurements

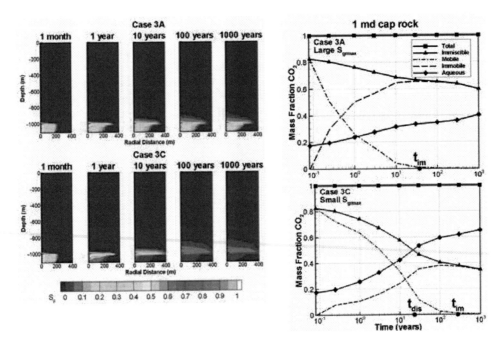

FIGURE 4. Example reservoir simulation shows fate of a plume of 900,000 tons of CO_2 injected beneath a poor quality cap rock (1-md permeability); Sg is the CO_2 saturation, mass of "mobile" plus "immobile" equals "free phase" CO_2 (from Doughty and Myer[iv]).

Need for Field Pilot Tests

Pilot programs provide regional knowledge base data, essential for large scale implementation. Pilots demonstrate best sequestration options, unique technologies and approaches, in the region. Pilots involve site-specific focus for testing technologies, defining costs, assessing leakage risks, gauging public acceptance, exercising regulatory requirements, and validating monitoring methods.

SUMMARY

The technology necessary to undertake CCS is available today. Its cost effectiveness is driven mostly by capture costs. The risks can be managed. Field testing is essential to gain experience. There are plenty of opportunities for innovation: Fossil power generation can be optimized for CCS. The basic physics of storage mechanisms must be understood. New monitoring approaches, with increased resolution are being developed. We must think beyond coal. A comprehensive approach must reach beyond coal plants to include, natural gas, industrial processes, fermentation processes (eg. biofuels), as well as linking with forest management.

AKNOWLDEGMENTS

This work was supported by the Assistant Secretary for Fossil Energy, Office of Coal and Power Systems, through the National Energy Technology Laboratory, of the U.S. Department of Energy under Contract No. DE-AC02-05CH1123.

REFERENCES

1. E. Burton, R. Myhre, L. Myer, and K. Birkinshaw. "Geologic Carbon Sequestration Strategies for California, The Assembly Bill 1925 Report to the California Legislature", California Energy Commission (2007).

2. E. Rubin, L. Myer, H. Coninck, J. Abanades, M. Akai, S. Benson, et al. "Carbon Dioxide Capture and Storage", Technical Summary of Intergovernmental Panel on Climate Change Special Report, Cambridge University Press, Cambridge, United Kingdom, and New York, NY, USA (2005).

3. L. Myer, C. Downey, J. Clinkenbeard, S. Thomas, S. Stevens, S. Benson, H. Zheng, H. Herzog, and B. Biediger. "Preliminary Geologic Characterization of West Coast States for Geologic Sequestration", report to Department of Energy, Contract No. DE-FC26-03NT41984 (2005).

4. C. Doughty and L. Myer. "Bounding calculations on geologic leakage of CO_2", in Science and Technology of Carbon Sequestration, B. McPherson and E. Sundquist, eds, American Geophysical Union, Washington DC, (2007).

Energy Storage for a Greener Grid

Imre Gyuk

Energy Storage Research Program
U.S. Department of Energy,
1000 Independence Avenue
Washington, DC 20585

Abstract. Energy Storage is an emerging technology with potential applications throughout the grid. Storage can provide power quality, frequency regulation, and renewable smoothing. It can increase asset utilization for generation, transmission, substations, and distribution. Storage can help in reducing peak loads, and making variable renewable energy more dispatchable. The article presents the technology of electrical storage, provides examples of different types of applications, and analyses the rationale for introducing storage. Storage makes the grid greener, more efficient, and more reliable.

CAVEAT

This paper is a report from the frontier. It is not intended to be a scientific review with profuse quotes and copious references. Instead, it is an account of the status and some of the rationale of the technologies of electrical energy storage and some of the recent applications on the electricity grid. This information changes weekly. It is gleaned from conferences, coffee breaks, phone calls, and site visits. The Energy Storage Program of the Department of Energy (DOE) is directly involved in a large number of the projects mentioned. The storage community is quite a tightly knit group and news of new projects and applications spreads quickly. But it is important to remember that a few years ago there were virtually no sizable projects in the U.S. or elsewhere. Most of the material presented here has developed in the last few years. Electrical energy storage is quite definitely an industry in the process of being born.

It is important to note that all opinions expressed in this paper are purely the author's and are not to be imputed to the U.S. Department of Energy.

INTRODUCTION

As a concept, energy storage has immediate appeal. It transfers energy through time, from generation to consumption, just as energy transmission transfers energy from one place to another. Indeed, in the early days of electricity the Leyden jar and Volta's 'columnar apparatus' were very much cutting edge technology. However, after the invention of the dynamo, AC generation, and the spectacular spread of continental

CP1044, *Physics of Sustainable Energy, Using Energy Efficiently and Producing It Renewably*
edited by D. Hafemeister, B. Levi, M. Levine, and P. Schwartz
2008 American Institute of Physics 978-0-7354-0572-1/08/$23.00

electric grids, storage of electricity became relatively unimportant. With one exception: the lead acid battery became an essential ingredient of the automobile. Due to large production volume this battery has become relatively inexpensive and fairly reliable. More recently energy storage has also become a critical component of electronic equipment such as portable computers, mobile phones, and i-pods. In this market, where cost is less crucial, the need for greater energy density, zero maintenance, and long cycle life has allowed advanced batteries such as lithium-ion and nickel-metal-hydride batteries to find wide application. Hybrid cars have become the next major application field for storage. Both footprint and price are important considerations here. The all-electric navy being planned by the military is looming on the horizon.

It is important to remember that there is a vast range for the power involved in the various applications of energy storage: From a few Watt for personal electronics, up to a hundred kW for hybrids, tens of MW for ships, and hundreds of MW for utility applications. It is, therefore, no surprise that different applications call for different storage technologies. In particular, among the requirements for utility storage, from half a MW to hundreds of MW, are price, price, and price! Of course reliability, lifetime, efficiency, environmental acceptability, and safety are also important.

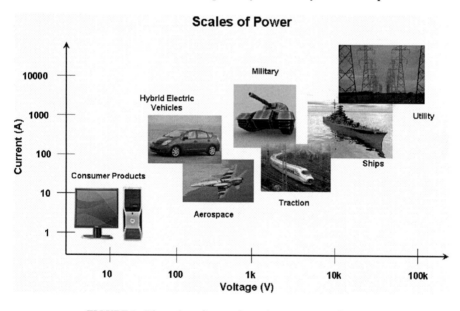

FIGURE 1. Dimension of power for various storage applications.

DRIVERS FOR THE MODERN GRID

Before we embark on a more detailed exploration of the role of storage in the utility industry, let us step back for a moment and consider some of the drivers facing the modern grid.

- **Digitization of Society.** From traffic lights to banking transactions, from telecommunication to high tech manufacturing – we have rapidly become a digital society. Unfortunately, digital systems are very sensitive to outages and cannot tolerate to be without power for even a cycle or two. Even lowered voltage by 20% or so proves fatal. By contrast, analog systems are forgiving and able to ride through such glitches. The resultant imperative for the grid is increased power quality.
- **Ecological Concern.** As a society we have finally realized that the earth is not infinite. If we exceed our boundaries we will indeed fall off! In particular, global warming has at last been accepted as real even by the denser strata of bureaucracy. And, although the U.S. has not signed the Kyoto protocol, stringent renewable mandates are being enacted by the States. But to realize its full potential and make it compatible with the grid renewable energy needs to become dispatchable.
- **Growth in Energy Consumption.** Energy is being devoured at an ever accelerated pace. Not only is the population growing, but our standard of living is continually increasing, bringing along higher per capita consumption of energy. Globally, the situation is even worse with populations growing faster due to improved health standards, and more and more regions of the world wishing to become accustomed to a lifestyle common in the West. In the U.S. the grid faces massive congestion as load and generation increase, requiring much better asset utilization to function smoothly.

Energy storage can contribute significantly to all these issues facing the modern grid. Storage can provide seamless continuity through brief outages, help in smoothing renewables and making them dispatchable, and provide better asset utilization for generation, transmission, and substations. Of course there is one more issue which has become pervasive in modern society and which I prefer not to discuss here: Security. Suffice it to say that any contribution towards resolving the issues facing the grid will also help with security and might even aid in diminishing the need for such security.

APPLICATIONS AND TECHNOLGIES

Following the scheme of challenges outlined above we find three broad application areas for energy storage. These range from short term "power" applications to long term "energy" applications. Since applications can be consumer / load oriented or utility / grid oriented, these gives us six broad application categories.

Power quality and digital reliability for the customer as well as voltage support and transient control on the grid require short term storage for cycles, seconds, or a few minutes. Intermediate storage applications include enabling load following for distributed resources such as fuel cells, smoothing of renewables, frequency regulations, and providing dispatch for a microgrid. Applications with a time scale of several hours include peak shaving to avoid demand charges for the customer as well as spinning reserve, dispatch of renewables, and mitigation of transmission congestion.

A fair number of technologies are available for energy storage. They are conveniently displayed in a chart with system power rating as the horizontal axis and discharge times as the vertical axis. Both scales are logarithmic.

The three application areas fall into diagonal bands. For power quality applications we find Li-Ion, flywheels, super capacitors, NiMetal-hydride, and super-

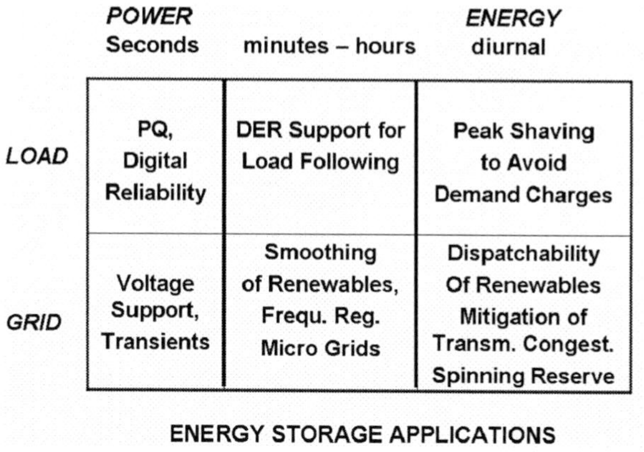

| | POWER | | ENERGY |
	Seconds	minutes – hours	diurnal
LOAD	PQ, Digital Reliability	DER Support for Load Following	Peak Shaving to Avoid Demand Charges
GRID	Voltage Support, Transients	Smoothing of Renewables, Frequ. Reg. Micro Grids	Dispatchability Of Renewables Mitigation of Transm. Congest. Spinning Reserve

ENERGY STORAGE APPLICATIONS

FIGURE 2. Schematic of storage applications.

conducting magnetic energy storage. In the band labeled bridging power there are the customary lead acid batteries, Ni-Cd, and various air batteries. For large energy applications we have the sodium sulfur (NaS) battery, and various flow batteries. Finally, for really large and long term applications we have pumped hydro and compressed air energy storage.

The total amount of electrical energy storage deployed in the world is not as yet overly large. There is some 110,000 MW of pumped hydro in North America, Europe, and the Far East. Compressed air amounts to 477MW. Battery storage deployment is much smaller but growing fast. It is instructive to inquire what these figures mean in terms of the total amount of electricity consumption. In the U.S. roughly 2% of electricity goes through storage. The numbers for Europe and Japan are 10% and 15% respectively. Need we ask which system is most prone to outages?

We shall now consider each of the three application areas separately and provide recent projects as examples. Of course it must be remembered that many applications could fall into either one of two categories. Boundaries are flexible. Moreover, perceptions change over time as well. Yesterday's small scale UPS suddenly morphs into a 20MW application.

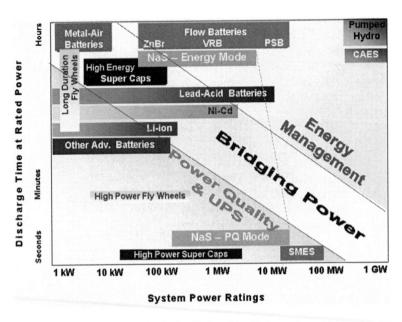

FIGURE 3. Storage technologies displayed by discharge time and power.

RELIABILITY AND POWER QUALITY

Reliability and power quality have become a necessity for modern digital society. No longer is quantity the only issue for electricity distribution but quality is becoming of equal importance. Which poses a problem: As a general principle, any stressed un-buffered non-linear system may be highly susceptible to collapse! The national grid is certainly complex. It is also quite non-linear. And, as we have seen, it is also largely un-buffered by storage.

Naturally this instability of the grid has economic consequences. A recent study by Joe Eto of LBL shows that the cost of outages to the U.S. amounts to some $79 billion per year. Compare this with the total U.S. cost of electricity of some $250 billion a year! One might well say that 33 cents of every electricity dollar is lost to the inefficiency of the grid.

The story gets even more interesting when one looks at the time signature of the outages. About $52 billion (about two third) of the loss is due to momentary interruptions of less than 5 minutes. Only $26 billion (about one third) are due to the spectacular huge outages like the one that knocked out half the East Coast in August of 2003. The reason for this vast economic impact of short term outages lies in the fact that an outage lasting only a few cycles can knock out an entire high tech manufacturing facility. Once a polymer extrusion factory is stopped, gunk starts to congeal in the dies and it may take an entire eight hour shift to repair the damage. An outage of a few cycles may lead to hours of downtime!

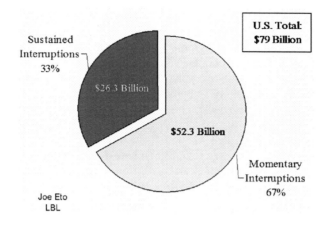

U.S. Total:
$79 Billion

Sustained
Interruptions
33%

$26.3 Billion

$52.3 Billion

Momentary
Interruptions
67%

Joe Eto
LBL

FIGURE 4. U.S. outage costs for sustained and momentary outages.

Number of "nines" is sometimes used as a rough measure of reliability. High tech industry would like to see nine nines of reliability: 0.999 999 999 parts out of 1.0. This computes to one cycle lost per year. Anything else runs into big money. Unfortunately the utilities can only provide about three nines – anything higher would substantially increase the cost of electricity to the general consumer. As a result, power quality control should be installed on the customer side of the meter. Energy storage is able to provide seamless continuity of power supply even for big customers. A system developed with DOE funding (Purewave by S&C), which resulted in an R&D 100 award, can protect facilities like microchip plants with 10MW or more for 30 seconds. After this, but no faster than 15 seconds, a genset can provide the factory with power.

FIGURE 5. A 10 MW Installation at a microchip plant.

On a larger scale, one of the world's biggest battery safeguards the transmission line from Anchorage to Fairbanks. Alaska has a single major transmission line, a single railroad, and a single major highway – all along the same corridor. And since most of the generation is in Anchorage, the line gets quite weak when it gets to Fairbanks. There were very numerous outages until Golden Valley Coop put in a 27 MW NiCd battery to give voltage support, prevent outages, and provide reactive power locally. During 2006 the battery responded to 82 events, preventing 311,000 member outages.

The take home message is that voltage fluctuations are expensive and require mitigation. Short term energy storage is an ideal tool to do this even on the large utility scale. Furthermore, outage costs represent a reduction in energy efficiency. Reducing them means that we get more productivity for a given amount of electricity generation.

FIGURE 6. A 40 MW NiCd battery in Fairbanks, Alaska.

FREQUENCY REGULATION, RENEWABLE SMOOTHING, AND MICROGRIDS

As a first example we will consider frequency regulation. In any given area, controlled by an independent system operator (ISO), load and generation are always slightly out of synch. For large changes the ISO brings new generators on the grid. But for small changes, frequency adjustment balances the grid. Participating generators will then slightly reduce or increase their output to bring the frequency back in line. This is an inefficient procedure because generators have to leave a reserve margin to participate in this ancillary service. It takes several minutes to adjust their output so that the response always lags behind and may actually be in the wrong direction. Changing output continually also has a negative effect on CO_2 emissions.

Because they respond instantaneously, flywheels can be twice as effective as regulation by fossil generation. 100MW of flywheels could eliminate 90% of all the frequency variation in California. Business models show that frequency regulation by flywheel storage is economically viable. A 20MW facility is now in the planning stage by Beacon Power. Best of all, flywheel regulation can result in an 80% reduction in CO_2 emission over present methods.

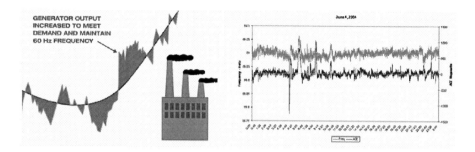

FIGURE 7. Schematic of frequency regulation (a) and actual frequency response (b).

FIGURE 8. Flywheel system assembly (a) and containerized 100kW system (b).

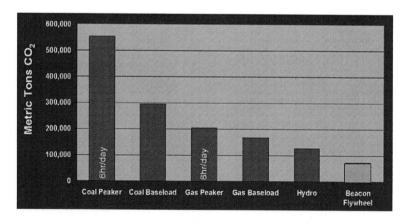

FIGURE 9. Comparisons of CO_2 emission for various modes of frequency regulation.

Substantial amounts of renewables are planned for most states in the U.S. This will bring about increased need for regulation. In California for example, with 20%

383

renewables planned for 2012, the need for regulation will increase by some 45%. It is unfortunately not true that small wind fluctuations will simply even out in the aggregate. Random variations can reinforce each other just as easily as they can cancel each other. If this increase has to be covered by fossil regulation resources it would in effect add a carbon footprint to renewables. With regulation by storage the renewable resources can stay green.

A wind smoothing project currently in progress is jointly supported by Bonneville Power Authority and DOE. Customers have registered numerous complaints due to the egregious power quality from a 48MW wind farm. The plan is to provide local regulation with a 10MW STATCOM supported by fast supercapacitors.

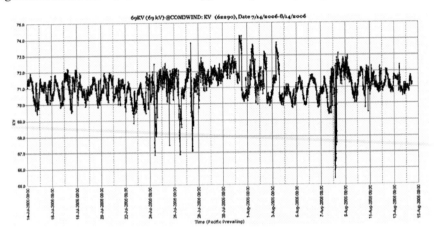

FIGURE 10. Measured voltage fluctuations during one month near the Condon wind farm.

FIGURE 11. Wind, hydro, gas turbine 1250kW microgrid in Palmdale, CA.

384

Microgrids are a concept that is attracting increasing interest among planners of residential communities, shopping centers, or office parks. The idea is to create a local network, incorporating distributed and renewable resources, and allowing the use of waste heat (CHP). The microgrid may be connected to the grid either for extra input or for export of green energy. During outages it may separate from the grid and function independently. Energy storage provides the ideal platform to serve as a controller and dispatcher for this shifting mosaic of loads and generation resources. As an example, CEC together with DOE are supporting the construction of a 1250kW microgrid at the Palmdale, CA, water treatment plant. The facility has multiple generation resources comprised of wind, hydro power, natural gas and diesel generators. 450kW of storage are provided by ultracapacitors for smoothing and backup.

PEAK SHAVING, ENERGY MANAGEMENT, UPGRADE DEFERRAL

It is well known that asset utilization for generation, transmission and distribution is not particularly good. Assets like substations or transmission lines have to be sized for peak demand with ample to spare for that extra hot day. A typical load duration curve shows that peak load is about 25% above the 95% load level. One quarter of the facility is devoted to maintain service during a 5% peak period. The goal of energy storage is to supply this peak load from energy stored during periods of least demand. A similar agenda is also pursued by demand management techniques.

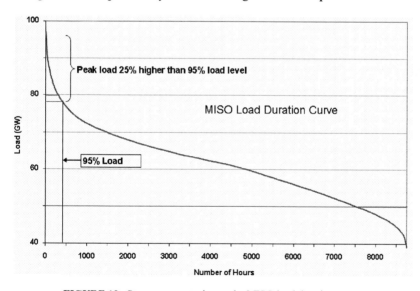

FIGURE 12. Storage opportunity on the MISO load duration curve.

Many utilities are already using energy storage in the form of pumped hydro. Hundreds of MW of off peak energy are used to pump water into an upper reservoir, to be released for electricity production during peak periods. This works very nicely, but unfortunately most suitable sites have already been utilized and the rest are unsuitable for environmental reasons. Moreover, historically pumped hydro had been closely linked with nuclear energy making it unacceptable to a certain part of the public. It is therefore worthwhile to explore storage by means of other technologies. Particularly promising candidates are flow batteries and the sodium sulfur battery.

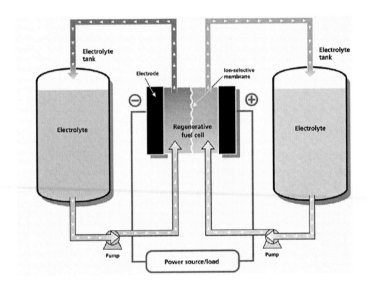

FIGURE 13. Schematic of flow battery.

Flow batteries basically consist of two electrolyte tanks, a pump, and an electrochemical cell. Energy is stored in the electrolytes, power is generated by the conversion cell which consists of two electrodes and a membrane between. For more energy make the tanks larger, for more power make the cell stack bigger. The main current technologies are vanadium redox batteries and ZnBr systems. Flow batteries tend to be less expensive than other batteries, but have a larger footprint.

FIGURE 14. Assembly of vanadium redox flow battery, tanks (a) and conversion cell stack (b).

As an example, a 250kW / 8hrs vanadium redox battery (by VRB Power Systems) has been installed in Castle Rock Utah where it can provide peak power for four hours. The town is at the end of a 209 mile distribution feeder and peak consumption has outgrown the capacity of the line. Upgrading the line would have involved considerable expense and taken some three years to complete. Furthermore, the town did not wish to use polluting diesel generators. Installation of storage proved to be most economical solution. The system has run un-manned for 5 years.

Sodium-sulfur (NAS) batteries were invented in the U.S., but perfected over the course of many years by NGK in Japan. It involves molten sulfur and sodium electrodes separated by a ceramic membrane. The technology has been extensively tested for safety and over 30 NAS systems are deployed in Japan. The first application in the U.S. was sponsored by American Electric Power (AEP) and DOE at a substation in West Virginia. The substation had been reaching the limit of its capacity and an upgrade was needed within a short time frame to handle the overload during peak periods. AEP decided on the alternate solution of installing energy storage. Due to its long operating experience the NaS battery was selected and a 1.2MW / 6 hour system was installed in June of 2006. Energy is stored at night when electricity is less expensive and released over a 6 hour period during peak load times. The system has performed flawlessly. Installation of storage has deferred a substation upgrade by 5 to 6 years. The main economic incentive comes from amortization of the deferral. There are also benefits from arbitrage. AEP has already committed itself to installing three more storage systems and intends to have 1000 MW on line by 2020

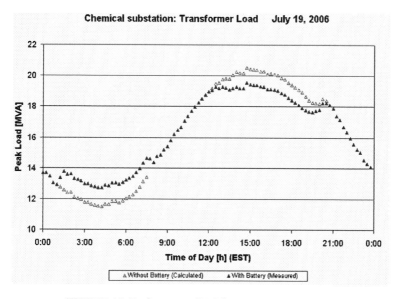

FIGURE 15. Performance of NaS battery at WV substation.

A similar system can also benefit the costumer directly. A 1MW / 6 hour NaS system is being installed at a Long Island natural gas refueling station serving 220 busses. Night time electricity will be stored to drive a 1,800 HP compressor system during the day. This relieves the local utility during peak load periods, simplifies operations at the plant, and decreases demand charges. The project is supported by the New York State Energy Research and Development Authority and DOE with cost shares from EPRI and a number of other utilities.

FIGURE 16. 1 MW/6hr NaS battery system and compressors at Long Island, NY bus depot.

RENEWABLE RESOURCES DISPATCH

The reality of global warming and the importance of reducing carbon footprints locally and globally appear to have become generally accepted. However, even to just maintain the 1990 level of greenhouse gas emissions, heroic measures are indicated. Short of relying on nuclear power, which is not a popular choice at the time, massive amounts of renewable resources are called for. Wind represents the primary renewable resource, but solar, and wave power may become important as well.

The problem is that these renewable resources are variable – even more so than the load they are intended to serve. There needs to be a buffer between them to make the system work smoothly. For small amounts of renewables the custom is to use the grid itself as a storage buffer. But there is a problem with this: every little excursion has to be accounted for by fossil fuel plants. They are needed for standby and for modulating the grid response. The result is that renewables lose a certain amount of their green quality and acquire a concomitant carbon footprint. Moreover, fossil generation plants operate less efficiently and produce higher greenhouse gas emissions when forced to ramp up and down than when operating at their design set point.

While the U.S. has not subscribed to the Kyoto protocol, individual states have developed ambitious mandates of their own for renewable resources. Representative numbers are: CA, 20% by 2010, NY, 24% by 2013, and TX, 5,800MW by 2015. It will be difficult to reach such levels of penetration without serious problems to the stability of the grid and continuity of service. What is needed is smoothing, protection against ramping, and eventually full dispatchability. Energy storage can contribute to these goals in an effective way. We have already seen that smoothing by flywheels (or super capacitors, or fast advanced batteries) is entirely feasible with a much smaller footprint than regulation by fossil fuel generators. Ramping is becoming a serious problem. It is entirely possible for the wind to die down over a large region and very rapidly. As an example, west Texas lost some 1,200MW of wind in about 10 minutes. Only massive load shedding could prevent a collapse. Extensive renewables deployment (say 30%) could lead to large numbers of standby fossil plants, considerable construction of new transmission capacity with bad asset utilization, and an unstable grid. Energy storage offers a much more elegant solution. Technologies are available and applications are beginning to move from MW, to tens, and even hundreds of MW.

A small example at first: a diesel, wind, battery hybrid generation plant was installed on King Island, Tasmania. There is 2,450 kW of wind power, backed up by 1,500 kW of diesel generators. This could cover the local load by continually varying the diesel to compensate the variations in wind power and load. An additional amount of 200 kW of storage for 4 hours provided a much better solution. A vanadium redox battery was used. The diesel is run flat during day time and turned off at night. This is a much more fuel efficient operation. Wind power is stored at night instead of spilling it when the load is minimal. During the day storage will make up the discrepancy between load and available wind plus diesel. Experience here, as well as in Alaska and Peru, shows that this mode of operation can lead to a 20% decrease in fuel consumption. It also decreases pollution and increases the life time of the generator. One might suggest that such a hybrid installation could be an excellent model for power generation for the one third of the world's population that is off grid and may well remain so.

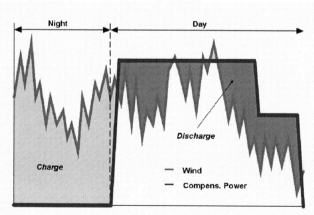

FIGURE 17. 34MW/7hr NaS battery system in Japan to dispatch a 51MW wind farm.

A larger storage venture will be completed in Japan during 2008. Japan has realized that its plans for 3,000 MW of wind by 2010 cannot be realized without appropriate storage buffering. The amount of dispatchable hydro, which currently compensates for fluctuations, will not be enough for such a large amount of renewables. To encourage the deployment of storage the Japanese government provides one third the construction cost of new storage. A substantial incentive! In addition, the day/night difference in electricity cost is relatively high allowing profitable arbitrage. The most recent storage installation is being built in Rokkasho in northern Japan to back up a 51 MW Wind farm with 34 MW / 7hr of NaS battery storage. The plan is to store all available wind during the night and then discharge flat during the day, thus making the wind truly dispatchable.

For yet larger amounts of storage we may want to turn to compressed air energy storage (CAES). In this technology air is compressed off peak and stored in salt domes, man made caverns, or deep aquifers. When extra energy is required during peak periods, air is released and fed directly into combustion turbines, eliminating the need for a compressor. While this does not eliminate the need for fuel, it increases the efficiency of the turbines substantially. There are two CAES units in existence – one in Germany (290MW) and one in Alabama (110MW). Both facilities use salt domes formed by solution mining. A 200 MW CAES facility is being planned in Iowa using deep aquifers. The unit could be used in a straightforward arbitrage mode. However, in view of 2000MW of wind at play in Iowa this CAES plant could play an important role as a renewables buffer. Wind energy, which otherwise might be spilled, can be used to compress air during the night. Besides producing electricity during peak periods, the plant can also be ready to take up the slack whenever a wind ramp occurs. This would eliminate the need for fossil fuel standby peaker plants.

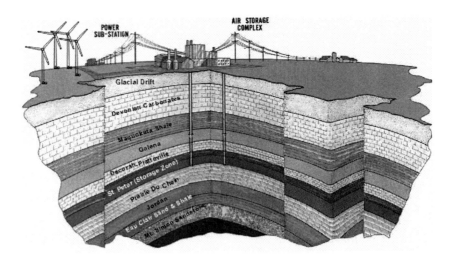

FIGURE 18. Schematic of proposed aquifer CAES plant in Iowa.

SUMMARY AND OUTLOOK

Energy storage offers a diverse portfolio of technologies for a wide spectrum of applications. It allows us to optimize operation of the grid to make the most of our increasingly precious resources. Energy storage can

- Provide power quality and digital reliability
- Provide voltage and frequency regulation
- Smooth renewables
- Allow better asset utilization for generation and transmission
- Provide relief to customers and utilities during peak load periods
- Provide spinning reserve and energy management to make renewables dispatchable

Advances in energy storage are an international concern. Besides the U.S., the European Union, Canada, Australia, and Japan have sizable storage efforts. Storage research centers are being created in the Basque Republic and in Saudi Arabia.

The U.S. Congress seriously considered energy storage in its Energy Bill of 2007. This is the first official recognition of the importance of storage by Congress. HR6-sec. 641 enjoins the Secretary of Energy to carry out "a research, development, and demonstration program ... in energy storage for electric drive vehicles, stationary applications, and electricity transmission and distribution". The bill authorizes some $260 million annually for basic and applied storage research, and for energy storage system demonstrations in applications treated in this paper. Of course this is only an authorization bill and carries no appropriated funding, but it will provide a basis for further Federal involvement.

Other emerging technologies have the potential of enhancing or augmenting storage. Smart grid concepts, for example, could link storage to demand response and enable aggregation of distributed storage. Plug-in hybrids and, eventually, electric vehicles add a whole new dimension by linking transportation to energy management.

Utilities are increasingly becoming involved in energy storage and states like California and New York continue to work with DOE in funding new projects. The investment community is becoming interested in providing venture capital for companies developing new technologies and funding ambitious large scale projects. The industry appears poised to move from single megawatt scale application to utility grade projects in the hundreds of megawatts. The goal is to make energy storage ubiquitous!

But a warning is in order: Energy Storage is a disruptive technology whose adoption will induce a paradigm shift in the entire utility industry. The electricity business will never be the same again.

ACKNOWLEDGEMENTS

The energy storage community consisting of manufacturers, developers, utility representatives, decision makers, as well as researchers and consultants is a dynamic but supportive group. Thanks are extended to all of them. Particular thanks to the storage team at Sandia, and to Pramod Kulkarni and Joe Sayer, my colleagues at the California Energy Commission and the New York State Energy Research and Development Authority. It is a pleasure working with all of you to make energy storage on the grid a reality.

WHERE TO LEARN MORE

Detailed information on energy storage technologies may be found in the voluminous EPRI/DOE Handbook of Energy Storage for Transmission and Distribution Applications. The Handbook together with a Wind Supplement is available free on the internet:

    http://my.epri.com/portal/server.pt?Abstract_id=000000000001001834
    http://my.epri.com/portal/server.pt?Abstract_id=000000000001008703

A good primer on the subject has been written by Richard Baxter, *Energy Storage, A Nontechnical Guide,* Tulsa, OK, Pennwell Corp (2006).

For proceedings of the Electricity Storage Association (ESA) and the Electric Energy Storage – Applications and Technology conference see their respective web sites:

    http://www.electricitystorage.org/
    http://www.sandia.gov/

APPENDICES

Energy and Environment Chronology

David Hafemeister

Physics Department
Cal Poly University
San Luis Obispo, CA 93407

1859
—Edwin Drake drills 21 meters for 500 barrels of oil at Titusville, Pennsylvania, to begin the *petroleum era*.
—The rechargeable lead–acid battery is developed by Gaston Plante.

1870
—John D. Rockefeller founds Standard Oil.

1879
—Thomas Edison and J. Swan, independently, invent the incandescent lightbulb.

1882
—First US coal-fired power plant lights up Manhattan by Thomas Edison using DC voltage.

1908
—Model–T Ford runs on gas or ethanol, getting 21 miles per gallon.

1911
—Standard Oil Company monopoly is broken by the Supreme Court.

1930s
—General Motors, Firestone and Standard Oil of California buy US electric streetcar systems across and replace them with buses.

1942
—December 2: Enrico Fermi's reactor goes critical at Stagg Field, University of Chicago, Illinois to begin the *nuclear era*.

1945
—August 15: Office of Price Administration (OPA) lifts gasoline rationing.

1946
—May 6: Division of Oil and Gas established in Department of Interior.
—May 21: US President Harry Truman orders US Government to take possession of coal mines during a strike.
—June 18: National Petroleum Council established.

1947
—January 1: Atomic Energy Commission begins operation.

—March 25: Coal-mine disaster kills 11 In Centralia, Illinois.
—June 16: Federal Power Commission authority extended to all natural gas producers.

1952
—December 5: Severe air pollution (0.7 ppm SO_x and particulates) kills 4000 in London in 4 days.

1953
—August 7: Congress gives US government jurisdiction of ocean floors beyond 3-mile boundary.
—December 8: US President Dwight Eisenhower delivers "Atoms for Peace" speech before the United Nations.

1954
—August 30: Atomic Energy Act of 1954 encourages peaceful use of nuclear energy.
—Bell Labs develops the silicon photovoltaic cell.

1956
—Federal Highway system begins at a cost of $129 billion.

1957
—King Hubbert correctly predicts US petroleum production peak between 1966–71, which happened in 1970 when lower forty-eight produced 9.1 Mbbl/day. Hubbert uses a finite resource in differential equations, but does not use economics.

1959
—March 10: Eisenhower limits oil imports to stimulate domestic production and refining capacity.

1962
—October 11: Congress authorizes the president to impose mandatory oil import quotas.

1963
—December 17: Clean Air Act provides assistance to states for air pollution research.

1965
—October 2: Water Quality Act establishes the Water Control Administration.
—October 20: Solid Waste Disposal Act provides assistance for study, collection, and disposal of solid wastes.
—November 9: First major power blackout covers northeast US.

1967
—November 21: Clean Air Act gives authority to Secretary of Health Education and Welfare to set auto emission standards.

1969
—January 1: National Environmental Policy Act (NEPA) establishes framework for Environmental Impact Statements and the Council of Environmental Quality (CEQ).
—January–February: Major oil spill from offshore drilling near Santa Barbara, CA.
—December 30: Oil depletion allowance reduced from 27.5% to 22%.

1970
—March 5: President Richard Nixon issues executive order requiring federal agencies to evaluate their activities under the National Environmental Policy Act.
—April 22: First "Earth Day" celebration.
—July 9: Nixon requests Congress to create Environmental Protection Agency (EPA).
—October 23: Merchant Marine Act Amendment provides subsidies for oil and liquified natural-gas tankers.

—December 24: Geothermal Steam Act authorizes leases for geothermal steam.
—Clean Air Act sets national air quality and auto emission standards.

1971
—July 23: Supreme Court decision on siting of nuclear power plant at Calvert Cliffs, Maryland, requires the Atomic Energy Commission to comply with the National Environmental Policy Act.

1972
—EPA bans DDT.
—Clean Water Act sets pollution standards for water.
—US and Canada agree to clean up the Great Lakes, source of 95% of US fresh water, used by 25 million persons.

1973
—June 29: White House Energy Policy Office created with former Governor John Love of Colorado as first "energy czar."
—October 17, 1973 to March 17, 1974: Organization of Petroleum Exporting Counties (OPEC) embargoes US and the Netherlands because of their support for Israel.
—November 7: Nixon creates Project Independence to end oil imports by 1980.
—November 27: Emergency Petroleum Allocation Act provides authority for oil allocations.
—December 4: Presidential Federal Energy Office created with William Simon as energy czar.
—December 15: Congress mandates daylight savings to save energy.
—December 28: Nixon signs Endangered Species Act.
—EPA begins phasing out leaded gasoline.
—EPA begins limits on factory pollution discharges.

1974
—June 22: Federal Energy Administration (FEA) authorized to order utilities and industry to convert from oil/gas to coal.
—September 3: Congress authorizes funds for geothermal energy and solar heating demonstrations.
—October 5: Congress repeals daylight savings time in winter to save energy.
—October 11: Energy Reorganization Act abolishes the AEC to create the Energy Research and Development Administration (ERDA, later DOE) and the Nuclear Regulatory Commission (NRC).
—December 31: Congress requires ERDA to submit an annual comprehensive energy plan.
—Arthur Rosenfeld, establishes research on buildings and energy at Lawrence Berkeley National Laboratory, to begin the *enhanced end-use efficiency energy era*. Rosenfeld was the last graduate student of Enrico Fermi, who began the *nuclear reactor era* in 1942. The LBL work was originally based on the 1974–5 study by the American Physical Society.

1975
—January 4: Congress establishes the 55-mph speed limit to save energy.
—March 17: Supreme Court rules that the states do not have jurisdiction over the outer continental shelf.
—October 29: ERDA dedicated its first wind power system at Sandusky, Ohio.
—December 22: Energy Policy and Conservation Act (EPCA) establishes prices for US crude oil, the Strategic Petroleum Reserve, emergency energy powers for the president, Corporate Average Fuel Economy (CAFE) standards of 27.5 mpg by 1985, and cost-effective, appliance energy standards.
—Introduction of catalytic converter mufflers for cars.

1976
—April 5: Congress authorizes production from naval petroleum reserves.
—August 14: Energy Conservation and Production Act (ECPA) creates incentives for conservation and renewables, funds weatherization for low income homes, establishes a program for energy standards for new buildings and establishes Solar Energy Research Institute (now the National Renewable Energy Laboratory) to begin the *renewable energy era*.

—Toxic Substance Control Act reduces environmental and human health risks. EPA begins phase out of PCBs.

1977
—April 7: President Jimmy Carter indefinitely defers reprocessing of nuclear fuel and stops construction of Clinch River Breeder Reactor.
—July 5: Solar Energy Research Institute (SERI) opens in Golden, Colorado; renamed in 1991 as the National Renewable Energy Laboratory (NREL).
—October 1: Department of Energy created from ERDA and the FEA. Jimmy Carter declares it is "the moral equivalent of war."

1978
—November 9: National Energy Act establishes weatherization grants for low income families, conservation programs for local governments, energy standards for consumer products, programs to convert utilities to coal, and energy tax credits.
—Buried, leaking containers found at Love Canal, NY; which was completed in 1998.

1979
—March 28: Accident at Three Mile Island nuclear power plant in PA effectively halts purchase of new reactors in the US.
—Spring: Second oil shock from Iran oil curtailments causes gasoline shortages.
—August 17: President Carter begins to lift price controls on domestic crude oil.
—November 3: US embassy in Iran seized by revolutionaries; Carter suspends oil imports from Iran on November 14.

1980
—April 2: Windfall profit tax on crude oil to give assistance for weatherization of homes of low income people.
—June 30: Energy Security Act creates the Synthetic Fuels Corporation and funds renewable energy projects.
—Tax on certain chemicals creates a "superfund" to pay for cleanup when responsible parties fail.

1981
—January 28: President Ronal Reagan completes price decontrol on domestic crude oil.
—National Research Council finds acid rain intensifying in northeast US.
—1981–1985: Reagan and Congress debate funding for conservation programs.

1982
—May 24: Reagan proposes transfer of Department of Energy functions to the Department of Commerce.

1983
—Clean-up of Chesapeake Bay begins.
—EPA encourages homeowners to test for radon.

1984
—December 3: Chemical disaster at Union Carbide plant in Bhopal, India, kills 3000 and partially disables 2700.

1985
—February 4: Department of the Interior reduces estimates for offshore oil (27 to 12 Bbbl) and gas (163 to 91 tcf).
—June 27: EPA modifies mileage test, lowering CAFE's 27.5 mpg by 2 mpg.
—June 28: EPA curbs tall smokestacks to avoid distant pollution.

—July 16: Appellate Court confirms EPCA's appliance standards by voiding DoE's "no-standard" standard. A key issue was the magnitude of the discount rate for future benefits.
—Antarctic Ozone Hole for stratospheric ozone discovered.

1986
—April 26: Accident at Soviet reactor in Chernobyl, Ukraine.
—August 21: Large release of carbon dioxide from the depths of Lake Nyos, Cameron, kill 1800.

1987
—Montreal Protocol for ozone protection bans chlorofluorocarbons.
—December 27: Congress approves Yucca Mountain as the only high-level nuclear waste repository under development.

1988
—Congress bans ocean dumping of sewage sludge and industrial waste.

1989
—March 24: Exxon Valdez spill of 0.3 million barrels of oil in Prince William Sound, Alaska.
—March 23: Stanley Pons and Martin Fleishman announce discovery of "cold fusion." Physicists do not believe them.

1991
—January-February: Iraq sets 700 oil fires in Kuwait in wake of Gulf War.
—Exxon pays $1 billion for Exxon Valdez spill.

1992
—EPA bans ocean dumping of sewage sludge.
—EPA launches Energy Star Program to identify energy efficient products.

1993
—EPA's Common Sense Initiative shifts OSHA regulations from pollution-based to industry-based.
—EPA research finds secondhand indoor cigarette smoke causes 3000 lung-cancer deaths per year to nonsmokers.

1994
—EPA launches Brownfields Program to restore abandoned city sites.

1995
—70% of US metropolitan areas that had unhealthy air in 1990 meet air quality standards in 1995.

1997
—EPA restricts particulate matter air emissions down to 2.5 μ diameter.
—Kyoto Protocol to limit greenhouse gases (GHG) is agreed in principle by most nations, including the US, but the details for trading allotments of GHG are not certain.

1999
—Minivans and sport utility vehicles (SUVs) will have the same emission standards as cars.
—EPA requires cars, SUVs, minivans and trucks to have the same standards by 2004, reducing SUV pollution by 77–95%.
—Honda sells the first commercial hybrid auto.

2000
—Most utilities buy combined-cycle natural gas turbines at 55–60% efficiency, a major supply side break through, establishing the *natural gas era*.
—California hit with rolling blackouts, stimulated by Enron maneuvers.

2001

—January: California's partial deregulation contributes to rolling blackouts caused by many factors.

—July 23: The Kyoto-Bonn Protocol on limiting GHG moved towards implementation, supported by 178 nations and the EU. Without support from the US, a former advocate turned critic, the future of the Kyoto process is in doubt. The US stated it would use voluntary caps on GHG emissions.

—Sales of minivan and SUVs equal sales of cars.

2002

—February: President Bush proposed a *voluntary* 18% carbon reduction by 2012 in terms of *greenhouse gas intensity*, which is the ratio of national total energy use of carbon fuels to GDP dollar.

—February 14: Secretary of Energy Spencer Abraham recommends approval of the Yucca Mountain, NV, geological repository for 77,000 tonnes of nuclear spent fuel, to begin in 2010.

—November: Bush Administration relaxes pollution standards on existing coal-fired power plants and gives managers of national forests more discretion to approve logging and commercial activities.

2003

—April 1: CAFE standards for light-truck/SUVs rise from 20.7 mpg to 21.0 in 2205, 21.6 in 2006 and 22.2 mpg in 2007.

—June 7: The 18-GW$_e$ Three Gorges Dam closes, to be completed by 2009, displacing 1.1 million people.

—August 11: Auto manufacturers no longer contest California's 2005–20 phase-in of low and zero emission vehicles.

—August 11: Gov. Michael Leavitt nominated to replace Gov. Christine Todd Whitman as EPA administrator.

—August 14: Electrical grid failure of 62 GW$_e$ darkens 8 states and Canada (previously on 11-9-65 and 7-13-77).

—August 27: President Bush gives emission exemptions from the Clean Air Act for rehabilitated power plants.

2004

—Natural gas prices double and supplies from Canada become more uncertain. Coal plant orders rise from 2 to 100 and utilities request site approval for 3 nuclear plants.

—July 9: US Court of Appeals in DC rules against the 10,000–year limit on radiation safety at Yucca Mountain. The court concluded that EPA must either issue a revised standard that is "consistent with" the NAS peak-dose standard "or return to Congress and seek legislative authority to deviate from the NAS report."

—September 24: The California Air Resources Board required automobiles to lower carbon emissions by 30%, to be phased in over 2009 and 2016. The Board claimed the extra cost would be $1000 per vehicle, but it would save $2500 in fuel. New York and other states claim they will follow suit.

—October 26: General Motors downsizes the Hummer from 6400 pounds (12 mpg) to 4700 pounds (16 city/20 highway).

—Kyoto climate change treaty goes into effect without US ratification.

2005

—January 19: A billion dollar LNG explosion at Skikda, Algeria clouds plans for 35 LNG projects in the US.

—February 16: The Kyoto Protocol enters into force after Russian ratification. The 120 nations that ratified emitted 61% of greenhouse gases by signatories, over the 55% threshold for ratification. The US, which emits 37%, did not ratify, while India and China did not sign. The carbon-trading allotments initially were selling for $10/tonne.

—August 29–30: Katrina, a category 3 hurricane, floods 80% of New Orleans, killing 1836 and costing $86 billion (2007$), the US's most expensive natural disaster.

2006

—Al Gore's documentary movie on the effects of global climate change, *An Unfortunate Truth*, heightens concerns on carbon dioxide emissions from burning fossil fuels. Greenland glaciers and the North Polar Cap melt faster than expected.

2007
—CAFE auto standards raised from 27.7 mpg to 35 mpg by 2020. California and other states lose in court to more the date forward on the grounds of cleaner air for their cities.

2008
—Gasoline hits $4.50 a gallon in the US, and twice that in many countries. SUV sales plummet.
—General Motors plans to introduce the Chevy Volt by about 2011 with an all–electric range of 40 miles with lithium ion batteries and an extended range from a 3–cycle engine. Toyota states it will follow, perhaps with nickel metal hydride batteries. EPRI points out that the un-used electricity at night can be put to good use on these proposals.
—Utilities propose to build some 30 nuclear power plants.
—China plans to build over 50 coal-fired power plants per year.
—British Petroleum gives Lawrence Berkeley Laboratory $500 million to develop cellulosic fuels from switch grass. Corn food price rises as corn–ethanol for autos increases.
—Polymer photovoltaics hit 5% efficiency, needing further research to enhance robustness.
—Refrigerators use less than 25% of former energy use. Movement towards green buildings expands, with the hope to have carbon–neutral buildings (with some local renewable energy).

Energy Outlook, 1980–2030

Energy Information Administration

Department of Energy
Washington, DC 20585

International Energy Outlook 2007

U.S. Annual Energy Outlook 2007

International Energy Outlook 2007

Figure 4. World Marketed Energy Use by Fuel Type, 1980-2030

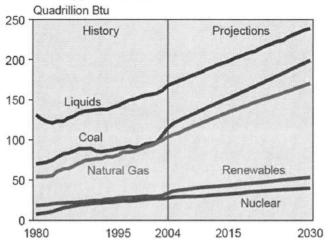

Sources: **History:** Energy Information Administration (EIA), *International Energy Annual 2004* (May-July 2006), web site www.eia.doe.gov/iea. **Projections:** EIA, System for the Analysis of Global Energy Markets (2007).

Figure 5. World Liquids Production, 2004-2030

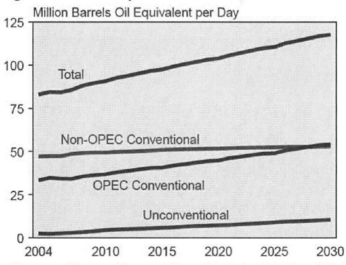

Sources: **History:** Energy Information Administration (EIA), *International Energy Annual 2004* (May-July 2006), web site www.eia.doe.gov/iea. **Projections:** EIA, System for the Analysis of Global Energy Markets (2007).

405

Figure 16. World Marketed Energy Consumption in Three Economic Growth Cases, 1980-2030

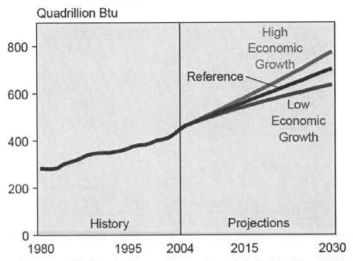

Sources: **History:** Energy Information Administration (EIA), *International Energy Annual 2004* (May-July 2006), web site www.eia.doe.gov/iea. **Projections:** EIA, System for the Analysis of Global Energy Markets (2007).

Figure 17. World Oil Prices in Three World Oil Price Cases, 1980-2030

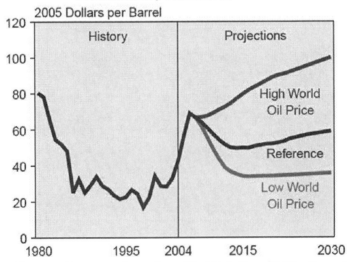

Source: Energy Information Administration (EIA), *Annual Energy Outlook 2007*, DOE/EIA-0383(2007) (Washington, DC, February 2007), web site www.eia.doe.gov/oiaf/aeo.

Figure 24. Energy Intensity by Region, 1980-2030

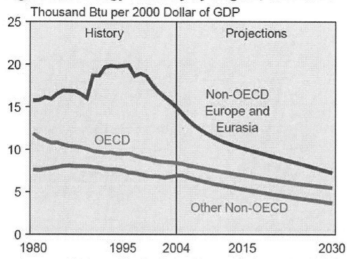

Thousand Btu per 2000 Dollar of GDP

Sources: **History:** Derived from Energy Information Administration (EIA), *International Energy Annual 2004* (May-July 2006), web site www.eia.doe.gov/iea. **Projections:** EIA, System for the Analysis of Global Energy Markets (2007).

Figure 36. OPEC and Non-OPEC Conventional and Unconventional Liquids Production, 1980-2030

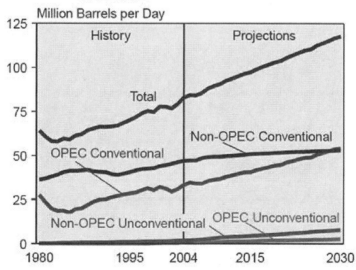

Million Barrels per Day

Sources: **1980-2004:** Energy Information Administration (EIA), *Short-Term Energy Outlook* (October 2006), and *International Energy Annual 2004* (May-July 2006), web site www.eia.doe.gov/iea.. **Projections:** EIA, System for the Analysis of Global Energy Markets (2007).

Figure 37. Cumulative World Production of Crude Oil and Lease Condensates in the Reference Case, 1980-2030

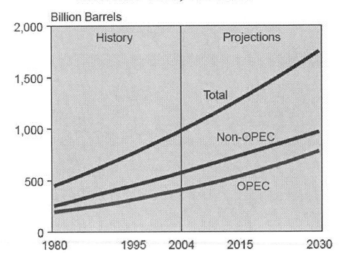

Sources: **1980-2004:** Energy Information Administration (EIA), *Short-Term Energy Outlook* (October 2006), and *International Energy Annual 2004* (May-July 2006), web site www.eia.doe.gov/iea. **Projections:** EIA, System for the Analysis of Global Energy Markets (2007).

Figure 38. World Crude Oil Reserves, 1980-2007

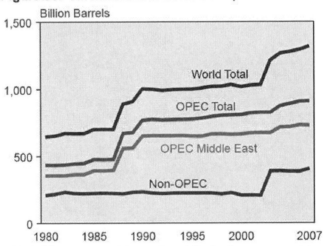

Note: Reserves include crude oil (including lease condensates) and natural gas plant liquids.

Sources: **1980-1993:** "Worldwide Oil and Gas at a Glance," *International Petroleum Encyclopedia* (Tulsa, OK: PennWell Publishing, various issues). **1994-2007:** *Oil & Gas Journal* (various issues).

Figure 42. World Natural Gas Reserves by Region, 1980-2007

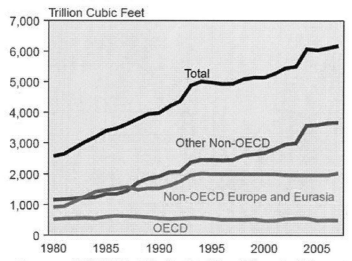

Sources: **1980-1993:** "Worldwide Oil and Gas at a Glance," *International Petroleum Encyclopedia* (Tulsa, OK: PennWell Publishing, various issues). **1994-2007:** *Oil & Gas Journal* (various issues).

Figure 46. U.S. Net Imports of Natural Gas by Source, 1990-2030

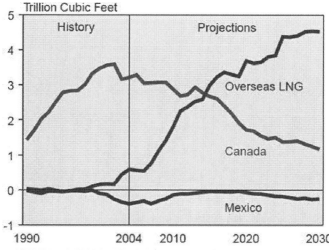

Sources: **History:** Energy Information Administration (EIA), *Annual Energy Review 2005*, DOE/EIA-0384(2005) (Washington, DC, August 2006), web site www.eia.doe.gov/emeu/aer. **Projections:** EIA, *Annual Energy Outlook 2007*, DOE/EIA-0383(2007) (Washington, DC, February 2007), web site www.eia.doe.gov/oiaf/aeo.

Figure 54. World Coal Consumption by Region, 1980-2030

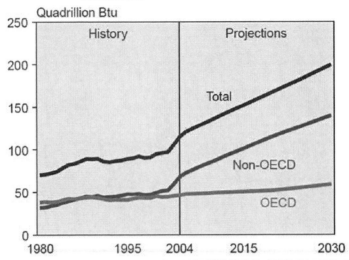

Sources: **History:** Energy Information Administration (EIA), *International Energy Annual 2004* (May-July 2006), web site www.eia.doe.gov/iea. **Projections:** EIA, System for the Analysis of Global Energy Markets (2007).

Figure 61. World Electric Power Generation by Region, 1980-2030

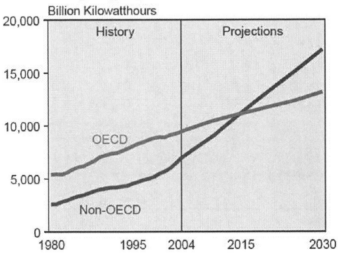

Sources: **History:** Energy Information Administration (EIA), *International Energy Annual 2004* (May-July 2006), web site www.eia.doe.gov/iea. **Projections:** EIA, System for the Analysis of Global Energy Markets (2007).

Figure 78. World Energy-Related Carbon Dioxide Emissions by Fuel Type, 1990-2030

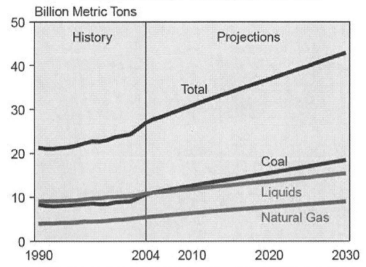

Sources: **History:** Energy Information Administration (EIA), *International Energy Annual 2004* (May-July 2006), web site www.eia.doe.gov/iea. **Projections:** EIA, System for the Analysis of Global Energy Markets (2007).

Figure 83. World Carbon Dioxide Emissions from Coal Combustion by Region, 1990-2030

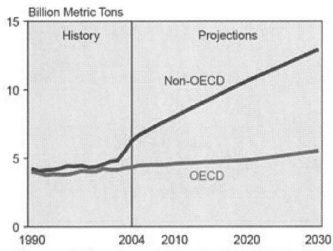

Sources: **History:** Energy Information Administration (EIA), *International Energy Annual 2004* (May-July 2006), web site www.eia.doe.gov/iea. **Projections:** EIA, System for the Analysis of Global Energy Markets (2007).

411

Figure 84. World Carbon Dioxide Emissions per Capita by Region, 1990-2030

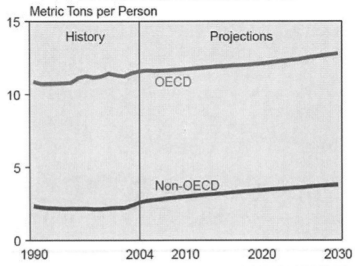

Sources: **History:** Energy Information Administration (EIA), *International Energy Annual 2004* (May-July 2006), web site www.eia.doe.gov/iea. **Projections:** EIA, System for the Analysis of Global Energy Markets (2007).

Figure 85. Carbon Dioxide Emissions and Gross Domestic Product per Capita by Region, 2004

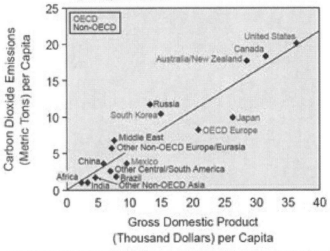

Source: Derived from Energy Information Administration, *International Energy Annual 2004* (May-July 2006), web site www.eia.doe.gov/iea.

412

Figure 1. Energy prices, 1980-2030 (2005 dollars per million Btu)

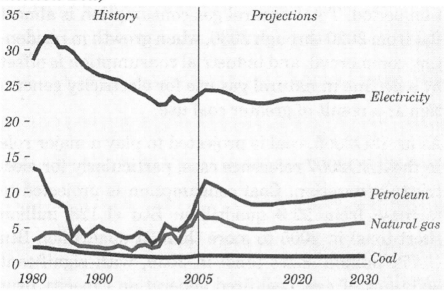

Figure 2. Delivered energy consumption by sector, 1980-2030 (quadrillion Btu)

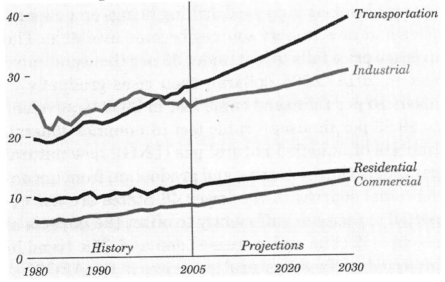

Figure 3. Energy consumption by fuel, 1980-2030 (quadrillion Btu)

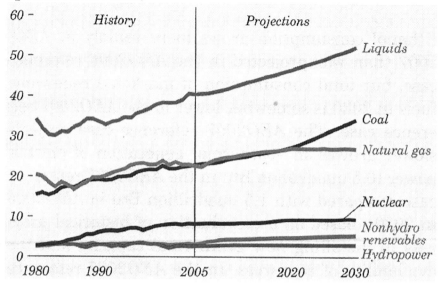

Figure 4. Energy use per capita and per dollar of gross domestic product, 1980-2030 (index, 1980 = 1)

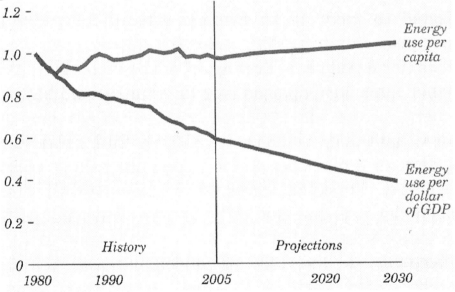

Figure 5. Electricity generation by fuel, 1980-2030 (billion kilowatthours)

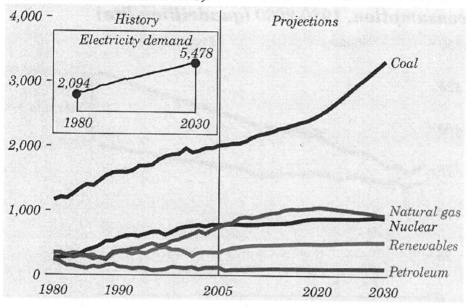

Figure 6. Total energy production and consumption, 1980-2030 (quadrillion Btu)

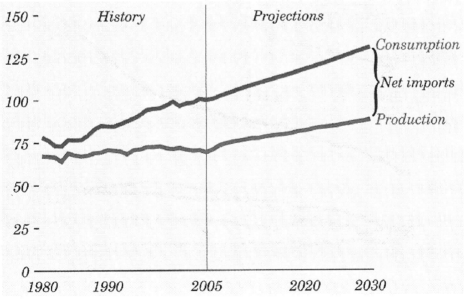

Figure 7. Energy production by fuel, 1980-2030 (quadrillion Btu)

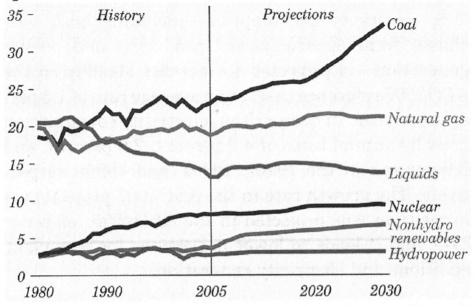

Figure 28. Energy expenditures as a share of gross domestic product, 1970-2030 (nominal expenditures as percent of nominal GDP)

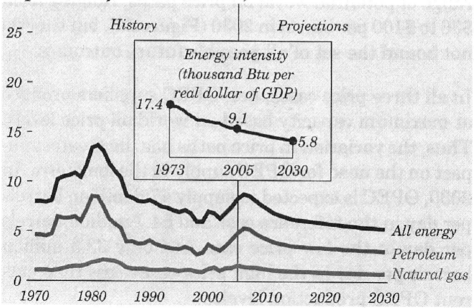

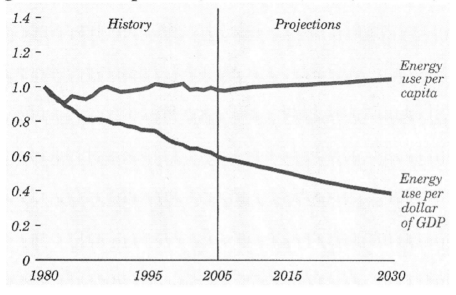

Figure 33. Energy use per capita and per dollar of gross domestic product, 1980-2030 (index, 1980 = 1)

Transportation Energy Use Is Expected To Increase

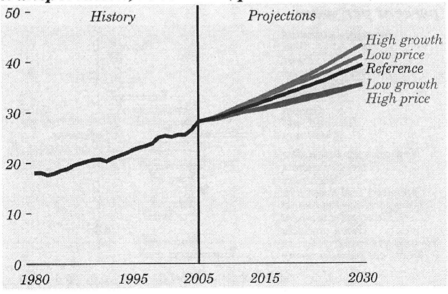

Figure 49. Delivered energy consumption for transportation, 1980-2030 (quadrillion Btu)

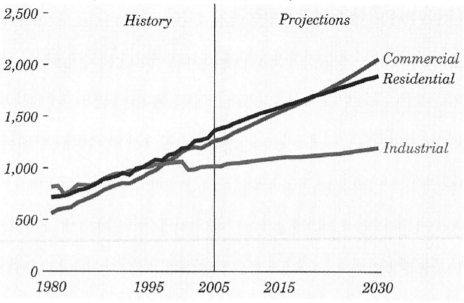

Figure 53. Annual electricity sales by sector, 1980-2030 (billion kilowatthours)

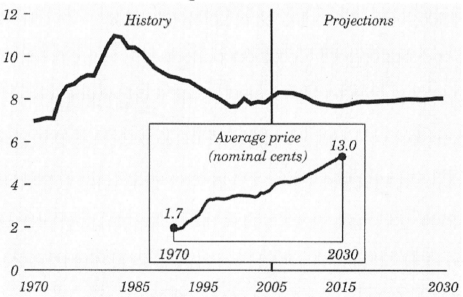

Figure 66. Average U.S. retail electricity prices, 1970-2030 (2005 cents per kilowatthour)

Diagram 1. Energy Flow, 2006
(Quadrillion Btu)

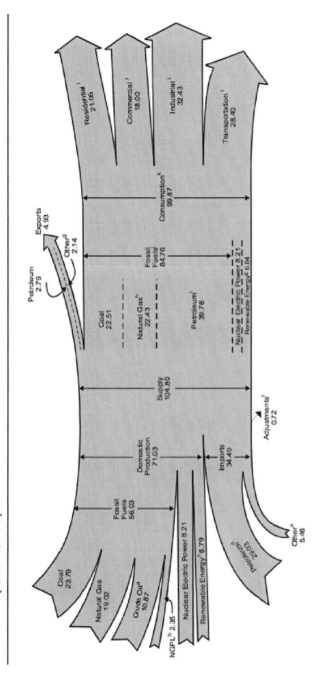

[a] Includes lease condensate.
[b] Natural gas plant liquids.
[c] Conventional hydroelectric power, biomass, geothermal, solar/PV, and wind.
[d] Crude oil and petroleum products. Includes imports into the Strategic Petroleum Reserve.
[e] Natural gas, coal, coal coke, fuel ethanol, and electricity.
[f] Stock changes, losses, gains, miscellaneous blending components, and unaccounted-for supply.
[g] Coal, natural gas, coal coke, and electricity.
[h] Natural gas only; excludes supplemental gaseous fuels.
[i] Petroleum products, including natural gas plant liquids, and crude oil burned as fuel.
[j] Includes 0.06 quadrillion Btu of coal coke net imports.
[k] Includes 0.06 quadrillion Btu of electricity net imports.
[l] Primary consumption, electricity retail sales, and electrical system energy losses, which are allocated to the end-use sectors in proportion to each sector's share of total electricity retail sales. See Note, "Electrical Systems Energy Losses," at end of Section 2.

Notes: • Data are preliminary. • Values are derived from source data prior to rounding for publication. • Totals may not equal sum of components due to independent rounding.

Sources: Tables 1.1, 1.2, 1.3, 1.4, and 2.1a.

419

Energy Units

Dimensional Prefixes

10	deka (da)	10^{-1}	deci (d)
10^2	hecto (h)	10^{-2}	centi (c)
10^3	kilo (k	10^{-3}	milli (m)
10^6	mega (M)	10^{-6}	micro (m)
10^9	giga (G)	10^{-9}	nano (n)
10^{12}	tera (T)	10^{-12}	pico (p)
10^{15}	peta (P)	10^{-15}	femto (f)
10^{18}	exa (E)	10^{-18}	atto (a)

Physics Constants

$a = e^2/\hbar c = 1/137.036$ (fine structure constant)

$c = 2.998 \times 10^8$ m/sec (speed of light in a vacuum)

$e = 1.60 \times 10^{-19}$ coulomb, C (electron/proton charge)

$1/4pe_o = 9.0 \times 10^9$ N-m^2/C^2

$e_o = 8.8 \times 10^{-12}$ C^2/N-m^2 (permittivity of space)

$m_o = 4p \times 10^{-7}$ N/A^2 (permeability of space)

$G = 6.67 \times 10^{-11}$ N-m^2/kg^2 (gravitational)

$g = 9.807$ m/sec^2 (acceleration of gravity at 45° latitude at sea level)

$h = 6.63 \times 10^{-34}$ J-sec $= 4.14 \times 10^{-15}$ eV-sec (Planck)

$\hbar = h/2p = 1.06 \times 10^{-34}$ J-sec $= 6.59 \times 10^{-14}$ eV-sec

$k_B = 1.38 \times 10^{-23}$ J/K $= 8.63 \times 10^5$ eV/K (Boltzman's constant)

$k_B T = 0.26$ eV $= 1/40$ eV (at room temperature 300K)

$m_e = 9.110 \times 10^{-31}$ kg (electron mass)

$m_p = 1.673 \times 10^{-27}$ kg $= 1.6726$ amu (proton mass)

$m_n = 1.675 \times 10^{-27}$ kg $= 1.6749$ amu, (neutron mass)

$m_e c^2 = 511$ keV, $m_p c^2 = 938.3$ MeV, $m_n c^2 = 939.6$ MeV

$N_A = 6.023 \times 10^{23}$ molecules/gram-mole (Avagodro's number)

volume of mole of gas at STP = 22.4 liter at 2.7×10^{19} molecules/cm^3

$R_{gas} = 8.31$ J/g-mole-K (ideal gas)

$s = 5.669 \times 10^{-8}$ J/m^2-K^4-sec (Stefan–Boltzman)

$p = 3.14159265358279 =$ the number of letters in the words in *How I need a drink, alcoholic of course, after the heavy lectures on quantum mechanics.*

Miscellaneous Useful Numbers

$a_o = \hbar^2/mc^2 = 0.053$ nm (Bohr radius)

$e^2/4pe_o = 1.44$ MeV-fermi (1 fm $= 10^{-15}$ m)

$e^2/8pe_o a_o = 13.6$ eV (hydrogen binding energy)

$E = hc/1 = 1.24/1$ (photon energy in eV with wavelength in m)

Wein $1 = 2.90 \times 10^6/T$ (blackbody maximum in nm with T in Kelvin)

$E = lm = 2$ eV (visible light from 6000 K at 0.6 m)

$E = hn = 0.1$ eV (IR at 300 K, 12 m)

band gap: Si (1.1 eV), Ge (0.7 eV)

1.3 fm $A^{1/3}$ (nuclear radius, A atomic number)

$c_V \gg 3R_{gas}$ = 24.9 J/g-mole-K (high temperature specific heat with $T \gg q_D$)

$R_{SI} = R_{English}/5.67$ and $U_{SI} = 5.67$ x $U_{English}$ (insulation)

0 K = –273.15°C = –459.7°F (absolute zero).

Length

1 mile (mi) = 5280 ft = 1.609 km

1 nautical mile = 1.86 km = 1.16 mi

1 m = 3.281 ft = 39.37 in

1 micron (m) = 10^{-6} m = 10^3 nm = 10^4 angstrom

1 inch (in) = 2.54 cm

1 fermi (fm) = 10^{-15} m

universe expanse 10^{26} m >> human height 1.8 m >> nuclear radius 10^{-15} m.

Area

1 m^2 = 10.8 ft^2

1 km^2 = 0.386 mi^2

1 acre = 43,560 ft^2 = 1 mi^2/640

1 hectare (ha) = 10^4 m^2 = 2.47 acres

1 barn (b) = 10^{-24} cm^2.

Volume

1 m^3 = 1000 liters = 264 US gallons (gal) = 35.3ft^3

1 $mile^3$ = 4.17 km^3

1 acre-foot = 43,560 ft^3 = 326,000 gal = 1234 m^3 = 0.1234 hectare-m

1 liter = 1000 cm^3 = 0.264 gal

1 bbl petroleum = 42 gal = 0.159 m^3.

Time

1 year = 365.25 days = 8766 hr = 3.154 x 10^7 sec » p x 10^7 sec

1 day = 86,400 sec

1 shake = 10^{-8} sec

1 age of universe = 4 x 10^{17} sec >> human 2 x 10^9 sec >> nuclear 10^{-23} sec ($2r/c$).

Mass

1 kg = 2.205 pounds (lb) = 32.3 ounces (oz)

1 lb = 16 oz = 453.6 g, 1 oz = 28.4 g

1 metric tonne (t) = 1000 kg = 1.102 tons

1 English ton = 2000 lb = 907.0 kg = 0.907 t

Force

1 newton (N) = 1 kg m/sec^2 = 10^5 dynes (dyn) = 0.22 lb

Pressure

1 bar (atm) = 76 cm Hg = 760 torr = 14.7 lb/in^2 = 10^5 pascal (1 N/m^2)

Energy/Heat
1 J = 1 W-sec = 1 calorie/4.2 = 1 kilocal/4200 = 6.242×10^{18} eV = 1 Btu/1055 = 0.738 ft-lb
1 eV = 1.602×10^{-19} eV
1 kWh = 3.6×10^6 J = 3412 Btu (electricity at h = 33% uses 10^4 Btu/kWh)
1 bbl crude petroleum = 5.8 MBtu = (42 gal)(138,000 Btu/gal)
1000 ft^3 (STP) natural gas = 100 terms = 1.03 MBtu = 1.09 GJ
1 trillion cubic feet (TCF) natural gas = 1.03 quads, 1 Gt coal = 27.8 quads
1 m^3 (STP) = 39 MJ, 1 TCF = 10^{12} ft^3
1 ton coal = 25.2 MBtu = 0.9 tonne coal
1 quad = 10^{15} Btu = 172 Mbbl = 0.97 TCF = 36 Mt coal = 292 G-kW_th = 1.05 EJ = $1.05 \, 10^{18}$ J
1 Gbbl = 5.8 quads
1 cubic foot natural gas at STP = 1000 Btu = 1 MJ
1 terawatt-year (TWyr) = 8.76×10^{12} kWh = 31.5 EJ
1 kWh/m^2 = 313 Btu/ft^2
1 kiloton TNT (kton) = 4.2×10^{12} J = 10^{12} calories
1 kg fission = 17 kton TNT (60 g/kton)
1 kg DT fusion = 85 kt (12 g/kton)
1 $MW_{thermal}$-day = 1 gram ^{235}U = 0.3 $MW_{electric}$-day.

Power
1 W = 1 J/sec = 1 N-M/sec = 1 kg-m^2/sec^3
1 hp = 550 ft-lb/sec = 0.746 kW = 746 J/sec
1 kW = 3412 Btu/hour
1 Mbbl/d = 0.365 Gbbl/year = 2.12 quads/yr = 71 GW_t(thermal)
US 100 quads/yr = 3400 GW_t = 47 Mbbl/d = 17 Gbbl/yr = 100 tcf/yr (equivalent)
US (280 M)/capita = 12 kW_t = 60 bbl/yr = 12.8 tonne/yr coal = 0.35 million ft^3/yr
1 lumen = 1/673 watt visible light
1 lux = 1 lumen/m^2, 1 foot candle = 1 lumen/ft^2 = 0.0929 lux.

Air
molecular weight (28.96)
density (1.293 kg/m^3)
sound speed (331.4 m/sec)
volume fraction (N_2/78%, O_2/21%, A/0.9%, H_2O/0.4%, CO_2/370 ppm)
specific heat (constant pressure, 1004 J/kg-K) and (constant volume, 720 J/kg-K)
viscosity (0.17 millipoise).

Water
density [1000 kg/m^3 (4 °C), 997 (25 °C), 958 (100 °C), 1025 (salt), 900 (ice)]
latent heat of fusion (333 kJ/kg), vaporization (2.26 MJ/kg)
specific heat water (4.2 kJ/kg-K), steam (100 °C, 2.01 kJ/kg-K), ice (2.1 kJ/kg-K)
viscosity (17.5 millipoise at 0°C, 2.8 at 100°C)
flow in Sverdup = 1 M m^3/sec
oceans (1350 x 10^{15} m^3), ice (29 x 10^{15} m^3), ground water (8.3 x 10^{15} m^3), lakes (0.13 x 10^{15} m^3).

Earth
radius R_E = 6357 km polar and 6378 km equatorial
area = 5.10 x 10^{14} km^2 (oceans 71%)
mass = 5.98 x 10^{24} kg
g' = 9.8 m/sec^2$(R_E/r)^2$
atmospheric pressure = 10^5 Pa e$^{-h/H}$, atmospheric height H = 8.1 km
atmospheric mass = 5.14 x 10^{18} kg with 1.3 x 10^{16} kg H$_2$O, oceanic mass = 1.4 x 10^{21} kg.

Sun
solar flux s_o = 1.367 kW/m^2 = 0.13 kW/ft^2 = 2.0 cal/minute-cm^2 = 435 Btu/ft^2-hr
24-hour average horizontal flux (40°N latitude) = 185 W/m^2
mass = 2 x 10^{30} kg
radius = 0.696 Mkm
distance to Earth = 150 Mkm.

Radiation (colloquial and SI units)
Rate of decay
1 curie (radiation of 1 g radium) = 1 Ci = 3.7 x 10^{10} decay/sec
1 bequerel (SI) = 1 Bq = 1 decay/sec.
Absorbed in air
1 Roentgen = 1 R = 87 ergs/g = 0.0087 J/kg.
Physical dose absorbed
1 rad = 100 erg/g = 0.01 J/kg
1 gray (SI) = 1 Gy = 1 J/kg = 100 rad.
Biological Dose Equivalent (absorbed dose times biological effectiveness Q)
x, g and e (Q = 1), n (Q = 5–20), alphas and fission fragments (Q = 20)
1 sievert (SI) = 1 Sv = 1 J/kg = 100 Rem
1 Rem = 0.01 J/kg = 0.1 Sv.
US annual average background dose (1990 BEIR-V) = 360 mRem (3.6 mSv):
radon (200 mRem), body radioactivity (39 mRem), medical x-rays (53 mRem)
cosmic radiation (31 mRem), sea level (28 mRem), Denver (81 mRem).

WWW Energy Sites

GENERAL SITES

American Geophysical Union: www.agu.org/sci_soc/
American Institute of Physics: www.aip.org/history/
American Physics Society: www.aps.org/public_affairs
APS Forum on Physics and Society: www.aps.org/units/fps
Congressional Legislation: //thomas.loc.gov
Congressional Budget Office: www.cbo.gov
Congressional Research Service: www.cnie.org/NLE/CRS
DOE Information Bridge: www.osti.gov/bridge/
DOE Labs: www.XX.gov; XX = anl, bnl, lanl, llnl, ornl, pnl, sandia, y12
Economic Report of the President: www.access.gpo.gov/eop
General Accounting Office: www.gao.gov
Government Printing Office: www.access.gpo.gov/su_docs/
National Academy Press: www.nap.edu
Office of Technology Assessment Legacy: www.wws.princeton.edu/~ota/
Science: www.sciencemag.org
White House: www.whitehouse.gov

ENERGY SITES

American Council for an Energy Efficient Economy: www.aceee.org
American Gas Association: www.aga.org
American Nuclear Society: www.ans.org
American Petroleum Institute: www.api.org
Ballard Fuel Cells: www.ballard.com
Bureau of Transportation Statistics: www.bts.gov/publications/
Clean Energies Future: www.ornl.gov/ORNL/Energy_Eff/CEF.htm
Davis Energy Group:www.davisenergy.com
DOE Efficiency/Renewable Energy (600 links): www.eren.doe.gov.
DOE Alternate Energy Vehicles: www.fleets.doe.gov
Efficient Windows: www.efficientwindows.org
Energy Information Agency: www.eia.doe.gov
Energy Star: www.energystar.gov
EPA fuel economy: www.fueleconomy.gov
First Solar: www.firstsolar.com
Fuel Cells: www.fuelcells.org

Hubbert: www.hubbertpeak.com
Hydrogen: www.hydrogenus.org or www.clean-air.org
Hydrogen Research: www.sc.doe.gov/bes/hydrogen.pdf
International Solar Energy Society: www.ises.org
Lawrence Berkeley National Lab: //enduse.lbl.gov
LBL Buildings: //eetd.lbl.gov/buildings.html
National Renewable Energy Laboratory: www.nrel.gov
National Transportation Statistics: www.bts.gov
Nuclear Power: //web.mit.edu/nuclearpower/
International Energy Agency: www.iea.org
Pacific Gas and Electric: www.pge.com/003_save_energy
Princeton: www.princeton.edu/~cees
Princeton Plasma Physics Lab simulations: //ippex.ppnl.gov
Rocky Mountain Institute: www.rmi.org
US Green Buildings Council: www.usgbc.org
Windpower: www.windpower.org

ENVIRONMENT SITES

Acid Rain: //bqs.usgs.gov/acidrain
Air Quality Management District: www.aqmd.gov
British Medical Journal: www.bmj.com
Chernobyl: www.ic-chernobyl.kiev.ua
Congressional Research Service: www.cnie.org/NLE/CRS
Dr. Everett Koop: www.drkoop.com
DOE Nuclear Waste: //cid.em.doe.gov
DOE Biology/Environment: www.sc.doe.gov/feature/biology_and_environment.htm
DOE Carbon Dioxide Information Analysis Center: //cdiac.esd.ornl.gov
Greening Earth Society: www.greeningearthsociety.org
Indiana Law: www.law.indiana.edu/v-lib/index.html
Intergovernmental Panel on Climate Change: www.ipcc.ch
Earth Data: //personal.cmich.edu/~Franc1m/homepage.htm
Environmental Protection Agency: www.epa.gov
Global Change Research Program: www.usgcrp.gov
NASA: //earthobservatory.nasa.gov
NASA Climate/Radiation: //climate.gsfc.nasa.gov
National Cancer Institute: www.nci.nih.gov
National Center for Atmospheric Research: www.ucar.edu
NOAA Climate: www.ncdc.noaa.gov
NSF: www.geo.nsf.gov/start.htm
Nuclear Waste Technology Review Board: www.nwtrb.gov
Michigan Radiation/Health: www.umich.edu/~radinfo
Pacific Institute: www.pacinst.org
US Geological Survey: www.usgs.gov
World Bank: //publications.worldbank.org
World Resources Institute: www.earthtrends.wri.org

Author Bio Briefs

Sam Arons is a member of Google's Green Business and Operations Strategy team, where his work focuses on Google's environmental initiatives, including RE<C (Renewable Energy Cheaper than Coal). Sam holds a Masters of Science in Energy and Resources from UC Berkeley. At UC Berkeley, he studied plug-in hybrid vehicles (PHEVs) as a policy tool for achieving greenhouse gas emission reductions and served as Co-Chair of the Chancellor's Advisory Committee on Sustainability.

Adam Brandt is a doctoral candidate in the Energy and Resources Group at the University of California, Berkeley. His academic interests include the depletion of conventional oil and the environmental impacts of substitutes for conventional oil. In particular, he is interested in understanding the environmental consequences of a transition to low-quality and unconventional hydrocarbon resources such as tar sands and oil shale.

Steven Chu is Director of the Lawrence Berkeley National Laboratory and Professor of Physics, Molecular and Cell Biology, UC–Berkeley, formerly he was at Stanford and Bell Laboratories. He co-chairs an Inter–Academy Council study "Lighting the Way: Toward a Sustainable Energy Future" and is active in the LBNL cellulosic bio–fuels program. Chu shared the 1997 Nobel Prize in Physics "for development of methods to cool and trap atoms with laser light." At Stanford, he helped start Bio-X, a multi-disciplinary initiative linking the physical and biological sciences with engineering and medicine.

George Crabtree is Senior Scientist at Argonne National Laboratory and he is Director of Argonne's Materials Science Division. This article is based on the conclusions contained in a report of the US Department of Energy Basic Energy Sciences Workshop on Solar Energy Utilization, April 18–21, 2005. Nathan Lewis served as the chair and George Crabtree served as co-chair of the workshop; they were the principal editors of the report.

Steven DenBaars is Professor of Materials and Co-Director of the Solid-State Lighting Center at the University of California Santa Barbara. From 1988-1991 he was a member of the technical staff at Hewlett-Packard's Opkoelectroncis Division, involved in the growth and fabrication of visible LEDs. Specific research interests include growth of wide-bandgap semiconductors (GaN based), and their application to Blue LEDs and lasers and high power electronic devices. This research has lead to the first US university demonstration of a Blue GaN laser diode, with related research on GaN growth and processing.

Darryl Dickerhoff is a Principal Research Associate at Lawrence Berkeley National Laboratory and has studied energy use in buildings since 1980. His emphasis has been in developing measurement techniques related to airflow including: infiltration, ventilation, and air leakage of the envelope and thermal distribution systems of both residential and commercial buildings. He is a collaborator with UC Berkeley's Center for the Built Environment research on underfloor ventilation systems.

Mark Duvall is Program Manager of the Electric Transportation Program at the Electric Power Research Institute. Prior to joining EPRI in 2001, he was Principal Development Engineer at the Hybrid Electric Vehicle Center, UC–Davis. He is currently focused on plug-in hybrid electric vehicle research, development and demonstration in collaboration with major automotive manufacturers, such as the EPRI-DaimlerChrysler PHEV Sprinter Van Program. He is also involved in advanced battery system development and testing, electric charging infrastructure, and environmental analysis of air quality and greenhouse gas emissions of plug-in hybrids and other electric transportation technologies.

Alex Farrell was Associate Professor in the Energy and Resources Group at UC– Berkeley and Director of the Transportation Sustainability Research Center. He was co-director of two studies for the state of California: *Managing Greenhouse Gases In California* and *A Low Carbon Fuel Standard for California*. Alex served on advisory committees for the National Academy of Engineering and the National Science Foundation. We were deeply shocked to learn that Alex died on April 13, 2008. We have lost a leader in energy matters and a friend to many of us. This book is dedicated to Alex's memory.

Clifford Federspiel is President of Federspiel Controls, LLC, which provides energy management solutions to the commercial buildings industry. Previously, he held an academic staff appointment at UC Berkeley, where he was affiliated with the Center for the Built Environment (CBE). At Berkeley, Federspiel managed several projects on the application of wireless sensor networks (motes) to building energy management. Prior to his appointment at UC Berkeley, he was a Senior Member of the Technical Staff at Johnson Controls.

Ashok Gadgil is Senior Scientist and Deputy Director (Strategic Planning) in the Environmental Energy Technologies Division at LBNL, and holds a concurrent appointment as Adjunct Professor of Energy and Resources at UC Berkeley. At LBNL, he leads research on airflow and pollutant transport in buildings, applying experimental and computational methods to understand complex flows and transport phenomena. At UC–Berkeley, his graduate course, "Technologies for Sustainable Communities," integrates a range of disciplines to understand why most technical attempts to help communities in the developing world fail and what it takes to help them succeed.

Imre Gyuk is Program Manager of DOE's Energy Storage Research Program, which funds work on a wide variety of technologies such as advanced batteries, flywheels, super-capacitors and compressed air energy storage. The storage applications include seamless continuity of power supply for high tech industry, frequency regulation, making renewables dispatchable, and energy management for congested distribution lines and substations. He has taught physics, civil engineering, and architecture at the University of Wisconsin and at Kuwait University. Diverse research interests have included work on elementary particles, groundwater flow, environmental architecture and cultural dynamics.

David Hafemeister is Physics Professor (emeritus) at California Polytechnic State University. He spent a dozen years in Washington at the US Senate, State Department, ACDA and National Academy of Sciences. He also did research at MIT, Princeton and Stanford universities and Argonne, Lawrence-Berkeley, and Los Alamos National Labs. He participated in the passage of the 1975 EPCA and 1976 ECPA energy laws. His energy lab at Cal Poly was dedicated to an inspirational physicist as "The Arthur Rosenfeld House Doctor Laboratory." His book, *Physics of Societal Issues: Calculations on National Security, Environment and Energy* quantifies some of the topics of this conference.

Danny Harvey is Professor in the Department of Geography, University of Toronto. His research interests include computer climate modeling, with particular emphasis on coupled climate-carbon cycle models and impacts of different future global energy scenarios. His interests have shifted to options to reduce emissions of greenhouse gases associated with energy use, with a particular emphasis on energy use and energy savings potential in buildings. His books include *Global Warming: The Hard Science, Climate and Global Environmental Change, A Handbook on Low-Energy Buildings and District Energy Systems.*

Jan Herbst is a Principal Researcher at the General Motors R&D Center. He has examined various approaches for the storage of hydrogen for automobiles. For his work on neodymium-iron-boron magnet materials he shared the 1986 APS International Prize for New Materials and he also received the John M. Campbell and Charles L McCuen Award from the GM Research Laboratories and the

Kettering Award from General Motors Corporation. His career research interests include magnetic, magnetostrictive and hydrogen storage materials.

Allan Hoffman is a senior analyst in DOE's Office of Energy Efficiency and Renewable Energy, working on renewable energy and water-energy issues. During 1974-1978 he served as Staff Scientist for the Senate Committee on Commerce, Science and Transportation, with responsibility for drafting legislation on automobile fuel economy standards (CAFE) and the formation of the White House Office of Science and Technology Policy. He subsequently held positions as Director/DOE Advanced Energy Systems Policy Division, Assistant Director for Industrial Programs/Mellon Institute Energy Productivity Center, Senior Analyst/ Office of Technology Assessment, Executive Director/Committee on Science, Engineering and Public Policy (NRC), and Associate and Acting Deputy Assistant Secretary for DOE's renewable electric programs. He has also served as U.S. Representative to and Vice Chairman of the International Energy Agency's Working Party on Renewable Energy.

Daniel Kammen is the Class of 1935 Distinguished Professor of Energy at University of California, Berkeley, where he holds appointments in the Energy and Resources Group, the Goldman School of Public Policy, and the Nuclear Engineering Department. Kammen is the founding director of the Renewable and Appropriate Energy Laboratory and Co-Director of the Berkeley Institute of the Environment. He was professor and Chair of Science, Technology and Environmental Policy at Princeton's in Woodrow Wilson School of Public and International Affairs, 1993–98. The focus of Kammen's work is on the science and policy of clean, renewable energy systems, energy efficiency, and the impacts of energy sources and technologies on development, particularly in Africa and Latin America. He was a coordinating lead author for the Intergovernmental Panel on Climate Change, which won the Nobel Peace Prize in 2007.

Lance Kim is a graduate student in the Department of Nuclear Engineering and the Goldman School of Public Policy at the University of California, Berkeley. His research interests include the application of multi-objective evolutionary algorithms to nuclear fuel cycle optimization, economic modeling, and probabilistic risk analysis (proliferation resistance, physical protection, and reactor safety). Prior to graduate school, he was a U.S. Support Program Fellow in the Department of Safeguards at the International Atomic Energy Agency and worked in risk analysis and advanced reactors at the U.S. Nuclear Regulatory Commission.

Barbara Goss Levi has spent most of her career as an editor for Physics Today magazine, writing news stories about current research. In the early 1980s she worked on issues of energy and arms control at Princeton University's Center for Energy and Environmental Studies. She has remained interested in those topics and has been an active member of the Forum on Physics and Society. Under the auspices of FPS, she has helped organize short courses on Energy Sources: *Conservation and Renewables* (1985), *Global Warming: Physics and Facts* (1991) and, this year, on *Physics of Sustainable Energy: Using Energy Efficiently and Producing It Renewably.*

Mark Levine was Director of the Environmental Division at Lawrence Berkeley National Laboratory from 1996-2006. The Division is a leader in research on buildings energy efficiency, indoor air quality, and various clean energy technologies. His major passion in the past two decades has involved analyzing and promoting energy efficiency in China, as Group Leader of the China Energy Group at LBNL. He is a board member of American Council for an Energy Efficient Economy, the Center for Clean Air Policy, Center for Resource Solutions, the California Clean Energy Fund and one in Asia. Dr. Levine had overall responsibility for the IPCC chapters on mitigating carbon emissions in buildings (2nd assessment report) and shared responsibility (4th assessment report) for which a Nobel Peace Prize was awarded.

Nathan Lewis is Professor of Chemistry at the California Institute of Technology in Pasadena, California, and he is Director of the Molecular Materials Research Center at Caltech's Beckman Institute. This article is based on the conclusions contained in a report of the US Department of Energy Basic Energy Sciences Workshop on Solar Energy Utilization, April 18–21, 2005. Nathan Lewis served as the chair and George Crabtree served as co-chair of the workshop; they were the principal editors of the report.

Michael McGehee is Assistant Professor of Materials Science and Engineering at Stanford University. His research interests are patterning materials at the nanometer length scale, semiconducting polymers, large area electronics and renewable energy. He has taught courses on nanotechnology, organic semiconductors, polymer science and solar cells. His PhD research was done on polymer lasers in the lab of Nobel Laureate Alan Heeger at University of California at Santa Barbara. This was followed with postdoctoral research on the self-assembly of organic-inorganic mesostructures. He won an NSF CAREER Award, a Dupont Young Professor Award, a Henry and Camille Dreyfus New Faculty Award, the 2007 Materials Research Society Outstanding Young Investigator Award and the Mohr Davidow Innovators Award.

Patrick McAuliffe is an advisor to Commissioner Arthur Rosenfeld of the California Energy Commission, focusing on wholesale and retail market structures, conservation, energy efficiency and demand response. He spent 25 years working in the electricity field. Initially, he worked on renewable energy and cogeneration, implementing legislation and guidebooks for potential developers and investors. Later, he worked as a resource planner, conducting integrated resource planning for the CEC to ascertain which combination of power plants and efficiency improvements made the most sense from cost, risk, and environmental impact perspectives. In the late 1990s, he worked on deregulation of the electricity industry, including development of market rules and assessments of various markets in the western United States with an emphasis on the relationship of pricing and system operation.

Jim McMahon is Head of the Energy Analysis Department, Co-chair of the Water Energy Technology Team and Leader of the Energy Efficiency Standards Group in the Environmental Energy Technology Division at Lawrence Berkeley National Laboratory. He is lead author of Chapter 9, "Buildings" for the North American carbon budget and implications for the global carbon cycle (2007). For 30 years, he has led or participated in research to 1) identify the feasibility and cost of engineering design changes that could increase energy or water efficiency for products used in buildings; 2) to analyze scenarios, quantifying economic impacts associated with adoption of these technologies; and 3) to assess potential impacts on key market elements including consumers, manufacturers, utility companies, government, the nation, and the environment.

Mark Mehos is Program Manager for the Concentrating Solar Power Program at the National Renewable Energy Laboratory since 2001. He has been the leader of the High Temperature Solar Thermal Team since 1998. He has managed and performed technical work within NREL's advanced optical materials, solar photocatalysis, and dish/Stirling research and development activities. The emphasis of the High Temperature Solar Thermal Team is the development of low-cost, high-performance, and high-reliability systems that use concentrated sunlight to generate power, with an emphasis on large multi-megawatt parabolic trough systems and kilowatt-scale concentrating photovoltaic systems. He currently is a participating member of New Mexico Governor Bill Richardson's Concentrating Solar Power Task Force, as well as the Solar Task Force for the Western Governors' Association Clean and Diversified Energy Initiative.

Larry Myer is Staff Scientist at Lawrence Berkeley National Laboratory, Earth Sciences Division, where he has conducted research in geophysics and geomechanics since 1981. He has led research activities in geologic sequestration since 1999. He holds a joint appointment with the California Energy Commission, Public Interest Energy Research Program, where he is Technical Director of the West Coast Regional Carbon Sequestration Partnership. The Partnership is evaluating carbon dioxide sequestration options and opportunities for the west coast of North America.

Per Peterson is Professor of Nuclear Engineering at UC–Berkeley. He has served as a National Science Foundation Presidential Young Investigator from 1990 to 1995, as chair of the Energy and Resources Group, an interdisciplinary graduate group, from 1998 to 2000; and as chair of the Nuclear Engineering Department from 2000 to 2005. His specific research interests focuses on topics in heat and mass transfer, fluid dynamics, and phase change. He has worked on problems in energy and environmental systems, including advanced reactors, inertial fusion, high-level nuclear waste processing, and nuclear materials management and security.

Lynn Price is a Research Scientist in the China Energy Group of the Energy Analysis Department, Environmental Energy Technologies Division of Lawrence Berkeley National Laboratory, where she has worked since 1990. Ms. Price has been a member of the Intergovernmental Panel on Climate Change since 1994, which won the Nobel Peace Prize in 2007. Most recently she was a lead author of the industrial sector chapter of Fourth Assessment Report on Mitigation of Climate Change. Ms. Price has provided project and technical leadership on numerous projects related to energy efficiency and greenhouse gas mitigation for the Energy Foundation, U.S. Environmental Protection Agency, the U.S. Department of Energy, the California Energy Commission, the World Bank, and the International Energy Agency.

Mike Robinson is the Deputy Director of the National Wind Technology Center at the National Renewable Energy Laboratory. He currently manages scientists and engineers conducting basic and applied research programs at the NWTC. Basic research activities include projects in rotor aerodynamics, inflow characterization and turbulence modeling, electrical generation and power electronics, controls, and structural dynamics. Applied research programs include research efforts in advanced component development, variable speed architectures, power systems and structural design codes. He maintains an active involvement in unsteady aerodynamics research including both field test and model development. Prior to joining NREL, Mike was the Associate Director of Engineering, Bioserve Space Technology at the University of Colorado.

Art Rosenfeld, formerly Professor of Physics at UC-Berkeley, formed the Center for Building Science at Lawrence Berkeley National Laboratory, which he led until 1994. From 1994 -1999 Dr. Rosenfeld served as Senior Advisor to DOE Assistant Secretary for Energy Efficiency and Renewable Energy. In 2000 California Governor Gray Davis appointed him Commissioner at the California Energy Commission, and he was re-appointed by Governor Arnold Schwarzenegger in 2005. He is responsible for the Public Interest Energy Research program for energy efficiency, including the California energy efficiency standards for buildings and for appliances. He is the Assigned Commissioner to collaborate with the Public Utilities Commission Proceeding on demand response, critical peak pricing, advanced metering, and energy efficiency programs, with a budget of $600 million/year. Rosenfeld received the APS Szilard Award for Physics in the Public Interest in 1986, the DOE Carnot Award for Energy Efficiency in 1993, and the DOE Enrico Fermi Award in 2006.

Marc Ross is Professor of Physics (emeritus) at the University of Michigan. His research is in environmental physics, especially energy use and its impacts and the reduction of those impacts through efficiency and conservation. Up to the mid-1980s his special focus was industrial energy use. Since the late 1980s he has focused on automobiles. Much of his current research concerns technologies to improve fuel economy. He has also analyzed the emissions of in-use cars. An important part of this has been creation and improvement of a model of fuel use and emissions as they depend on driving. Recently he has been studying the relationships between traffic safety, vehicle mass and other variables in order to evaluate effects of changes in vehicle mass and design on traffic fatalities.

Peter Schwartz is an associate professor at Cal Poly University. After 10 years of nanotechnology research he changed his research to *energy sustainability*. Schwartz spent the 2006-7 year at UC Berkeley's Energy and Resources Group. His research interests include Concentrated Solar Power, Electric Transportation, and Economic Analysis of Increased Efficiency and Renewable Energy Substitution. The later includes work with Santa Barbara's Community Environmental Council. At Cal Poly, Schwartz is teaching courses on "Energy, Society and Environment" and "Appropriate Technology for Impoverished Communities."

Stephen Selkowitz is Department Head of the Building Technologies Department, Lawrence Berkeley National Laboratory, where he manages a building science R&D program encompassing Windows,

Daylighting and Lighting Systems Research, Simulation Tools, Commercial Building Performance, and Demand Response Research. He has over 30 years of experience in the field of building energy performance and sustainable design, with projects that range from basic materials research to near term demonstrations of emerging technologies to research that will enable a new generation of "zero energy" or "carbon-neutral" buildings. He is a frequent invited speaker on the topic of building energy efficiency, and is the author of over 170 publications and holds 2 patents.

Max Sherman is Leader of the Energy Performance of Buildings Group at Lawrence Berkeley National Laboratory and Senior Scientist. Sherman invented the blower door and discovered ways to tighten air ducting in buildings. Sherman is a Distinguished Fellow and former member of the Board of Directors of the American Society of Heating, Refrigerating, and Air Conditioning Engineers (ASHRAE). He chaired the committee that wrote the nation's first standard on residential ventilation, ASHRAE 62.2.

Robert H. Socolow, Professor of Mechanical and Aerospace Engineering at Princeton University, teaches in both the School of Engineering and Applied Science and the Woodrow Wilson School of Public and International Affairs. With ecologist, Stephen Pacala, Socolow leads the University's Carbon Mitigation Initiative. His research focuses on technology and policy for fossil fuels under climate constraints. He was the co-leader of the American Physical Society's 1974 Summer Study, "Efficient Use of Energy," held in Princeton. He was awarded the 2003 Leo Szilard Lectureship Award by the American Physical Society: "For leadership in establishing energy and environmental problems as legitimate research fields for physicists, and for demonstrating that these broadly defined problems can be addressed with the highest scientific standards."

Venkat Srinivasan is Staff Scientist at LBNL where he contributes towards solving the multitude of problems that prevent Li-ion batteries from being used in Hybrid Electric and Plug-in Hybrid Vehicles. In addition, he serves as the technical manager of the Batteries for Advanced Transportation Technologies (BATT) program. He has been involved with projects in lead-acid, alkaline, Ni-MH, and Li-ion batteries. His present interest is primarily in the area of energy storage devices (batteries and capacitors) where he uses both theoretical and experimental techniques to understand their behavior. His approach spans both the fundamental and the applied aspects of their operation.

Robert Thresher is the NREL Wind Energy Research Fellow and Director of the National Wind Technology Center at the National Renewable Energy Laboratory. He was a professor in Mechanical Engineering at Oregon State University in the 1970's and 1980's, where he taught courses in Applied Mechanics and initiated pioneering research in the mechanics of wind energy systems. He joined NREL in 1984 and has provided leadership for the growth of NREL's wind program at its inception to its current level of $30 million/year. He also serves as a member of the Advisory Panel on Ocean Energy Technologies for the Electric Power Research Institute's, testifying in 2005 before the US Senate's Energy Committee on the R&D on the need to accelerate the use of Wind and Ocean Technologies on the US Outer Continental Shelf. He was recognized as the Person of the Year by the American Wind Energy Association in1997.

Paul Veers is a Distinguished Member of the Technical Staff at Sandia National Laboratories where he has worked in Engineering Sciences from 1980 through 1987 and since then has worked in the Wind Energy Technology Department. He has conducted research on various aspects of wind systems including atmospheric turbulence simulation, fatigue analysis, reliability, structural dynamics, aeroelastic tailoring, and the evaluation of design safety factors. Paul is currently the Chief Editor for *Wind Energy*, an international journal for progress and applications in wind power, and is a past Associate Editor for the *Journal of Solar Energy Engineering, Transactions of the ASME*.

Iain Walker is a scientist at Lawrence Berkeley National Laboratory. His current work focuses on Zero/Low Energy Homes and HVAC systems in residential buildings through field and laboratory evaluations, modeling and simulation activities, and standards setting. Walker is the Executive Editor and Secretary of the board of *Home Energy* magazine. He is the task group leader for ASTM standards committees on building and duct system air leakage and sealant longevity. For ASHRAE, Walker serves on Standards committees for indoor air quality, weather, moisture design, and equipment air leakage. He also serves on the ASHRAE Research Activities Committee and the Building Performance Institute Technical Standards Committee.

Tom Wenzel is a Research Scientist at Lawrence–Berkeley National Laboratory, working in transportation energy and environmental policy analysis. He has been analyzing in-use vehicle emission data since 1992 and is a national expert on the evaluation of state vehicle emission inspection and maintenance programs. He has assisted in the development of a model for predicting modal emissions of in-use vehicles. Since 2000 he has analyzed the effect on occupant safety by reducing vehicle mass in order to improve fuel economy.

James Woolaway has 25 years experience in the design, fabrication, testing, and system integration of infrared focal plane arrays (IR-FPA) and infrared camera systems. He has designed, or contributed to the design, of over 100 advanced sensors including the first IR-FPA containing on-chip analog to digital conversion, the first windowing IR-FPA, and the first IR-FPA containing a neural network. In 1980 while still a student, Jim worked at the Natural Sciences Laboratory at the University of Groningen in the Netherlands as a Research Assistant for the Hyperfine Interactions Group. Jim was a founder of Indigo Systems Corporation and is now the Vice President of Technology at FLIR Systems Corporation in Santa Barbara California.

Craig Wray is a Mechanical Engineer at Lawrence Berkeley National Laboratory. Over the past 24 years, he has conducted research on building energy use, heat transfer, ventilation, smoke transport, simulation theory, and commissioning. Much of his work has directly supported national-scale energy efficiency and technology transfer activities such as the Canadian R-2000 and Energuide for Houses programs. Public and proprietary tools that incorporate his work have been used to generate performance ratings and retrofit plans for more than a quarter million houses. He currently serves as Chair of the ASHRAE Technical Activities Committee and was the former Section Head of the ASHRAE Load Calculations and Energy Requirements Technical Committee.

Sustainable Energy Participants

Paul Albertus
James Alessio
Daisy Allen
Ingrid Anderson
Laura Meadow Anderson
Steven Anton
Jerry Artz
Susan Amrose
Jessy Baker
John Bates
Luis J. Bautista
Nina Beck
Thomas Becker
Calvin Berggren
Rex Berney
Fabrizio Bisetti
David Blackman
Janice Blumenkrantz
Stephen Blumenkrantz
Nate Bluemenkrantz
David Boals
Zoe Boekelheide
Vincent Bouchiat
Jason Bowers
Sarah Boyd
David Braun
Jennifer Burney
Alex Byrne
Jerome Carman
Craig Carr
James Carroll
Christian Casillas
Nick Cizek
Alison Chaiken
Candace Chan
Kwang-Ping Cheng
Steve Chu
Nicole Clark
Richard Cohen
Joanne Cohn

Eugene Commins
Steve Connor
Pierce Corden
Ingrid Cotores
Paul Craig
Bruce Craver
Doug Crowder
Lifeng Cui
Yi Cui
Tanja Cuk
Michelangelo D'Agostino
Kassie Dallavis
Gregory Davis
Robert Dibble
Paul Debevec
Steve DenBaars
Jessie Denver
David Dixon
David Dobereiner
Alexsandar Donev
Mark Duvall
Kuniyoshi Ebina
Robert Echols
Elizabeth Edwards
Dirk Englund
Gerald Enrenstein
Doug Epperson
Ann Erickson
Lara Ettenson
Bruce Failor
Andre Faraon
Isaac Fenrenbach
Michael Fitzsimmons
Leah Fletcher
Kate Foreman
Craig Foster
David Foster
John Fox
Chris France
Ilya Fushman

Jim Gates
Oliver Gessner
Dimitrios Giannakis
David Gibbs
Judith Gibbs
Maxsim Gibiansky
Kohl Gill
Matt Golden
Sam Golding
Peter Gollon
Rafael Gomez-Sjoberg
Dan Goncher
Chris Gould
Karen Graul
Kim Griest
Imre Gyuk
Siina Haapanen
Nathan Haese
David Hafemeister
Wilson Hago
Shaul Hanany
Jonathan Hardis
Laurence Hardwick
Jim Harris
Danny Harvey
Hugh Haskell
Richard Haskell
David Heppe
Jan Herbst
Frank Hicks
Ananda Hirsch
Bettina Hodel
Eric Hoke
Ching-Mei Hsu
Rafael Jaramillo
Daniel Jarvis
Erik Jonsson
Stanley Jonsson
Kyungseon Joo
Michael Jura

Michael P. Jura
Jade Juhl
Tina Kaarsberg
Dan Kammen
Aidan Kelleher
Andreas Kemp
Na Young Kim
Sang Kim
Randy Knight
Ryoichi Komiyama
Lena Fitting Kourkoutis
Richard Kriske
Mari Kryder
Theodore Lavine
Justin Lawson
Ilan Levi
Barb Levi
Mark Levine
Herb Lin
Jamie Link
Roger Longden
William Love
Ben MacBride
Massimo Maniaci
Jeffrey Marque
Takechio Maruyama
Daria Mashnik
Phil Matheson
Laila Mattos
Erin McCowen
James McDonough
Michael McGehee
Chris McGuinness
Ron McKnight
Jim Mcmahon
Mark Mehos
Michael Meserve
Fletcher Miller
Brady Mills
Dan Milstein
Jonathan Mingle
Vivek Mohta
Jeremy Monat
Thomas Mosier

Richard Muller
Tom Murphy
Larry Myer
John Neal
James Neff
Joshua Nollenberg
Brian Nordstrom
Zack Norwood
Tyler Otto
Jennifer Ouellette
Feryal Ozel
Tapan Parekh
Sarah Peach
Michael Pelizzari
Hailen Peng
Dan Petersen
Per Peterson
Kevin Price
Lynn Price
John Poling
Stephen Portis
Mike Preiner
Daniel Ratner
Jayarami Reddy
Corinne Reich-Weiser
Chris Rivest
Yvonne Rodriguez
Robert Rohde
Art Rosenfeld
Ridah Sabouni
Michael Sarahan
Edina Sarajic
Wayne Saslow
Daniel Saunders
Samveg Saxena
Fred Schlachter
David Schoen
Peter Schwartz
Steve Selkowitz
Robert Shanbrom
Matthew Sharp
Pete Shoemaker
Navin Sivanondam
John Smedley

Rob Socolow
Sofia Soltys
Daniel Soto
Venkat Srinivasan
David Stamper
Duncan Stewart
John Paul Strachan
Elizabeth Stoltzfus
Allen Sussman
Patricia Terry
Julia Thompson
Neil Thompson
Nellie Tong
Robert Thresher
Dale Tutaj
Kaito Umemura
Andrew Van Blarigan
Patrick Varilly
Sam Vigil
Miren Vizeaino
James Waite
Leonard Wall
Jigang Wang
Robert Ward
Mark Waters
Max Wei
Tom Wenzel
James White
Martin White
Richard Wiener
Brian Wilfley
Dean Wilkening
Adam Winter
Jesse Wodin
Philip Wolf
Gordon Wozniak
Craig Wray
Ralph Wuerker
Lily Yang
Yuan Yang

AUTHOR INDEX

A

Akbari, H., 192
Arons, S., 235

B

Brandt, A., 235

C

Chu, S., 266
Crabtree, G. W., 309

D

DenBaars, S., 141
Dickerhoff, D. J., 149

F

Farrell, A., 235
Federspiel, C. C., 149

G

Gadgil, A., 176
Goh, C., 322
Gyuk, I., 376

H

Hafemeister, D., 88, 395
Harvey, L. D. D., 67
Herbst, J. F., 297
Hoffman, A. R., 55

K

Kammen, D. M., 49
Kim, L., 360

L

Levine, M. D., 15
Lewis, N. S., 309

M

McAuliffe, P., 3
McGehee, M. D., 322
McMahon, J. E., 124
Mehos, M., 331
Meier, A., 209
Myer, L. R., 366

P

Peterson, P. F., 360
Price, L., 163

R

Robinson, M., 340
Rosenfeld, A. H., 3
Ross, M., 251

S

Selkowitz, S., 112
Sherman, M. H., 149
Socolow, R. H., 28
Srinivasan, V., 283

T

Thresher, R., 340

437